Mathematical Methods in Quantum Mechanics

With Applications
to Schrödinger Operators,
Second Edition

Mathematical Methods in Quantum Mechanics

With Applications to Schrödinger Operators, Second Edition

Gerald Teschl

Graduate Studies
in Mathematics
Volume 157

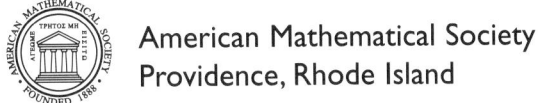

American Mathematical Society
Providence, Rhode Island

Mathematical Methods in Quantum Mechanics

With Applications to Schrödinger Operators, Second Edition

Gerald Teschl

Graduate Studies in Mathematics

Volume 157

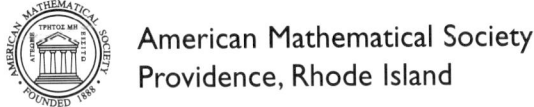

American Mathematical Society
Providence, Rhode Island

EDITORIAL COMMITTEE

Dan Abramovich
Daniel S. Freed
Rafe Mazzeo (Chair)
Gigliola Staffilani

2010 *Mathematics Subject Classification.* Primary 81-01, 81Qxx, 46-01, 34Bxx, 47B25.

For additional information and updates on this book, visit
www.ams.org/bookpages/gsm-157

Library of Congress Cataloging-in-Publication Data
Teschl, Gerald, 1970–
  Mathematical methods in quantum mechanics : with applications to Schrödinger operators / Gerald Teschl.– Second edition.
    pages cm. — (Graduate studies in mathematics ; volume 157)
  Includes bibliographical references and index.
  ISBN 978-1-4704-1704-8 (alk. paper)
  1. Schrödinger operator. 2. Quantum theory—Mathematics. I. Title.
QC174.17.S3T47 2014
530.1201′51—dc23
                                                                                          2014019123

Copying and reprinting. Individual readers of this publication, and nonprofit libraries acting for them, are permitted to make fair use of the material, such as to copy select pages for use in teaching or research. Permission is granted to quote brief passages from this publication in reviews, provided the customary acknowledgment of the source is given.

Republication, systematic copying, or multiple reproduction of any material in this publication is permitted only under license from the American Mathematical Society. Permissions to reuse portions of AMS publication content are handled by Copyright Clearance Center's RightsLink® service. For more information, please visit: http://www.ams.org/rightslink.

Send requests for translation rights and licensed reprints to reprint-permission@ams.org.

Excluded from these provisions is material for which the author holds copyright. In such cases, requests for permission to reuse or reprint material should be addressed directly to the author(s). Copyright ownership is indicated on the copyright page, or on the lower right-hand corner of the first page of each article within proceedings volumes.

© 2014 by the American Mathematical Society. All rights reserved.
The American Mathematical Society retains all rights
except those granted to the United States Government.
Printed in the United States of America.

∞ The paper used in this book is acid-free and falls within the guidelines
established to ensure permanence and durability.
Visit the AMS home page at http://www.ams.org/

10 9 8 7 6 5 4 3 2 1   19 18 17 16 15 14

To Susanne, Simon, and Jakob

Contents

Preface   xi

Part 0. Preliminaries

Chapter 0. A first look at Banach and Hilbert spaces   3
- §0.1. Warm up: Metric and topological spaces   3
- §0.2. The Banach space of continuous functions   14
- §0.3. The geometry of Hilbert spaces   21
- §0.4. Completeness   26
- §0.5. Bounded operators   27
- §0.6. Lebesgue L^p spaces   30
- §0.7. Appendix: The uniform boundedness principle   38

Part 1. Mathematical Foundations of Quantum Mechanics

Chapter 1. Hilbert spaces   43
- §1.1. Hilbert spaces   43
- §1.2. Orthonormal bases   45
- §1.3. The projection theorem and the Riesz lemma   49
- §1.4. Orthogonal sums and tensor products   52
- §1.5. The C^* algebra of bounded linear operators   54
- §1.6. Weak and strong convergence   55
- §1.7. Appendix: The Stone–Weierstraß theorem   59

Chapter 2. Self-adjointness and spectrum   63

§2.1.	Some quantum mechanics	63
§2.2.	Self-adjoint operators	66
§2.3.	Quadratic forms and the Friedrichs extension	76
§2.4.	Resolvents and spectra	83
§2.5.	Orthogonal sums of operators	89
§2.6.	Self-adjoint extensions	91
§2.7.	Appendix: Absolutely continuous functions	95

Chapter 3. The spectral theorem — 99
- §3.1. The spectral theorem — 99
- §3.2. More on Borel measures — 112
- §3.3. Spectral types — 118
- §3.4. Appendix: Herglotz–Nevanlinna functions — 120

Chapter 4. Applications of the spectral theorem — 131
- §4.1. Integral formulas — 131
- §4.2. Commuting operators — 135
- §4.3. Polar decomposition — 138
- §4.4. The min-max theorem — 140
- §4.5. Estimating eigenspaces — 142
- §4.6. Tensor products of operators — 143

Chapter 5. Quantum dynamics — 145
- §5.1. The time evolution and Stone's theorem — 145
- §5.2. The RAGE theorem — 150
- §5.3. The Trotter product formula — 155

Chapter 6. Perturbation theory for self-adjoint operators — 157
- §6.1. Relatively bounded operators and the Kato–Rellich theorem — 157
- §6.2. More on compact operators — 160
- §6.3. Hilbert–Schmidt and trace class operators — 163
- §6.4. Relatively compact operators and Weyl's theorem — 170
- §6.5. Relatively form-bounded operators and the KLMN theorem — 174
- §6.6. Strong and norm resolvent convergence — 179

Part 2. Schrödinger Operators

Chapter 7. The free Schrödinger operator — 187
- §7.1. The Fourier transform — 187

§7.2.	Sobolev spaces	194
§7.3.	The free Schrödinger operator	197
§7.4.	The time evolution in the free case	199
§7.5.	The resolvent and Green's function	201

Chapter 8. Algebraic methods — 207
- §8.1. Position and momentum — 207
- §8.2. Angular momentum — 209
- §8.3. The harmonic oscillator — 212
- §8.4. Abstract commutation — 214

Chapter 9. One-dimensional Schrödinger operators — 217
- §9.1. Sturm–Liouville operators — 217
- §9.2. Weyl's limit circle, limit point alternative — 223
- §9.3. Spectral transformations I — 231
- §9.4. Inverse spectral theory — 238
- §9.5. Absolutely continuous spectrum — 242
- §9.6. Spectral transformations II — 245
- §9.7. The spectra of one-dimensional Schrödinger operators — 250

Chapter 10. One-particle Schrödinger operators — 257
- §10.1. Self-adjointness and spectrum — 257
- §10.2. The hydrogen atom — 258
- §10.3. Angular momentum — 261
- §10.4. The eigenvalues of the hydrogen atom — 265
- §10.5. Nondegeneracy of the ground state — 272

Chapter 11. Atomic Schrödinger operators — 275
- §11.1. Self-adjointness — 275
- §11.2. The HVZ theorem — 278

Chapter 12. Scattering theory — 283
- §12.1. Abstract theory — 283
- §12.2. Incoming and outgoing states — 286
- §12.3. Schrödinger operators with short range potentials — 289

Part 3. Appendix

Appendix A. Almost everything about Lebesgue integration — 295
- §A.1. Borel measures in a nutshell — 295

§A.2.	Extending a premeasure to a measure	303
§A.3.	Measurable functions	307
§A.4.	How wild are measurable objects?	309
§A.5.	Integration — Sum me up, Henri	312
§A.6.	Product measures	319
§A.7.	Transformation of measures and integrals	322
§A.8.	Vague convergence of measures	328
§A.9.	Decomposition of measures	331
§A.10.	Derivatives of measures	334

Bibliographical notes	341
Bibliography	345
Glossary of notation	349
Index	353

Preface

Overview

The present text was written for my course *Schrödinger Operators* held at the University of Vienna in winter 1999, summer 2002, summer 2005, and winter 2007. It gives a brief but rather self-contained introduction to the mathematical methods of quantum mechanics with a view towards applications to Schrödinger operators. The applications presented are highly selective; as a result, many important and interesting items are not touched upon.

Part 1 is a stripped-down introduction to spectral theory of unbounded operators where I try to introduce only those topics which are needed for the applications later on. This has the advantage that you will (hopefully) not get drowned in results which are never used again before you get to the applications. In particular, I am not trying to present an encyclopedic reference. Nevertheless I still feel that the first part should provide a solid background covering many important results which are usually taken for granted in more advanced books and research papers.

My approach is built around the spectral theorem as the central object. Hence I try to get to it as quickly as possible. Moreover, I do not take the detour over bounded operators but I go straight for the unbounded case. In addition, existence of spectral measures is established via the Herglotz rather than the Riesz representation theorem since this approach paves the way for an investigation of spectral types via boundary values of the resolvent as the spectral parameter approaches the real line.

Part 2 starts with the free Schrödinger equation and computes the free resolvent and time evolution. In addition, I discuss position, momentum, and angular momentum operators via algebraic methods. This is usually found in any physics textbook on quantum mechanics, with the only difference being that I include some technical details which are typically not found there. Then there is an introduction to one-dimensional models (Sturm–Liouville operators) including generalized eigenfunction expansions (Weyl–Titchmarsh theory) and subordinacy theory from Gilbert and Pearson. These results are applied to compute the spectrum of the hydrogen atom, where again I try to provide some mathematical details not found in physics textbooks. Further topics are nondegeneracy of the ground state, spectra of atoms (the HVZ theorem), and scattering theory (the Enß method).

Prerequisites

I assume some previous experience with Hilbert spaces and bounded linear operators which should be covered in any basic course on functional analysis. However, while this assumption is reasonable for mathematics students, it might not always be for physics students. For this reason there is a preliminary chapter reviewing all necessary results (including proofs). In addition, there is an appendix (again with proofs) providing all necessary results from measure theory.

Literature

The present book is highly influenced by the four volumes of Reed and Simon [49]–[52] (see also [16]) and by the book by Weidmann [70] (an extended version of which has recently appeared in two volumes [72], [73], however, only in German). Other books with a similar scope are, for example, [16], [17], [21], [26], [28], [30], [48], [57], [63], and [65]. For those who want to know more about the physical aspects, I can recommend the classical book by Thirring [68] and the visual guides by Thaller [66], [67]. Further information can be found in the bibliographical notes at the end.

Reader's guide

There is some intentional overlap among Chapter 0, Chapter 1, and Chapter 2. Hence, provided you have the necessary background, you can start reading in Chapter 1 or even Chapter 2. Chapters 2 and 3 are key chapters, and you should study them in detail (except for Section 2.6 which can be skipped on first reading). Chapter 4 should give you an idea of how the spectral theorem is used. You should have a look at (e.g.) the first section, and you can come back to the remaining ones as needed. Chapter 5 contains two key results from quantum dynamics: Stone's theorem and the RAGE theorem. In particular, the RAGE theorem shows the connections between long-time behavior and spectral types. Finally, Chapter 6 is again of central importance and should be studied in detail.

The chapters in the second part are mostly independent of each other except for Chapter 7, which is a prerequisite for all others except for Chapter 9.

If you are interested in one-dimensional models (Sturm–Liouville equations), Chapter 9 is all you need.

If you are interested in atoms, read Chapter 7, Chapter 10, and Chapter 11. In particular, you can skip the separation of variables (Sections 10.3 and 10.4, which require Chapter 9) method for computing the eigenvalues of the hydrogen atom, if you are happy with the fact that there are countably many which accumulate at the bottom of the continuous spectrum.

If you are interested in scattering theory, read Chapter 7, the first two sections of Chapter 10, and Chapter 12. Chapter 5 is one of the key prerequisites in this case.

2nd edition

Several people have sent me valuable feedback and pointed out misprints since the appearance of the first edition. All of these comments are of course taken into account. Moreover, numerous small improvements were made throughout. Chapter 3 has been reworked, and I hope that it is now more accessible to beginners. Also some proofs in Section 9.4 have been simplified (giving slightly better results at the same time). Finally, the appendix on measure theory has also grown a bit: I have added several examples and some material around the change of variables formula and integration of radial functions.

Updates

The AMS is hosting a web page for this book at

http://www.ams.org/bookpages/gsm-157/

where updates, corrections, and other material may be found, including a link to material on my own web site:

> http://www.mat.univie.ac.at/~gerald/ftp/book-schroe/

Acknowledgments

I would like to thank Volker Enß for making his lecture notes [20] available to me. Many colleagues and students have made useful suggestions and pointed out mistakes in earlier drafts of this book, in particular: Kerstin Ammann, Jörg Arnberger, Chris Davis, Fritz Gesztesy, Maria Hoffmann-Ostenhof, Zhenyou Huang, Helge Krüger, Katrin Grunert, Wang Lanning, Daniel Lenz, Christine Pfeuffer, Roland Möws, Arnold L. Neidhardt, Serge Richard, Harald Rindler, Alexander Sakhnovich, Robert Stadler, Johannes Temme, Karl Unterkofler, Joachim Weidmann, Rudi Weikard, and David Wimmesberger.

My thanks for pointing out mistakes in the first edition go to: Erik Makino Bakken, Alexander Beigl, Stephan Bogendörfer, Søren Fournais, Semra Demirel-Frank, Katrin Grunert, Jason Jo, Helge Krüger, Oliver Leingang, Serge Richard, Gerardo González Robert, Bob Sims, Oliver Skocek, Robert Stadler, Fernando Torres-Torija, Gerhard Tulzer, Hendrik Vogt, and David Wimmesberger.

If you also find an error or if you have comments or suggestions (no matter how small), please let me know.

I have been supported by the Austrian Science Fund (FWF) during much of this writing, most recently under grant Y330.

Gerald Teschl

Vienna, Austria
April 2014

Gerald Teschl
Fakultät für Mathematik
Oskar-Morgenstern-Platz 1
Universität Wien
1090 Wien, Austria

E-mail: Gerald.Teschl@univie.ac.at
URL: http://www.mat.univie.ac.at/~gerald/

Part 0

Preliminaries

Chapter 0

A first look at Banach and Hilbert spaces

I assume that the reader has some basic familiarity with measure theory and functional analysis. For convenience, some facts needed from Banach and L^p spaces are reviewed in this chapter. A crash course in measure theory can be found in Appendix A. If you feel comfortable with terms like *Lebesgue L^p spaces*, *Banach space*, or *bounded linear operator*, you can skip this entire chapter. However, you might want to at least browse through it to refresh your memory.

0.1. Warm up: Metric and topological spaces

Before we begin, I want to recall some basic facts from metric and topological spaces. I presume that you are familiar with these topics from your calculus course. As a general reference I can warmly recommend Kelly's classical book [33].

A **metric space** is a space X together with a distance function $d : X \times X \to \mathbb{R}$ such that

(i) $d(x, y) \geq 0$,
(ii) $d(x, y) = 0$ if and only if $x = y$,
(iii) $d(x, y) = d(y, x)$,
(iv) $d(x, z) \leq d(x, y) + d(y, z)$ (**triangle inequality**).

If (ii) does not hold, d is called a **pseudometric**. Moreover, it is straightforward to see the **inverse triangle inequality** (Problem 0.1)

$$\tag{0.1} |d(x, y) - d(z, y)| \leq d(x, z).$$

Example. Euclidean space $\mathbb{R}^n$ together with $d(x,y) = (\sum_{k=1}^n (x_k - y_k)^2)^{1/2}$ is a metric space and so is $\mathbb{C}^n$ together with $d(x,y) = (\sum_{k=1}^n |x_k - y_k|^2)^{1/2}$. ⋄

The set

$$(0.2) \qquad B_r(x) = \{y \in X | d(x,y) < r\}$$

is called an **open ball** around x with radius $r > 0$. A point x of some set U is called an **interior point** of U if U contains some ball around x. If x is an interior point of U, then U is also called a **neighborhood** of x. A point x is called a **limit point** of U (also **accumulation** or **cluster point**) if $(B_r(x) \setminus \{x\}) \cap U \neq \emptyset$ for every ball around x. Note that a limit point x need not lie in U, but U must contain points arbitrarily close to x. A point x is called an **isolated point** of U if there exists a neighborhood of x not containing any other points of U. A set which consists only of isolated points is called a **discrete set**. If any neighborhood of x contains at least one point in U and at least one point not in U, then x is called a **boundary point** of U. The set of all boundary points of U is called the boundary of U and denoted by ∂U.

Example. Consider $\mathbb{R}$ with the usual metric and let $U = (-1,1)$. Then every point $x \in U$ is an interior point of U. The points $[-1,1]$ are limit points of U, and the points $\{-1,+1\}$ are boundary points of U. ⋄

A set consisting only of interior points is called **open**. The family of open sets $\mathcal{O}$ satisfies the properties

(i) $\emptyset, X \in \mathcal{O}$,

(ii) $O_1, O_2 \in \mathcal{O}$ implies $O_1 \cap O_2 \in \mathcal{O}$,

(iii) $\{O_\alpha\} \subseteq \mathcal{O}$ implies $\bigcup_\alpha O_\alpha \in \mathcal{O}$.

That is, $\mathcal{O}$ is closed under finite intersections and arbitrary unions.

In general, a space X together with a family of sets $\mathcal{O}$, the open sets, satisfying (i)–(iii), is called a **topological space**. The notions of interior point, limit point, and neighborhood carry over to topological spaces if we replace open ball by open set.

There are usually different choices for the topology. Two not too interesting examples are the **trivial topology** $\mathcal{O} = \{\emptyset, X\}$ and the **discrete topology** $\mathcal{O} = \mathfrak{P}(X)$ (the powerset of X). Given two topologies $\mathcal{O}_1$ and $\mathcal{O}_2$ on X, $\mathcal{O}_1$ is called **weaker** (or **coarser**) than $\mathcal{O}_2$ if and only if $\mathcal{O}_1 \subseteq \mathcal{O}_2$.

Example. Note that different metrics can give rise to the same topology. For example, we can equip $\mathbb{R}^n$ (or $\mathbb{C}^n$) with the Euclidean distance $d(x,y)$

0.1. Warm up: Metric and topological spaces

as before or we could also use

$$\tilde{d}(x,y) = \sum_{k=1}^{n} |x_k - y_k|. \tag{0.3}$$

Then

$$\frac{1}{\sqrt{n}} \sum_{k=1}^{n} |x_k| \le \sqrt{\sum_{k=1}^{n} |x_k|^2} \le \sum_{k=1}^{n} |x_k| \tag{0.4}$$

shows $B_{r/\sqrt{n}}(x) \subseteq \tilde{B}_r(x) \subseteq B_r(x)$, where B, $\tilde{B}$ are balls computed using d, $\tilde{d}$, respectively. ◇

Example. We can always replace a metric d by the bounded metric

$$\tilde{d}(x,y) = \frac{d(x,y)}{1 + d(x,y)} \tag{0.5}$$

without changing the topology (since the family of open balls does not change: $B_\delta(x) = \tilde{B}_{\delta/(1+\delta)}(x)$). ◇

Every subspace Y of a topological space X becomes a topological space of its own if we call $O \subseteq Y$ open if there is some open set $\tilde{O} \subseteq X$ such that $O = \tilde{O} \cap Y$. This natural topology $\mathcal{O} \cap Y$ is known as the **relative topology** (also **subspace**, **trace** or **induced topology**).

Example. The set $(0,1] \subseteq \mathbb{R}$ is not open in the topology of $X = \mathbb{R}$, but it is open in the relative topology when considered as a subset of $Y = [-1, 1]$. ◇

A family of open sets $\mathcal{B} \subseteq \mathcal{O}$ is called a **base** for the topology if for each x and each neighborhood $U(x)$, there is some set $O \in \mathcal{B}$ with $x \in O \subseteq U(x)$. Since an open set O is a neighborhood of every one of its points, it can be written as $O = \bigcup_{O \supseteq \tilde{O} \in \mathcal{B}} \tilde{O}$ and we have

Lemma 0.1. *If $\mathcal{B} \subseteq \mathcal{O}$ is a base for the topology, then every open set can be written as a union of elements from $\mathcal{B}$.*

If there exists a countable base, then X is called **second countable**.

Example. By construction, the open balls $B_{1/n}(x)$ are a base for the topology in a metric space. In the case of $\mathbb{R}^n$ (or $\mathbb{C}^n$) it even suffices to take balls with rational center, and hence $\mathbb{R}^n$ (as well as $\mathbb{C}^n$) is second countable. ◇

A topological space is called a **Hausdorff space** if for two different points there are always two disjoint neighborhoods.

Example. Any metric space is a Hausdorff space: Given two different points x and y, the balls $B_{d/2}(x)$ and $B_{d/2}(y)$, where $d = d(x,y) > 0$, are disjoint neighborhoods (a pseudometric space will not be Hausdorff). ◇

The complement of an open set is called a **closed set**. It follows from de Morgan's rules that the family of closed sets $\mathcal{C}$ satisfies

(i) $\emptyset, X \in \mathcal{C}$,
(ii) $C_1, C_2 \in \mathcal{C}$ implies $C_1 \cup C_2 \in \mathcal{C}$,
(iii) $\{C_\alpha\} \subseteq \mathcal{C}$ implies $\bigcap_\alpha C_\alpha \in \mathcal{C}$.

That is, closed sets are closed under finite unions and arbitrary intersections.

The smallest closed set containing a given set U is called the **closure**

$$(0.6) \qquad \overline{U} = \bigcap_{C \in \mathcal{C}, U \subseteq C} C,$$

and the largest open set contained in a given set U is called the **interior**

$$(0.7) \qquad U^\circ = \bigcup_{O \in \mathcal{O}, O \subseteq U} O.$$

It is not hard to see that the closure satisfies the following axioms (**Kuratowski closure axioms**):

(i) $\overline{\emptyset} = \emptyset$,
(ii) $U \subset \overline{U}$,
(iii) $\overline{\overline{U}} = \overline{U}$,
(iv) $\overline{U \cup V} = \overline{U} \cup \overline{V}$.

In fact, one can show that they can equivalently be used to define the topology by observing that the closed sets are precisely those which satisfy $\overline{A} = A$.

We can define interior and limit points as before by replacing the word ball by open set. Then it is straightforward to check

Lemma 0.2. *Let X be a topological space. Then the interior of U is the set of all interior points of U, and the closure of U is the union of U with all limit points of U.*

Example. The closed ball

$$(0.8) \qquad \bar{B}_r(x) = \{y \in X | d(x,y) \leq r\}$$

is a closed set (check that its complement is open). But in general we have only

$$(0.9) \qquad \overline{B_r(x)} \subseteq \bar{B}_r(x)$$

since an isolated point y with $d(x,y) = r$ will not be a limit point. In $\mathbb{R}^n$ (or $\mathbb{C}^n$) we have of course equality. ◇

A sequence $(x_n)_{n=1}^\infty \subseteq X$ is said to **converge** to some point $x \in X$ if $d(x, x_n) \to 0$. We write $\lim_{n \to \infty} x_n = x$ as usual in this case. Clearly the limit is unique if it exists (this is not true for a pseudometric).

0.1. Warm up: Metric and topological spaces

Every convergent sequence is a **Cauchy sequence**; that is, for every $\varepsilon > 0$ there is some $N \in \mathbb{N}$ such that

$$(0.10) \qquad d(x_n, x_m) \leq \varepsilon, \qquad n, m \geq N.$$

If the converse is also true, that is, if every Cauchy sequence has a limit, then X is called **complete**.

Example. Both $\mathbb{R}^n$ and $\mathbb{C}^n$ are complete metric spaces. ◇

Note that in a metric space x is a limit point of U if and only if there exists a sequence $(x_n)_{n=1}^\infty \subseteq U \setminus \{x\}$ with $\lim_{n \to \infty} x_n = x$. Hence U is closed if and only if for every convergent sequence the limit is in U. In particular,

Lemma 0.3. *A closed subset of a complete metric space is again a complete metric space.*

Note that convergence can also be equivalently formulated in topological terms: A sequence x_n converges to x if and only if for every neighborhood U of x there is some $N \in \mathbb{N}$ such that $x_n \in U$ for $n \geq N$. In a Hausdorff space the limit is unique.

A set U is called **dense** if its closure is all of X, that is, if $\overline{U} = X$. A metric space is called **separable** if it contains a countable dense set.

Lemma 0.4. *A metric space is separable if and only if it is second countable as a topological space.*

Proof. From every dense set we get a countable base by considering open balls with rational radii and centers in the dense set. Conversely, from every countable base we obtain a dense set by choosing an element from each element of the base. □

Lemma 0.5. *Let X be a separable metric space. Every subset Y of X is again separable.*

Proof. Let $A = \{x_n\}_{n \in \mathbb{N}}$ be a dense set in X. The only problem is that $A \cap Y$ might contain no elements at all. However, some elements of A must be at least arbitrarily close: Let $J \subseteq \mathbb{N}^2$ be the set of all pairs (n, m) for which $B_{1/m}(x_n) \cap Y \neq \emptyset$ and choose some $y_{n,m} \in B_{1/m}(x_n) \cap Y$ for all $(n, m) \in J$. Then $B = \{y_{n,m}\}_{(n,m) \in J} \subseteq Y$ is countable. To see that B is dense, choose $y \in Y$. Then there is some sequence x_{n_k} with $d(x_{n_k}, y) < 1/k$. Hence $(n_k, k) \in J$ and $d(y_{n_k, k}, y) \leq d(y_{n_k, k}, x_{n_k}) + d(x_{n_k}, y) \leq 2/k \to 0$. □

Next, we come to functions $f : X \to Y$, $x \mapsto f(x)$. We use the usual conventions $f(U) = \{f(x) | x \in U\}$ for $U \subseteq X$ and $f^{-1}(V) = \{x | f(x) \in V\}$ for $V \subseteq Y$. The set $\text{Ran}(f) = f(X)$ is called the **range** of f, and X is called the **domain** of f. A function f is called **injective** if for each $y \in Y$ there is at most one $x \in X$ with $f(x) = y$ (i.e., $f^{-1}(\{y\})$ contains at most one point) and **surjective** or **onto** if $\text{Ran}(f) = Y$. A function f which is both injective and surjective is called **bijective**.

A function f between metric spaces X and Y is called continuous at a point $x \in X$ if for every $\varepsilon > 0$ we can find a $\delta > 0$ such that

$$(0.11) \qquad d_Y(f(x), f(y)) \leq \varepsilon \quad \text{if} \quad d_X(x,y) < \delta.$$

If f is continuous at every point, it is called **continuous**.

Lemma 0.6. *Let X be a metric space. The following are equivalent:*
  (i) *f is continuous at x (i.e., (0.11) holds).*
  (ii) *$f(x_n) \to f(x)$ whenever $x_n \to x$.*
  (iii) *For every neighborhood V of $f(x)$, $f^{-1}(V)$ is a neighborhood of x.*

Proof. (i) $\Rightarrow$ (ii) is obvious. (ii) $\Rightarrow$ (iii): If (iii) does not hold, there is a neighborhood V of $f(x)$ such that $B_\delta(x) \not\subseteq f^{-1}(V)$ for every δ. Hence we can choose a sequence $x_n \in B_{1/n}(x)$ such that $f(x_n) \notin f^{-1}(V)$. Thus $x_n \to x$ but $f(x_n) \not\to f(x)$. (iii) $\Rightarrow$ (i): Choose $V = B_\varepsilon(f(x))$ and observe that by (iii), $B_\delta(x) \subseteq f^{-1}(V)$ for some δ. $\square$

The last item implies that f is continuous if and only if the inverse image of every open set is again open (equivalently, the inverse image of every closed set is closed). If the image of every open set is open, then f is called **open**. A bijection f is called a **homeomorphism** if both f and its inverse f^{-1} are continuous. Note that if f is a bijection, then f^{-1} is continuous if and only if f is open.

In a topological space, (iii) is used as the definition for continuity. However, in general (ii) and (iii) will no longer be equivalent unless one uses generalized sequences, so-called nets, where the index set $\mathbb{N}$ is replaced by arbitrary directed sets.

The **support** of a function $f : X \to \mathbb{C}^n$ is the closure of all points x for which $f(x)$ does not vanish; that is,

$$(0.12) \qquad \text{supp}(f) = \overline{\{x \in X | f(x) \neq 0\}}.$$

If X and Y are metric spaces, then $X \times Y$ together with

$$(0.13) \qquad d((x_1, y_1), (x_2, y_2)) = d_X(x_1, x_2) + d_Y(y_1, y_2)$$

is a metric space. A sequence (x_n, y_n) converges to (x, y) if and only if $x_n \to x$ and $y_n \to y$. In particular, the projections onto the first $(x, y) \mapsto x$, respectively, onto the second $(x, y) \mapsto y$, coordinate are continuous. Moreover, if X and Y are complete, so is $X \times Y$.

0.1. Warm up: Metric and topological spaces

In particular, by the inverse triangle inequality (0.1),

$$(0.14) \qquad |d(x_n, y_n) - d(x, y)| \leq d(x_n, x) + d(y_n, y),$$

we see that $d : X \times X \to \mathbb{R}$ is continuous.

Example. If we consider $\mathbb{R} \times \mathbb{R}$, we do not get the Euclidean distance of $\mathbb{R}^2$ unless we modify (0.13) as follows:

$$(0.15) \qquad \tilde{d}((x_1, y_1), (x_2, y_2)) = \sqrt{d_X(x_1, x_2)^2 + d_Y(y_1, y_2)^2}.$$

As noted in our previous example, the topology (and thus also convergence/continuity) is independent of this choice. ◇

If X and Y are just topological spaces, the **product topology** is defined by calling $O \subseteq X \times Y$ open if for every point $(x, y) \in O$ there are open neighborhoods U of x and V of y such that $U \times V \subseteq O$. In other words, the products of open sets form a basis of the product topology. In the case of metric spaces this clearly agrees with the topology defined via the product metric (0.13).

A **cover** of a set $Y \subseteq X$ is a family of sets $\{U_\alpha\}$ such that $Y \subseteq \bigcup_\alpha U_\alpha$. A cover is called open if all U_α are open. Any subset of $\{U_\alpha\}$ which still covers Y is called a **subcover**.

Lemma 0.7 (Lindelöf). *If X is second countable, then every open cover has a countable subcover.*

Proof. Let $\{U_\alpha\}$ be an open cover for Y, and let $\mathcal{B}$ be a countable base. Since every U_α can be written as a union of elements from $\mathcal{B}$, the set of all $B \in \mathcal{B}$ which satisfy $B \subseteq U_\alpha$ for some α form a countable open cover for Y. Moreover, for every B_n in this set we can find an α_n such that $B_n \subseteq U_{\alpha_n}$. By construction, $\{U_{\alpha_n}\}$ is a countable subcover. □

A subset $K \subset X$ is called **compact** if every open cover has a finite subcover. A set is called **relatively compact** if its closure is compact.

Lemma 0.8. *A topological space is compact if and only if it has the **finite intersection property**: The intersection of a family of closed sets is empty if and only if the intersection of some finite subfamily is empty.*

Proof. By taking complements, to every family of open sets there is a corresponding family of closed sets and vice versa. Moreover, the open sets are a cover if and only if the corresponding closed sets have empty intersection. □

Lemma 0.9. *Let X be a topological space.*

  (i) *The continuous image of a compact set is compact.*
  (ii) *Every closed subset of a compact set is compact.*

(iii) If X is Hausdorff, every compact set is closed.

(iv) The product of finitely many compact sets is compact.

(v) The finite union of compact sets is again compact.

(vi) If X is Hausdorff, any intersection of compact sets is again compact.

Proof. (i) Observe that if $\{O_\alpha\}$ is an open cover for $f(Y)$, then $\{f^{-1}(O_\alpha)\}$ is one for Y.

(ii) Let $\{O_\alpha\}$ be an open cover for the closed subset Y (in the induced topology). Then there are open sets $\tilde{O}_\alpha$ with $O_\alpha = \tilde{O}_\alpha \cap Y$ and $\{\tilde{O}_\alpha\} \cup \{X \backslash Y\}$ is an open cover for X which has a finite subcover. This subcover induces a finite subcover for Y.

(iii) Let $Y \subseteq X$ be compact. We show that $X \backslash Y$ is open. Fix $x \in X \backslash Y$ (if $Y = X$, there is nothing to do). By the definition of Hausdorff, for every $y \in Y$ there are disjoint neighborhoods $V(y)$ of y and $U_y(x)$ of x. By compactness of Y, there are $y_1, \ldots, y_n$ such that the $V(y_j)$ cover Y. But then $U(x) = \bigcap_{j=1}^n U_{y_j}(x)$ is a neighborhood of x which does not intersect Y.

(iv) Let $\{O_\alpha\}$ be an open cover for $X \times Y$. For every $(x, y) \in X \times Y$ there is some $\alpha(x, y)$ such that $(x, y) \in O_{\alpha(x,y)}$. By definition of the product topology there is some open rectangle $U(x, y) \times V(x, y) \subseteq O_{\alpha(x,y)}$. Hence for fixed x, $\{V(x, y)\}_{y \in Y}$ is an open cover of Y. Hence there are finitely many points $y_k(x)$ such that the $V(x, y_k(x))$ cover Y. Set $U(x) = \bigcap_k U(x, y_k(x))$. Since finite intersections of open sets are open, $\{U(x)\}_{x \in X}$ is an open cover and there are finitely many points x_j such that the $U(x_j)$ cover X. By construction, the $U(x_j) \times V(x_j, y_k(x_j)) \subseteq O_{\alpha(x_j, y_k(x_j))}$ cover $X \times Y$.

(v) Note that a cover of the union is a cover for each individual set and the union of the individual subcovers is the subcover we are looking for.

(vi) Follows from (ii) and (iii) since an intersection of closed sets is closed. □

As a consequence we obtain a simple criterion when a continuous function is a homeomorphism.

Corollary 0.10. *Let X and Y be topological spaces with X compact and Y Hausdorff. Then every continuous bijection $f : X \to Y$ is a homeomorphism.*

Proof. It suffices to show that f maps closed sets to closed sets. By (ii) every closed set is compact, by (i) its image is also compact, and by (iii) it is also closed. □

0.1. Warm up: Metric and topological spaces

A subset $K \subset X$ is called **sequentially compact** if every sequence from K has a convergent subsequence. In a metric space, compact and sequentially compact are equivalent.

Lemma 0.11. *Let X be a metric space. Then a subset is compact if and only if it is sequentially compact.*

Proof. Suppose X is compact and let x_n be a sequence which has no convergent subsequence. Then $K = \{x_n\}$ has no limit points and is hence compact by Lemma 0.9 (ii). For every n there is a ball $B_{\varepsilon_n}(x_n)$ which contains only finitely many elements of K. However, finitely many suffice to cover K, a contradiction.

Conversely, suppose X is sequentially compact and let $\{O_\alpha\}$ be some open cover which has no finite subcover. For every $x \in X$ we can choose some $\alpha(x)$ such that if $B_r(x)$ is the largest ball contained in $O_{\alpha(x)}$, then either $r \geq 1$ or there is no β with $B_{2r}(x) \subset O_\beta$ (show that this is possible). Now choose a sequence x_n such that $x_n \notin \bigcup_{m<n} O_{\alpha(x_m)}$. Note that by construction the distance $d = d(x_m, x_n)$ to every successor of x_m is either larger than 1 or the ball $B_{2d}(x_m)$ will not fit into any of the O_α.

Now let y be the limit of some convergent subsequence and fix some $r \in (0,1)$ such that $B_r(y) \subseteq O_{\alpha(y)}$. Then this subsequence must eventually be in $B_{r/5}(y)$, but this is impossible since if $d = d(x_{n_1}, x_{n_2})$ is the distance between two consecutive elements of this subsequence, then $B_{2d}(x_{n_1})$ cannot fit into $O_{\alpha(y)}$ by construction whereas on the other hand $B_{2d}(x_{n_1}) \subseteq B_{4r/5}(a) \subseteq O_{\alpha(y)}$. $\square$

In a metric space, a set is called **bounded** if it is contained inside some ball. Note that compact sets are always bounded since Cauchy sequences are bounded (show this!). In $\mathbb{R}^n$ (or $\mathbb{C}^n$) the converse also holds.

Theorem 0.12 (Heine–Borel). *In $\mathbb{R}^n$ (or $\mathbb{C}^n$) a set is compact if and only if it is bounded and closed.*

Proof. By Lemma 0.9 (ii) and (iii) it suffices to show that a closed interval in $I \subseteq \mathbb{R}$ is compact. Moreover, by Lemma 0.11, it suffices to show that every sequence in $I = [a,b]$ has a convergent subsequence. Let x_n be our sequence and divide $I = [a, \frac{a+b}{2}] \cup [\frac{a+b}{2}, b]$. Then at least one of these two intervals, call it I_1, contains infinitely many elements of our sequence. Let $y_1 = x_{n_1}$ be the first one. Subdivide I_1 and pick $y_2 = x_{n_2}$, with $n_2 > n_1$ as before. Proceeding like this, we obtain a Cauchy sequence y_n (note that by construction $I_{n+1} \subseteq I_n$ and hence $|y_n - y_m| \leq \frac{b-a}{n}$ for $m \geq n$). $\square$

By Lemma 0.11 this is equivalent to

Theorem 0.13 (Bolzano–Weierstraß)**.** *Every bounded infinite subset of $\mathbb{R}^n$ (or $\mathbb{C}^n$) has at least one limit point.*

Combining Theorem 0.12 with Lemma 0.9 (i) we also obtain the **extreme value theorem**.

Theorem 0.14 (Weierstraß)**.** *Let X be compact. Every continuous function $f : X \to \mathbb{R}$ attains its maximum and minimum.*

A metric space for which the Heine–Borel theorem holds is called **proper**. Lemma 0.9 (ii) shows that X is proper if and only if every closed ball is compact. Note that a proper metric space must be complete (since every Cauchy sequence is bounded). A topological space is called **locally compact** if every point has a compact neighborhood. Clearly a proper metric space is locally compact.

The **distance** between a point $x \in X$ and a subset $Y \subseteq X$ is

$$(0.16) \qquad \mathrm{dist}(x, Y) = \inf_{y \in Y} d(x, y).$$

Note that x is a limit point of Y if and only if $\mathrm{dist}(x, Y) = 0$.

Lemma 0.15. *Let X be a metric space. Then*

$$(0.17) \qquad |\mathrm{dist}(x, Y) - \mathrm{dist}(z, Y)| \le d(x, z).$$

In particular, $x \mapsto \mathrm{dist}(x, Y)$ is continuous.

Proof. Taking the infimum in the triangle inequality $d(x, y) \le d(x, z) + d(z, y)$ shows $\mathrm{dist}(x, Y) \le d(x, z) + \mathrm{dist}(z, Y)$. Hence $\mathrm{dist}(x, Y) - \mathrm{dist}(z, Y) \le d(x, z)$. Interchanging x and z shows $\mathrm{dist}(z, Y) - \mathrm{dist}(x, Y) \le d(x, z)$. □

Lemma 0.16 (Urysohn)**.** *Suppose C_1 and C_2 are disjoint closed subsets of a metric space X. Then there is a continuous function $f : X \to [0, 1]$ such that f is zero on C_2 and one on C_1.*

If X is locally compact and C_1 is compact, one can choose f with compact support.

Proof. To prove the first claim, set $f(x) = \frac{\mathrm{dist}(x, C_2)}{\mathrm{dist}(x, C_1) + \mathrm{dist}(x, C_2)}$. For the second claim, observe that there is an open set O such that $\overline{O}$ is compact and $C_1 \subset O \subset \overline{O} \subset X \setminus C_2$. In fact, for every $x \in C_1$, there is a ball $B_\varepsilon(x)$ such that $\overline{B_\varepsilon(x)}$ is compact and $\overline{B_\varepsilon(x)} \subset X \setminus C_2$. Since C_1 is compact, finitely many of them cover C_1 and we can choose the union of those balls to be O. Now replace C_2 by $X \setminus O$. □

Note that Urysohn's lemma implies that a metric space is **normal**; that is, for any two disjoint closed sets C_1 and C_2, there are disjoint open sets

0.1. Warm up: Metric and topological spaces

O_1 and O_2 such that $C_j \subseteq O_j$, $j = 1, 2$. In fact, choose f as in Urysohn's lemma and set $O_1 = f^{-1}([0, 1/2))$, respectively, $O_2 = f^{-1}((1/2, 1])$.

Lemma 0.17. *Let X be a locally compact metric space. Suppose K is a compact set and $\{O_j\}_{j=1}^n$ is an open cover. Then there is a **partition of unity** for K subordinate to this cover; that is, there are continuous functions $h_j : X \to [0, 1]$ such that h_j has compact support contained in O_j and*

$$(0.18) \qquad \sum_{j=1}^n h_j(x) \leq 1$$

with equality for $x \in K$.

Proof. For every $x \in K$ there is some ε and some j such that $\overline{B_\varepsilon(x)} \subseteq O_j$. By compactness of K, finitely many of these balls cover K. Let K_j be the union of those balls which lie inside O_j. By Urysohn's lemma there are continuous functions $g_j : X \to [0, 1]$ such that $g_j = 1$ on K_j and $g_j = 0$ on $X \setminus O_j$. Now set

$$h_j = g_j \prod_{k=1}^{j-1} (1 - g_k).$$

Then $h_j : X \to [0, 1]$ has compact support contained in O_j and

$$\sum_{j=1}^n h_j(x) = 1 - \prod_{j=1}^n (1 - g_j(x))$$

shows that the sum is one for $x \in K$, since $x \in K_j$ for some j implies $g_j(x) = 1$ and causes the product to vanish. □

Problem 0.1. *Show that $|d(x, y) - d(z, y)| \leq d(x, z)$.*

Problem 0.2. *Show the **quadrangle inequality** $|d(x, y) - d(x', y')| \leq d(x, x') + d(y, y')$.*

Problem 0.3. *Show that the closure satisfies the Kuratowski closure axioms.*

Problem 0.4. *Show that the closure and interior operators are dual in the sense that*
$$X \setminus \overline{A} = (X \setminus A)^\circ \qquad \text{and} \qquad X \setminus A^\circ = \overline{(X \setminus A)}.$$
(Hint: De Morgan's laws.)

Problem 0.5. *Let $U \subseteq V$ be subsets of a metric space X. Show that if U is dense in V and V is dense in X, then U is dense in X.*

Problem 0.6. *Show that every open set $O \subseteq \mathbb{R}$ can be written as a countable union of disjoint intervals. (Hint: Let $\{I_\alpha\}$ be the set of all maximal open subintervals of O; that is, $I_\alpha \subseteq O$ and there is no other subinterval of O which contains I_α. Then this is a cover of disjoint open intervals which has a countable subcover.)*

Problem 0.7. *Show that the boundary of A is given by $\partial A = \overline{A}\backslash A^\circ$.*

0.2. The Banach space of continuous functions

Now let us have a first look at Banach spaces by investigating the set of continuous functions $C(I)$ on a compact interval $I = [a,b] \subset \mathbb{R}$. Since we want to handle complex models, we will always consider complex-valued functions!

One way of declaring a distance, well-known from calculus, is the **maximum norm**:

$$\|f\|_\infty = \max_{x \in I} |f(x)|. \tag{0.19}$$

It is not hard to see that with this definition $C(I)$ becomes a normed vector space:

A **normed vector space** X is a vector space X over $\mathbb{C}$ (or $\mathbb{R}$) with a nonnegative function (the **norm**) $\|.\|$ such that

- $\|f\| > 0$ for $f \neq 0$ (**positive definiteness**),
- $\|\alpha f\| = |\alpha| \|f\|$ for all $\alpha \in \mathbb{C}$, $f \in X$ (**positive homogeneity**), and
- $\|f + g\| \leq \|f\| + \|g\|$ for all $f, g \in X$ (**triangle inequality**).

If positive definiteness is dropped from the requirements, one calls $\|.\|$ a **seminorm**.

From the triangle inequality we also get the **inverse triangle inequality** (Problem 0.8)

$$|\|f\| - \|g\|| \leq \|f - g\|. \tag{0.20}$$

Once we have a norm, we have a **distance** $d(f,g) = \|f - g\|$ and hence we know when a sequence of vectors f_n **converges** to a vector f. We will write $f_n \to f$ or $\lim_{n\to\infty} f_n = f$, as usual, in this case. Moreover, a mapping $F : X \to Y$ between two normed spaces is called **continuous** if $f_n \to f$ implies $F(f_n) \to F(f)$. In fact, the norm, vector addition, and multiplication by scalars are continuous (Problem 0.9).

In addition to the concept of convergence we have also the concept of a **Cauchy sequence** and hence the concept of completeness: A normed space is called **complete** if every Cauchy sequence has a limit. A complete normed space is called a **Banach space**.

Example. The space $\ell^1(\mathbb{N})$ of all complex-valued sequences $a = (a_j)_{j=1}^\infty$ for which the norm

$$\|a\|_1 = \sum_{j=1}^\infty |a_j| \tag{0.21}$$

0.2. The Banach space of continuous functions

is finite is a Banach space.

To show this, we need to verify three things: (i) $\ell^1(\mathbb{N})$ is a vector space that is closed under addition and scalar multiplication, (ii) $\|.\|_1$ satisfies the three requirements for a norm, and (iii) $\ell^1(\mathbb{N})$ is complete.

First of all, observe

$$(0.22) \quad \sum_{j=1}^{k} |a_j + b_j| \leq \sum_{j=1}^{k} |a_j| + \sum_{j=1}^{k} |b_j| \leq \|a\|_1 + \|b\|_1$$

for every finite k. Letting $k \to \infty$, we conclude that $\ell^1(\mathbb{N})$ is closed under addition and that the triangle inequality holds. That $\ell^1(\mathbb{N})$ is closed under scalar multiplication together with homogeneity as well as definiteness are straightforward. It remains to show that $\ell^1(\mathbb{N})$ is complete. Let $a^n = (a_j^n)_{j=1}^{\infty}$ be a Cauchy sequence; that is, for given $\varepsilon > 0$ we can find an N_ε such that $\|a^m - a^n\|_1 \leq \varepsilon$ for $m, n \geq N_\varepsilon$. This implies in particular $|a_j^m - a_j^n| \leq \varepsilon$ for every fixed j. Thus a_j^n is a Cauchy sequence for fixed j and, by completeness of $\mathbb{C}$, it has a limit: $\lim_{n \to \infty} a_j^n = a_j$. Now consider

$$(0.23) \quad \sum_{j=1}^{k} |a_j^m - a_j^n| \leq \varepsilon$$

and take $m \to \infty$:

$$(0.24) \quad \sum_{j=1}^{k} |a_j - a_j^n| \leq \varepsilon.$$

Since this holds for all finite k, we even have $\|a - a^n\|_1 \leq \varepsilon$. Hence $(a - a^n) \in \ell^1(\mathbb{N})$ and since $a^n \in \ell^1(\mathbb{N})$, we finally conclude $a = a^n + (a - a^n) \in \ell^1(\mathbb{N})$. By our estimate $\|a - a^n\|_1 \leq \varepsilon$, our candidate a is indeed the limit of a^n. ⋄

Example. The previous example can be generalized by considering the space $\ell^p(\mathbb{N})$ of all complex-valued sequences $a = (a_j)_{j=1}^{\infty}$ for which the norm

$$(0.25) \quad \|a\|_p = \left(\sum_{j=1}^{\infty} |a_j|^p \right)^{1/p}, \quad p \in [1, \infty),$$

is finite. By $|a_j + b_j|^p \leq 2^p \max(|a_j|, |b_j|)^p = 2^p \max(|a_j|^p, |b_j|^p) \leq 2^p(|a_j|^p + |b_j|^p)$ it is a vector space, but the triangle inequality is only easy to see in the case $p = 1$. (It is also not hard to see that it fails for $p < 1$, which explains our requirement $p \geq 1$. See also Problem 0.17.)

To prove it we need the elementary inequality (Problem 0.12)

$$(0.26) \quad \alpha^{1/p} \beta^{1/q} \leq \frac{1}{p}\alpha + \frac{1}{q}\beta, \quad \frac{1}{p} + \frac{1}{q} = 1, \quad \alpha, \beta \geq 0,$$

which implies **Hölder's inequality**

(0.27) $$\|ab\|_1 \leq \|a\|_p \|b\|_q$$

for $x \in \ell^p(\mathbb{N})$, $y \in \ell^q(\mathbb{N})$. In fact, by homogeneity of the norm it suffices to prove the case $\|a\|_p = \|b\|_q = 1$. But this case follows by choosing $\alpha = |a_j|^p$ and $\beta = |b_j|^q$ in (0.26) and summing over all j.

Now using $|a_j + b_j|^p \leq |a_j| \, |a_j + b_j|^{p-1} + |b_j| \, |a_j + b_j|^{p-1}$, we obtain from Hölder's inequality (note $(p-1)q = p$)

$$\|a+b\|_p^p \leq \|a\|_p \|(a+b)^{p-1}\|_q + \|b\|_p \|(a+b)^{p-1}\|_q$$
$$= (\|a\|_p + \|b\|_p) \|(a+b)\|_p^{p-1}.$$

Hence ℓ^p is a normed space. That it is complete can be shown as in the case $p = 1$ (Problem 0.13). ◇

Example. The space $\ell^\infty(\mathbb{N})$ of all complex-valued bounded sequences $a = (a_j)_{j=1}^\infty$ together with the norm

(0.28) $$\|a\|_\infty = \sup_{j \in \mathbb{N}} |a_j|$$

is a Banach space (Problem 0.14). Note that with this definition, Hölder's inequality (0.27) remains true for the cases $p = 1$, $q = \infty$ and $p = \infty$, $q = 1$. The reason for the notation is explained in Problem 0.16. ◇

Example. Every closed subspace of a Banach space is again a Banach space. For example, the space $c_0(\mathbb{N}) \subset \ell^\infty(\mathbb{N})$ of all sequences converging to zero is a closed subspace. In fact, if $a \in \ell^\infty(\mathbb{N}) \setminus c_0(\mathbb{N})$, then $\liminf_{j \to \infty} |a_j| \geq \varepsilon > 0$ and thus $\|a - b\|_\infty \geq \varepsilon$ for every $b \in c_0(\mathbb{N})$. ◇

Now what about convergence in the space $C(I)$? A sequence of functions $f_n(x)$ converges to f if and only if

(0.29) $$\lim_{n \to \infty} \|f - f_n\|_\infty = \lim_{n \to \infty} \sup_{x \in I} |f_n(x) - f(x)| = 0.$$

That is, in the language of real analysis, f_n converges uniformly to f. Now let us look at the case where f_n is only a Cauchy sequence. Then $f_n(x)$ is clearly a Cauchy sequence of real numbers for every fixed $x \in I$. In particular, by completeness of $\mathbb{C}$, there is a limit $f(x)$ for each x. Thus we get a limiting function $f(x)$. Moreover, letting $m \to \infty$ in

(0.30) $$|f_m(x) - f_n(x)| \leq \varepsilon \qquad \forall m, n > N_\varepsilon,\ x \in I,$$

we see

(0.31) $$|f(x) - f_n(x)| \leq \varepsilon \qquad \forall n > N_\varepsilon,\ x \in I;$$

0.2. The Banach space of continuous functions

that is, $f_n(x)$ converges uniformly to $f(x)$. However, up to this point we do not know whether it is in our vector space $C(I)$, that is, whether it is continuous. Fortunately, there is a well-known result from real analysis which tells us that the uniform limit of continuous functions is again continuous: Fix $x \in I$ and $\varepsilon > 0$. To show that f is continuous we need to find a δ such that $|x - y| < \delta$ implies $|f(x) - f(y)| < \varepsilon$. Pick n so that $\|f_n - f\|_\infty < \varepsilon/3$ and δ so that $|x - y| < \delta$ implies $|f_n(x) - f_n(y)| < \varepsilon/3$. Then $|x - y| < \delta$ implies

$$|f(x) - f(y)| \leq |f(x) - f_n(x)| + |f_n(x) - f_n(y)| + |f_n(y) - f(y)| < \frac{\varepsilon}{3} + \frac{\varepsilon}{3} + \frac{\varepsilon}{3} = \varepsilon$$

as required. Hence $f(x) \in C(I)$ and thus every Cauchy sequence in $C(I)$ converges. Or, in other words,

Theorem 0.18. *$C(I)$ with the maximum norm is a Banach space.*

Next we want to look at *countable bases*. To this end we introduce a few definitions first.

The set of all finite linear combinations of a set of vectors $\{u_n\} \subset X$ is called the **span** of $\{u_n\}$ and denoted by $\mathrm{span}\{u_n\}$. A set of vectors $\{u_n\} \subset X$ is called **linearly independent** if every finite subset is. If $\{u_n\}_{n=1}^N \subset X$, $N \in \mathbb{N} \cup \{\infty\}$, is countable, we can throw away all elements which can be expressed as linear combinations of the previous ones to obtain a subset of linearly independent vectors which have the same span.

We will call a countable set of vectors $\{u_n\}_{n=1}^N \subset X$ a **Schauder basis** if every element $f \in X$ can be uniquely written as a countable linear combination of the basis elements:

$$(0.32) \qquad f = \sum_{n=1}^N c_n u_n, \qquad c_n = c_n(f) \in \mathbb{C},$$

where the sum has to be understood as a limit if $N = \infty$ (the sum is not required to converge unconditionally). Since we have assumed the coefficients $c_n(f)$ to be uniquely determined, the vectors are necessarily linearly independent.

Example. The set of vectors δ^n, with $\delta_n^n = 1$ and $\delta_m^n = 0$, $n \neq m$, is a Schauder basis for the Banach space $\ell^1(\mathbb{N})$.

Let $a = (a_j)_{j=1}^\infty \in \ell^1(\mathbb{N})$ be given and set $a^n = \sum_{j=1}^n a_j \delta^j$. Then

$$\|a - a^n\|_1 = \sum_{j=n+1}^\infty |a_j| \to 0$$

since $a_j^n = a_j$ for $1 \le j \le n$ and $a_j^n = 0$ for $j > n$. Hence

$$a = \sum_{j=1}^{\infty} a_j \delta^j$$

and $\{\delta^n\}_{n=1}^{\infty}$ is a Schauder basis (linear independence is left as an exercise). ◇

A set whose span is dense is called **total**, and if we have a countable total set, we also have a countable dense set (consider only linear combinations with rational coefficients — show this). A normed vector space containing a countable dense set is called **separable**.

Example. Every Schauder basis is total and thus every Banach space with a Schauder basis is separable (the converse is not true). In particular, the Banach space $\ell^1(\mathbb{N})$ is separable. ◇

While we will not give a Schauder basis for $C(I)$, we will at least show that it is separable. In order to prove this, we need a lemma first.

Lemma 0.19 (Smoothing)**.** *Let $u_n(x)$ be a sequence of nonnegative continuous functions on $[-1, 1]$ such that*

$$(0.33) \qquad \int_{|x| \le 1} u_n(x) dx = 1 \quad \text{and} \quad \int_{\delta \le |x| \le 1} u_n(x) dx \to 0, \quad \delta > 0.$$

(In other words, u_n has mass one and concentrates near $x = 0$ as $n \to \infty$.)

Then for every $f \in C[-\frac{1}{2}, \frac{1}{2}]$ which vanishes at the endpoints, $f(-\frac{1}{2}) = f(\frac{1}{2}) = 0$, we have that

$$(0.34) \qquad f_n(x) = \int_{-1/2}^{1/2} u_n(x-y) f(y) dy$$

converges uniformly to $f(x)$.

Proof. Since f is uniformly continuous, for given ε we can find a $\delta < 1/2$ (independent of x) such that $|f(x) - f(y)| \le \varepsilon$ whenever $|x - y| \le \delta$. Moreover, we can choose n such that $\int_{\delta \le |y| \le 1} u_n(y) dy \le \varepsilon$. Now abbreviate $M = \max_{x \in [-1/2, 1/2]} \{1, |f(x)|\}$ and note

$$|f(x) - \int_{-1/2}^{1/2} u_n(x-y) f(x) dy| = |f(x)| \, |1 - \int_{-1/2}^{1/2} u_n(x-y) dy| \le M\varepsilon.$$

In fact, either the distance of x to one of the boundary points $\pm\frac{1}{2}$ is smaller than δ and hence $|f(x)| \le \varepsilon$ or otherwise $[-\delta, \delta] \subset [x - 1/2, x + 1/2]$ and the difference between one and the integral is smaller than ε.

0.2. The Banach space of continuous functions

Using this, we have

$$|f_n(x) - f(x)| \le \int_{-1/2}^{1/2} u_n(x-y)|f(y) - f(x)|dy + M\varepsilon$$

$$= \int_{|y|\le 1/2, |x-y|\le \delta} u_n(x-y)|f(y) - f(x)|dy$$

$$+ \int_{|y|\le 1/2, |x-y|\ge \delta} u_n(x-y)|f(y) - f(x)|dy + M\varepsilon$$

(0.35) $$\le \varepsilon + 2M\varepsilon + M\varepsilon = (1+3M)\varepsilon,$$

which proves the claim. □

Note that f_n will be as smooth as u_n, hence the title smoothing lemma. Moreover, f_n will be a polynomial if u_n is. The same idea is used to approximate noncontinuous functions by smooth ones (of course the convergence will no longer be uniform in this case).

Now we are ready to show:

Theorem 0.20 (Weierstraß)**.** *Let I be a compact interval. Then the set of polynomials is dense in $C(I)$.*

Proof. Let $f(x) \in C(I)$ be given. By considering $f(x) - f(a) - \frac{f(b)-f(a)}{b-a}(x-a)$ it is no loss to assume that f vanishes at the boundary points. Moreover, without restriction, we only consider $I = [-\frac{1}{2}, \frac{1}{2}]$ (why?).

Now the claim follows from Lemma 0.19 using

$$u_n(x) = \frac{1}{I_n}(1-x^2)^n,$$

where

$$I_n = \int_{-1}^{1}(1-x^2)^n dx = \frac{n}{n+1}\int_{-1}^{1}(1-x)^{n-1}(1+x)^{n+1}dx$$

$$= \cdots = \frac{n!}{(n+1)\cdots(2n+1)}2^{2n+1} = \frac{(n!)^2 2^{2n+1}}{(2n+1)!} = \frac{n!}{\frac{1}{2}(\frac{1}{2}+1)\cdots(\frac{1}{2}+n)}.$$

Indeed, the first part of (0.33) holds by construction, and the second part follows from the elementary estimate

$$\frac{2}{2n+1} \le I_n < 2.$$

□

Corollary 0.21. *$C(I)$ is separable.*

However, $\ell^\infty(\mathbb{N})$ is not separable (Problem 0.15)!

Problem 0.8. *Show that $|\|f\| - \|g\|| \le \|f - g\|$.*

Problem 0.9. Let X be a Banach space. Show that the norm, vector addition, and multiplication by scalars are continuous. That is, if $f_n \to f$, $g_n \to g$, and $\alpha_n \to \alpha$, then $\|f_n\| \to \|f\|$, $f_n + g_n \to f + g$, and $\alpha_n g_n \to \alpha g$.

Problem 0.10. Let X be a Banach space. Show that $\sum_{j=1}^{\infty} \|f_j\| < \infty$ implies that
$$\sum_{j=1}^{\infty} f_j = \lim_{n \to \infty} \sum_{j=1}^{n} f_j$$
exists. The series is called **absolutely convergent** in this case.

Problem 0.11. While $\ell^1(\mathbb{N})$ is separable, it still has room for an uncountable set of linearly independent vectors. Show this by considering vectors of the form
$$a^{\alpha} = (1, \alpha, \alpha^2, \dots), \qquad \alpha \in (0, 1).$$
(Hint: Take n such vectors and cut them off after $n+1$ terms. If the cut-off vectors are linearly independent, so are the original ones. Recall the Vandermonde determinant.)

Problem 0.12. Prove (0.26). (Hint: Take logarithms on both sides.)

Problem 0.13. Show that $\ell^p(\mathbb{N})$ is a separable Banach space.

Problem 0.14. Show that $\ell^{\infty}(\mathbb{N})$ is a Banach space.

Problem 0.15. Show that $\ell^{\infty}(\mathbb{N})$ is not separable. (Hint: Consider sequences which take only the value one and zero. How many are there? What is the distance between two such sequences?)

Problem 0.16. Show that if $a \in \ell^{p_0}(\mathbb{N})$ for some $p_0 \in [1, \infty)$, then $a \in \ell^p(\mathbb{N})$ for $p \geq p_0$ and
$$\lim_{p \to \infty} \|a\|_p = \|a\|_{\infty}.$$

Problem 0.17. Formally extend the definition of $\ell^p(\mathbb{N})$ to $p \in (0, 1)$. Show that $\|.\|_p$ does not satisfy the triangle inequality. However, show that it is a **quasinormed space**; that is, it satisfies all requirements for a normed space except for the triangle inequality which is replaced by
$$\|a + b\| \leq K(\|a\| + \|b\|)$$
with some constant $K \geq 1$. Show, in fact,
$$\|a + b\|_p \leq 2^{1/p - 1}(\|a\|_p + \|b\|_p), \qquad p \in (0, 1).$$
Moreover, show that $\|.\|_p^p$ satisfies the triangle inequality in this case, but of course it is no longer homogeneous (but at least you can get an honest metric $d(a, b) = \|a - b\|_p^p$ which gives rise to the same topology). (Hint: Show $\alpha + \beta \leq (\alpha^p + \beta^p)^{1/p} \leq 2^{1/p - 1}(\alpha + \beta)$ for $0 < p < 1$ and $\alpha, \beta \geq 0$.)

0.3. The geometry of Hilbert spaces

So it looks like $C(I)$ has all the properties we want. However, there is still one thing missing: How should we define orthogonality in $C(I)$? In Euclidean space, two vectors are called **orthogonal** if their scalar product vanishes, so we would need a scalar product:

Suppose $\mathfrak{H}$ is a vector space. A map $\langle .,..\rangle : \mathfrak{H} \times \mathfrak{H} \to \mathbb{C}$ is called a **sesquilinear form** if it is conjugate linear in the first argument and linear in the second; that is,

$$
(0.36) \quad \begin{aligned} \langle \alpha_1 f_1 + \alpha_2 f_2, g \rangle &= \alpha_1^* \langle f_1, g \rangle + \alpha_2^* \langle f_2, g \rangle, \\ \langle f, \alpha_1 g_1 + \alpha_2 g_2 \rangle &= \alpha_1 \langle f, g_1 \rangle + \alpha_2 \langle f, g_2 \rangle, \end{aligned} \quad \alpha_1, \alpha_2 \in \mathbb{C},
$$

where '*' denotes complex conjugation. A sesquilinear form satisfying the requirements

(i) $\langle f, f \rangle > 0$ for $f \neq 0$ \quad (positive definite),
(ii) $\langle f, g \rangle = \langle g, f \rangle^*$ \quad (symmetry)

is called an **inner product** or **scalar product**. Associated with every scalar product is a norm

$$(0.37) \qquad \|f\| = \sqrt{\langle f, f \rangle}.$$

Only the triangle inequality is nontrivial. It will follow from the Cauchy–Schwarz inequality below. Until then, just regard (0.37) as a convenient short hand notation.

The pair $(\mathfrak{H}, \langle .,..\rangle)$ is called an **inner product space**. If $\mathfrak{H}$ is complete (with respect to the norm (0.37)), it is called a **Hilbert space**.

Example. Clearly $\mathbb{C}^n$ with the usual scalar product

$$(0.38) \qquad \langle a, b \rangle = \sum_{j=1}^{n} a_j^* b_j$$

is a (finite dimensional) Hilbert space. ◇

Example. A somewhat more interesting example is the Hilbert space $\ell^2(\mathbb{N})$, that is, the set of all complex-valued sequences

$$(0.39) \qquad \left\{ (a_j)_{j=1}^{\infty} \Big| \sum_{j=1}^{\infty} |a_j|^2 < \infty \right\}$$

with scalar product

$$(0.40) \qquad \langle a, b \rangle = \sum_{j=1}^{\infty} a_j^* b_j.$$

(Show that this is in fact a separable Hilbert space — Problem 0.13.) ◇

A vector $f \in \mathfrak{H}$ is called **normalized** or a **unit vector** if $\|f\| = 1$. Two vectors $f, g \in \mathfrak{H}$ are called **orthogonal** or **perpendicular** ($f \perp g$) if $\langle f, g \rangle = 0$ and **parallel** if one is a multiple of the other.

If f and g are orthogonal, we have the **Pythagorean theorem**:

$$\|f + g\|^2 = \|f\|^2 + \|g\|^2, \quad f \perp g, \tag{0.41}$$

which is one line of computation (do it!).

Suppose u is a unit vector. Then the projection of f in the direction of u is given by

$$f_\| = \langle u, f \rangle u, \tag{0.42}$$

and $f_\perp$, defined via

$$f_\perp = f - \langle u, f \rangle u, \tag{0.43}$$

is perpendicular to u since $\langle u, f_\perp \rangle = \langle u, f - \langle u, f \rangle u \rangle = \langle u, f \rangle - \langle u, f \rangle \langle u, u \rangle = 0$.

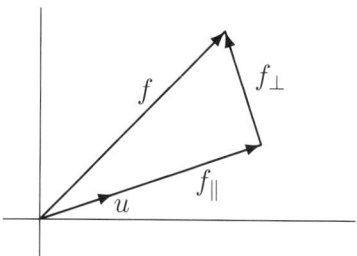

Taking any other vector parallel to u, we obtain from (0.41)

$$\|f - \alpha u\|^2 = \|f_\perp + (f_\| - \alpha u)\|^2 = \|f_\perp\|^2 + |\langle u, f \rangle - \alpha|^2 \tag{0.44}$$

and hence $f_\| = \langle u, f \rangle u$ is the unique vector parallel to u which is closest to f.

As a first consequence we obtain the **Cauchy–Schwarz–Bunjakowski** inequality:

Theorem 0.22 (Cauchy–Schwarz–Bunjakowski). *Let $\mathfrak{H}_0$ be an inner product space. Then for every $f, g \in \mathfrak{H}_0$ we have*

$$|\langle f, g \rangle| \leq \|f\| \|g\| \tag{0.45}$$

with equality if and only if f and g are parallel.

Proof. It suffices to prove the case $\|g\| = 1$. But then the claim follows from $\|f\|^2 = |\langle g, f \rangle|^2 + \|f_\perp\|^2$. □

Note that the Cauchy–Schwarz inequality implies that the scalar product is continuous in both variables; that is, if $f_n \to f$ and $g_n \to g$, we have $\langle f_n, g_n \rangle \to \langle f, g \rangle$.

0.3. The geometry of Hilbert spaces

As another consequence we infer that the map $\|.\|$ is indeed a norm. In fact,

(0.46) $$\|f + g\|^2 = \|f\|^2 + \langle f, g \rangle + \langle g, f \rangle + \|g\|^2 \leq (\|f\| + \|g\|)^2.$$

But let us return to $C(I)$. Can we find a scalar product which has the maximum norm as associated norm? Unfortunately the answer is no! The reason is that the maximum norm does not satisfy the parallelogram law (Problem 0.20).

Theorem 0.23 (Jordan–von Neumann). *A norm is associated with a scalar product if and only if the* **parallelogram law**

(0.47) $$\|f + g\|^2 + \|f - g\|^2 = 2\|f\|^2 + 2\|g\|^2$$

holds.

In this case the scalar product can be recovered from its norm by virtue of the **polarization identity**

(0.48) $$\langle f, g \rangle = \frac{1}{4} \left(\|f + g\|^2 - \|f - g\|^2 + \mathrm{i}\|f - \mathrm{i}g\|^2 - \mathrm{i}\|f + \mathrm{i}g\|^2 \right).$$

Proof. If an inner product space is given, verification of the parallelogram law and the polarization identity is straightforward (Problem 0.22).

To show the converse, we define

$$s(f, g) = \frac{1}{4} \left(\|f + g\|^2 - \|f - g\|^2 + \mathrm{i}\|f - \mathrm{i}g\|^2 - \mathrm{i}\|f + \mathrm{i}g\|^2 \right).$$

Then $s(f, f) = \|f\|^2$ and $s(f, g) = s(g, f)^*$ are straightforward to check. Moreover, another straightforward computation using the parallelogram law shows

$$s(f, g) + s(f, h) = 2s(f, \frac{g + h}{2}).$$

Now choosing $h = 0$ (and using $s(f, 0) = 0$) shows $s(f, g) = 2s(f, \frac{g}{2})$ and thus $s(f, g) + s(f, h) = s(f, g + h)$. Furthermore, by induction we infer $\frac{m}{2^n} s(f, g) = s(f, \frac{m}{2^n} g)$; that is, $\alpha s(f, g) = s(f, \alpha g)$ for every positive rational α. By continuity (which follows from the triangle inequality for $\|.\|$) this holds for all $\alpha > 0$ and $s(f, -g) = -s(f, g)$, respectively, $s(f, \mathrm{i}g) = \mathrm{i}s(f, g)$, finishes the proof. $\square$

Note that the parallelogram law and the polarization identity even hold for sesquilinear forms (Problem 0.22).

But how do we define a scalar product on $C(I)$? One possibility is

(0.49) $$\langle f, g \rangle = \int_a^b f^*(x) g(x) \, dx.$$

The corresponding inner product space is denoted by $\mathcal{L}^2_{cont}(I)$. Note that we have

$$\|f\| \leq \sqrt{|b-a|} \|f\|_\infty \tag{0.50}$$

and hence the maximum norm is stronger than the $\mathcal{L}^2_{cont}$ norm.

Suppose we have two norms $\|.\|_1$ and $\|.\|_2$ on a vector space X. Then $\|.\|_2$ is said to be **stronger** than $\|.\|_1$ if there is a constant $m > 0$ such that

$$\|f\|_1 \leq m\|f\|_2. \tag{0.51}$$

It is straightforward to check the following.

Lemma 0.24. *If $\|.\|_2$ is stronger than $\|.\|_1$, then every $\|.\|_2$ Cauchy sequence is also a $\|.\|_1$ Cauchy sequence.*

Hence if a function $F : X \to Y$ is continuous in $(X, \|.\|_1)$, it is also continuous in $(X, \|.\|_2)$, and if a set is dense in $(X, \|.\|_2)$, it is also dense in $(X, \|.\|_1)$.

In particular, $\mathcal{L}^2_{cont}$ is separable. But is it also complete? Unfortunately the answer is no:

Example. Take $I = [0,2]$ and define

$$f_n(x) = \begin{cases} 0, & 0 \leq x \leq 1 - \frac{1}{n}, \\ 1 + n(x-1), & 1 - \frac{1}{n} \leq x \leq 1, \\ 1, & 1 \leq x \leq 2. \end{cases} \tag{0.52}$$

Then $f_n(x)$ is a Cauchy sequence in $\mathcal{L}^2_{cont}$, but there is no limit in $\mathcal{L}^2_{cont}$! Clearly the limit should be the step function which is 0 for $0 \leq x < 1$ and 1 for $1 \leq x \leq 2$, but this step function is discontinuous (Problem 0.25)! ⋄

This shows that in infinite dimensional vector spaces, different norms will give rise to different convergent sequences! In fact, the key to solving problems in infinite dimensional spaces is often finding the right norm! This is something which cannot happen in the finite dimensional case.

Theorem 0.25. *If X is a finite dimensional vector space, then all norms are equivalent. That is, for any two given norms $\|.\|_1$ and $\|.\|_2$, there are positive constants m_1 and m_2 such that*

$$\frac{1}{m_2}\|f\|_1 \leq \|f\|_2 \leq m_1\|f\|_1. \tag{0.53}$$

Proof. Since equivalence of norms is an equivalence relation (check this!) we can assume that $\|.\|_2$ is the usual Euclidean norm. Moreover, we can choose an orthogonal basis u_j, $1 \le j \le n$, such that $\|\sum_j \alpha_j u_j\|_2^2 = \sum_j |\alpha_j|^2$. Let $f = \sum_j \alpha_j u_j$. Then by the triangle and Cauchy–Schwarz inequalities,

$$\|f\|_1 \le \sum_j |\alpha_j| \|u_j\|_1 \le \sqrt{\sum_j \|u_j\|_1^2} \, \|f\|_2$$

and we can choose $m_2 = \sqrt{\sum_j \|u_j\|_1}$.

In particular, if f_n is convergent with respect to $\|.\|_2$, it is also convergent with respect to $\|.\|_1$. Thus $\|.\|_1$ is continuous with respect to $\|.\|_2$ and attains its minimum $m > 0$ on the unit sphere (which is compact by the Heine–Borel theorem, Theorem 0.12). Now choose $m_1 = 1/m$. □

Problem 0.18. Show that the norm in a Hilbert space satisfies $\|f + g\| = \|f\| + \|g\|$ if and only if $f = \alpha g$, $\alpha \ge 0$, or $g = 0$.

Problem 0.19 (Generalized parallelogram law). Show that, in a Hilbert space,

$$\sum_{1 \le j < k \le n} \|x_j - x_k\|^2 + \|\sum_{1 \le j \le n} x_j\|^2 = n \sum_{1 \le j \le n} \|x_j\|^2.$$

The case $n = 2$ is (0.47).

Problem 0.20. Show that the maximum norm on $C[0, 1]$ does not satisfy the parallelogram law.

Problem 0.21. In a Banach space, the unit ball is convex by the triangle inequality. A Banach space X is called **uniformly convex** if for every $\varepsilon > 0$ there is some δ such that $\|x\| \le 1$, $\|y\| \le 1$, and $\|\frac{x+y}{2}\| \ge 1 - \delta$ imply $\|x - y\| \le \varepsilon$.

Geometrically this implies that if the average of two vectors inside the closed unit ball is close to the boundary, then they must be close to each other.

Show that a Hilbert space is uniformly convex and that one can choose $\delta(\varepsilon) = 1 - \sqrt{1 - \frac{\varepsilon^2}{4}}$. Draw the unit ball for $\mathbb{R}^2$ for the norms $\|x\|_1 = |x_1| + |x_2|$, $\|x\|_2 = \sqrt{|x_1|^2 + |x_2|^2}$, and $\|x\|_\infty = \max(|x_1|, |x_2|)$. Which of these norms makes $\mathbb{R}^2$ uniformly convex?

(Hint: For the first part, use the parallelogram law.)

Problem 0.22. Suppose $\mathfrak{Q}$ is a vector space. Let $s(f, g)$ be a sesquilinear form on $\mathfrak{Q}$ and $q(f) = s(f, f)$ the associated quadratic form. Prove the **parallelogram law**

(0.54) $$q(f + g) + q(f - g) = 2q(f) + 2q(g)$$

and the **polarization identity**

(0.55) $$s(f,g) = \frac{1}{4}\left(q(f+g) - q(f-g) + \mathrm{i}\,q(f-\mathrm{i}g) - \mathrm{i}\,q(f+\mathrm{i}g)\right).$$

Show that $s(f,g)$ is symmetric if and only if $q(f)$ is real-valued.

Problem 0.23. *A sesquilinear form is called* **bounded** *if*

$$\|s\| = \sup_{\|f\|=\|g\|=1} |s(f,g)|$$

is finite. Similarly, the associated quadratic form q is **bounded** *if*

$$\|q\| = \sup_{\|f\|=1} |q(f)|$$

is finite. Show

$$\|q\| \le \|s\| \le 2\|q\|.$$

(Hint: Use the parallelogram law and the polarization identity from the previous problem.)

Problem 0.24. *Suppose $\mathfrak{Q}$ is a vector space. Let $s(f,g)$ be a sesquilinear form on $\mathfrak{Q}$ and $q(f) = s(f,f)$ the associated quadratic form. Show that the Cauchy–Schwarz inequality*

(0.56) $$|s(f,g)| \le q(f)^{1/2} q(g)^{1/2}$$

holds if $q(f) \ge 0$.

(Hint: Consider $0 \le q(f + \alpha g) = q(f) + 2\operatorname{Re}(\alpha\, s(f,g)) + |\alpha|^2 q(g)$ and choose $\alpha = t\, s(f,g)^*/|s(f,g)|$ with $t \in \mathbb{R}$.)

Problem 0.25. *Prove the claims made about f_n, defined in (0.52), in the last example.*

0.4. Completeness

Since $\mathcal{L}^2_{cont}$ is not complete, how can we obtain a Hilbert space from it? Well, the answer is simple: take the **completion**.

If X is an (incomplete) normed space, consider the set of all Cauchy sequences $\tilde{X}$. Call two Cauchy sequences equivalent if their difference converges to zero and denote by $\bar{X}$ the set of all equivalence classes. It is easy to see that $\bar{X}$ (and $\tilde{X}$) inherit the vector space structure from X. Moreover,

Lemma 0.26. *If x_n is a Cauchy sequence, then $\|x_n\|$ converges.*

Consequently, the norm of a Cauchy sequence $(x_n)_{n=1}^\infty$ can be defined by $\|(x_n)_{n=1}^\infty\| = \lim_{n\to\infty} \|x_n\|$ and is independent of the equivalence class (show this!). Thus $\bar{X}$ is a normed space ($\tilde{X}$ is not! Why?).

Theorem 0.27. *$\bar{X}$ is a Banach space containing X as a dense subspace if we identify $x \in X$ with the equivalence class of all sequences converging to x.*

Proof. (Outline) It remains to show that $\bar{X}$ is complete. Let $\xi_n = [(x_{n,j})_{j=1}^\infty]$ be a Cauchy sequence in $\bar{X}$. Then it is not hard to see that $\xi = [(x_{j,j})_{j=1}^\infty]$ is its limit. □

Let me remark that the completion $\bar{X}$ is unique. More precisely, every other complete space which contains X as a dense subset is isomorphic to $\bar{X}$. This can for example be seen by showing that the identity map on X has a unique extension to $\bar{X}$ (compare Theorem 0.29 below).

In particular, it is no restriction to assume that a normed vector space or an inner product space is complete. However, in the important case of $\mathcal{L}^2_{cont}$, it is somewhat inconvenient to work with equivalence classes of Cauchy sequences and hence we will give a different characterization using the Lebesgue integral later.

Problem 0.26. *Provide a detailed proof of Theorem 0.27.*

0.5. Bounded operators

A linear map A between two normed spaces X and Y will be called a **(linear) operator**

$$(0.57) \qquad A : \mathfrak{D}(A) \subseteq X \to Y.$$

The linear subspace $\mathfrak{D}(A)$ on which A is defined is called the **domain** of A and is usually required to be dense. The **kernel** (also **null space**)

$$(0.58) \qquad \mathrm{Ker}(A) = \{f \in \mathfrak{D}(A) | Af = 0\} \subseteq X$$

and **range**

$$(0.59) \qquad \mathrm{Ran}(A) = \{Af | f \in \mathfrak{D}(A)\} = A\mathfrak{D}(A) \subseteq Y$$

are defined as usual. The operator A is called **bounded** if the operator norm

$$(0.60) \qquad \|A\| = \sup_{f \in \mathfrak{D}(A), \|f\|_X = 1} \|Af\|_Y$$

is finite.

By construction, a bounded operator is Lipschitz continuous,

$$(0.61) \qquad \|Af\|_Y \leq \|A\| \|f\|_X, \qquad f \in \mathfrak{D}(A),$$

and hence continuous. The converse is also true:

Theorem 0.28. *An operator A is bounded if and only if it is continuous.*

Proof. Suppose A is continuous but not bounded. Then there is a sequence of unit vectors u_n such that $\|Au_n\| \geq n$. Then $f_n = \frac{1}{n} u_n$ converges to 0 but $\|Af_n\| \geq 1$ does not converge to 0. $\square$

In particular, if X is finite dimensional, then every operator is bounded. Note that in general one and the same operation might be bounded (i.e. continuous) or unbounded, depending on the norm chosen.

Example. Consider the vector space of differentiable functions $X = C^1[0,1]$ and equip it with the norm (cf. Problem 0.29)

$$\|f\|_{\infty,1} = \max_{x \in [0,1]} |f(x)| + \max_{x \in [0,1]} |f'(x)|.$$

Let $Y = C[0,1]$ and observe that the differential operator $A = \frac{d}{dx} : X \to Y$ is bounded since

$$\|Af\|_\infty = \max_{x \in [0,1]} |f'(x)| \leq \max_{x \in [0,1]} |f(x)| + \max_{x \in [0,1]} |f'(x)| = \|f\|_{\infty,1}.$$

However, if we consider $A = \frac{d}{dx} : \mathfrak{D}(A) \subseteq Y \to Y$ defined on $\mathfrak{D}(A) = C^1[0,1]$, then we have an unbounded operator. Indeed, choose

$$u_n(x) = \sin(n\pi x)$$

which is normalized, $\|u_n\|_\infty = 1$, and observe that

$$Au_n(x) = u_n'(x) = n\pi \cos(n\pi x)$$

is unbounded, $\|Au_n\|_\infty = n\pi$. Note that $\mathfrak{D}(A)$ contains the set of polynomials and thus is dense by the Weierstraß approximation theorem (Theorem 0.20). $\diamond$

If A is bounded and densely defined, it is no restriction to assume that it is defined on all of X.

Theorem 0.29 (B.L.T. theorem). *Let $A : \mathfrak{D}(A) \subseteq X \to Y$ be a bounded linear operator and let Y be a Banach space. If $\mathfrak{D}(A)$ is dense, there is a unique (continuous) extension of A to X which has the same operator norm.*

Proof. Since a bounded operator maps Cauchy sequences to Cauchy sequences, this extension can only be given by

$$\overline{A}f = \lim_{n \to \infty} Af_n, \quad f_n \in \mathfrak{D}(A), \quad f \in X.$$

To show that this definition is independent of the sequence $f_n \to f$, let $g_n \to f$ be a second sequence and observe

$$\|Af_n - Ag_n\| = \|A(f_n - g_n)\| \leq \|A\| \|f_n - g_n\| \to 0.$$

Since for $f \in \mathfrak{D}(A)$ we can choose $f_n = f$, we see that $\overline{A}f = Af$ in this case; that is, $\overline{A}$ is indeed an extension. From continuity of vector addition and

0.5. Bounded operators

scalar multiplication it follows that $\overline{A}$ is linear. Finally, from continuity of the norm we conclude that the operator norm does not increase. □

The set of all bounded linear operators from X to Y is denoted by $\mathfrak{L}(X,Y)$. If $X = Y$, we write $\mathfrak{L}(X,X) = \mathfrak{L}(X)$. An operator in $\mathfrak{L}(X,\mathbb{C})$ is called a **bounded linear functional**, and the space $X^* = \mathfrak{L}(X,\mathbb{C})$ is called the dual space of X.

Example. Consider $X = C(I)$. Then for every $t_0 \in I$ the point evaluation $\ell_{t_0}(x) = x(t_0)$ is a bounded linear functional. ◇

Theorem 0.30. *The space $\mathfrak{L}(X,Y)$ together with the operator norm (0.60) is a normed space. It is a Banach space if Y is.*

Proof. That (0.60) is indeed a norm is straightforward. If Y is complete and A_n is a Cauchy sequence of operators, then $A_n f$ converges to an element g for every f. Define a new operator A via $Af = g$. By continuity of the vector operations, A is linear and by continuity of the norm $\|Af\| = \lim_{n\to\infty} \|A_n f\| \leq (\lim_{n\to\infty} \|A_n\|)\|f\|$, it is bounded. Furthermore, given $\varepsilon > 0$, there is some N such that $\|A_n - A_m\| \leq \varepsilon$ for $n, m \geq N$ and thus $\|A_n f - A_m f\| \leq \varepsilon \|f\|$. Taking the limit $m \to \infty$, we see $\|A_n f - A f\| \leq \varepsilon \|f\|$; that is, $A_n \to A$. □

The Banach space of bounded linear operators $\mathfrak{L}(X)$ even has a multiplication given by composition. Clearly this multiplication satisfies

$$(0.62) \quad (A+B)C = AC + BC, \quad A(B+C) = AB + BC, \quad A, B, C \in \mathfrak{L}(X)$$

and

$$(0.63) \quad (AB)C = A(BC), \quad \alpha(AB) = (\alpha A)B = A(\alpha B), \quad \alpha \in \mathbb{C}.$$

Moreover, it is easy to see that we have

$$(0.64) \quad \|AB\| \leq \|A\|\|B\|.$$

In other words, $\mathfrak{L}(X)$ is a so-called **Banach algebra**. However, note that our multiplication is not commutative (unless X is one-dimensional). We even have an **identity**, the identity operator $\mathbb{I}$ satisfying $\|\mathbb{I}\| = 1$.

Problem 0.27. *Consider $X = \mathbb{C}^n$ and let $A : X \to X$ be a matrix. Equip X with the norm (show that this is a norm)*

$$\|x\|_\infty = \max_{1 \leq j \leq n} |x_j|$$

and compute the operator norm $\|A\|$ with respect to this matrix in terms of the matrix entries. Do the same with respect to the norm

$$\|x\|_1 = \sum_{1 \leq j \leq n} |x_j|.$$

Problem 0.28. Show that the integral operator

$$(Kf)(x) = \int_0^1 K(x,y) f(y) dy,$$

where $K(x,y) \in C([0,1] \times [0,1])$, defined on $\mathfrak{D}(K) = C[0,1]$, is a bounded operator both in $X = C[0,1]$ (max norm) and $X = \mathcal{L}^2_{cont}(0,1)$.

Problem 0.29. Let I be a compact interval. Show that the set of differentiable functions $C^1(I)$ becomes a Banach space if we set $\|f\|_{\infty,1} = \max_{x \in I} |f(x)| + \max_{x \in I} |f'(x)|$.

Problem 0.30. Show that $\|AB\| \leq \|A\|\|B\|$ for every $A, B \in \mathfrak{L}(X)$. Conclude that the multiplication is continuous: $A_n \to A$ and $B_n \to B$ imply $A_n B_n \to AB$.

Problem 0.31. Let $A \in \mathfrak{L}(X)$ be a bijection. Show

$$\|A^{-1}\|^{-1} = \inf_{f \in X, \|f\|=1} \|Af\|.$$

Problem 0.32. Let

$$f(z) = \sum_{j=0}^{\infty} f_j z^j, \qquad |z| < R,$$

be a convergent power series with convergence radius $R > 0$. Suppose A is a bounded operator with $\|A\| < R$. Show that

$$f(A) = \sum_{j=0}^{\infty} f_j A^j$$

exists and defines a bounded linear operator (cf. Problem 0.10).

0.6. Lebesgue L^p spaces

For this section, some basic facts about the Lebesgue integral are required. The necessary background can be found in Appendix A. To begin with, Sections A.1, A.3, and A.5 will be sufficient.

We fix some σ-finite measure space (X, Σ, μ) and denote by $\mathcal{L}^p(X, d\mu)$, $1 \leq p$, the set of all complex-valued measurable functions for which

(0.65) $$\|f\|_p = \left(\int_X |f|^p d\mu \right)^{1/p}$$

is finite. First of all, note that $\mathcal{L}^p(X, d\mu)$ is a vector space, since $|f+g|^p \leq 2^p \max(|f|, |g|)^p = 2^p \max(|f|^p, |g|^p) \leq 2^p(|f|^p + |g|^p)$. Of course our hope is that $\mathcal{L}^p(X, d\mu)$ is a Banach space. However, there is a small technical problem (recall that a property is said to hold almost everywhere if the set where it fails to hold is contained in a set of measure zero):

0.6. Lebesgue L^p spaces

Lemma 0.31. *Let f be measurable. Then*

(0.66)
$$\int_X |f|^p \, d\mu = 0$$

if and only if $f(x) = 0$ almost everywhere with respect to μ.

Proof. Observe that we have $A = \{x | f(x) \neq 0\} = \bigcup_n A_n$, where $A_n = \{x \mid |f(x)| > \frac{1}{n}\}$. If $\int_X |f|^p d\mu = 0$, we must have $\mu(A_n) = 0$ for every n and hence $\mu(A) = \lim_{n \to \infty} \mu(A_n) = 0$.

Conversely, we have $\int_X |f|^p \, d\mu = \int_A |f|^p \, d\mu = 0$ since $\mu(A) = 0$ implies $\int_A s \, d\mu = 0$ for every simple function and thus for any integrable function by definition of the integral. □

Note that the proof also shows that if f is not 0 almost everywhere, there is an $\varepsilon > 0$ such that $\mu(\{x \mid |f(x)| \geq \varepsilon\}) > 0$.

Example. Let λ be the Lebesgue measure on $\mathbb{R}$. Then the characteristic function of the rationals $\chi_\mathbb{Q}$ is zero a.e. (with respect to λ).

Let Θ be the Dirac measure centered at 0. Then $f(x) = 0$ a.e. (with respect to Θ) if and only if $f(0) = 0$. ◇

Thus $\|f\|_p = 0$ only implies $f(x) = 0$ for almost every x, but not for all! Hence $\|.\|_p$ is not a norm on $\mathcal{L}^p(X, d\mu)$. The way out of this misery is to identify functions which are equal almost everywhere: Let

(0.67) $\qquad \mathcal{N}(X, d\mu) = \{f \mid f(x) = 0 \; \mu\text{-almost everywhere}\}.$

Then $\mathcal{N}(X, d\mu)$ is a linear subspace of $\mathcal{L}^p(X, d\mu)$ and we can consider the quotient space

(0.68) $\qquad L^p(X, d\mu) = \mathcal{L}^p(X, d\mu) / \mathcal{N}(X, d\mu).$

If $d\mu$ is the Lebesgue measure on $X \subseteq \mathbb{R}^n$, we simply write $L^p(X)$. Observe that $\|f\|_p$ is well-defined on $L^p(X, d\mu)$.

Even though the elements of $L^p(X, d\mu)$ are, strictly speaking, equivalence classes of functions, we will still call them functions for notational convenience. However, note that for $f \in L^p(X, d\mu)$ the value $f(x)$ is not well-defined (unless there is a continuous representative and different continuous functions are in different equivalence classes, e.g., in the case of Lebesgue measure).

With this modification we are back in business since $L^p(X, d\mu)$ turns out to be a Banach space. We will show this in the following sections.

But before that, let us also define $L^\infty(X, d\mu)$. It should be the set of bounded measurable functions $B(X)$ together with the sup norm. The only problem is that if we want to identify functions equal almost everywhere, the supremum is no longer independent of the representative in the equivalence class. The solution is the **essential supremum**

$$\|f\|_\infty = \inf\{C \,|\, \mu(\{x|\,|f(x)|>C\})=0\}. \tag{0.69}$$

That is, C is an essential bound if $|f(x)| \leq C$ almost everywhere and the essential supremum is the infimum over all essential bounds.

Example. If λ is the Lebesgue measure, then the essential sup of $\chi_\mathbb{Q}$ with respect to λ is 0. If Θ is the Dirac measure centered at 0, then the essential sup of $\chi_\mathbb{Q}$ with respect to Θ is 1 (since $\chi_\mathbb{Q}(0) = 1$, and $x = 0$ is the only point which counts for Θ). $\diamond$

As before, we set

$$L^\infty(X, d\mu) = B(X)/\mathcal{N}(X, d\mu) \tag{0.70}$$

and observe that $\|f\|_\infty$ is independent of the equivalence class.

If you wonder where the ∞ comes from, have a look at Problem 0.33.

As a preparation for proving that L^p is a Banach space, we will need Hölder's inequality, which plays a central role in the theory of L^p spaces. In particular, it will imply Minkowski's inequality, which is just the triangle inequality for L^p.

Theorem 0.32 (Hölder's inequality)**.** *Let p and q be* **dual** *indices; that is,*

$$\frac{1}{p} + \frac{1}{q} = 1 \tag{0.71}$$

with $1 \leq p \leq \infty$. If $f \in L^p(X, d\mu)$ and $g \in L^q(X, d\mu)$, then $fg \in L^1(X, d\mu)$ and

$$\|fg\|_1 \leq \|f\|_p \|g\|_q. \tag{0.72}$$

Proof. The case $p = 1$, $q = \infty$ (respectively $p = \infty$, $q = 1$) follows directly from the properties of the integral and hence it remains to consider $1 < p, q < \infty$.

First of all, it is no restriction to assume $\|f\|_p = \|g\|_q = 1$. Then, using (0.26) with $\alpha = |f|^p$ and $\beta = |g|^q$ and integrating over X gives

$$\int_X |fg|\,d\mu \leq \frac{1}{p}\int_X |f|^p d\mu + \frac{1}{q}\int_X |g|^q d\mu = 1$$

and finishes the proof. $\square$

As a consequence we also get

Theorem 0.33 (Minkowski's inequality)**.** *Let $f, g \in L^p(X, d\mu)$. Then*

$$\|f + g\|_p \leq \|f\|_p + \|g\|_p. \tag{0.73}$$

0.6. Lebesgue L^p spaces

Proof. Since the cases $p = 1, \infty$ are straightforward, we only consider $1 < p < \infty$. Using $|f + g|^p \leq |f||f + g|^{p-1} + |g||f + g|^{p-1}$, we obtain from Hölder's inequality (note $(p-1)q = p$)

$$\|f + g\|_p^p \leq \|f\|_p \|(f + g)^{p-1}\|_q + \|g\|_p \|(f + g)^{p-1}\|_q$$
$$= (\|f\|_p + \|g\|_p)\|(f + g)\|_p^{p-1}.$$

□

This shows that $L^p(X, d\mu)$ is a normed vector space. Finally, it remains to show that $L^p(X, d\mu)$ is complete.

Theorem 0.34. *The space $L^p(X, d\mu)$, $1 \leq p \leq \infty$, is a Banach space.*

Proof. We begin with the case $1 \leq p < \infty$. Suppose f_n is a Cauchy sequence. It suffices to show that some subsequence converges (show this). Hence we can drop some terms such that

$$\|f_{n+1} - f_n\|_p \leq \frac{1}{2^n}.$$

Now consider $g_n = f_n - f_{n-1}$ (set $f_0 = 0$). Then

$$G(x) = \sum_{k=1}^{\infty} |g_k(x)|$$

is in L^p. This follows from

$$\left\|\sum_{k=1}^{n} |g_k|\right\|_p \leq \sum_{k=1}^{n} \|g_k\|_p \leq \|f_1\|_p + 1$$

using the monotone convergence theorem. In particular, $G(x) < \infty$ almost everywhere and the sum

$$\sum_{n=1}^{\infty} g_n(x) = \lim_{n \to \infty} f_n(x)$$

is absolutely convergent for those x. Now let $f(x)$ be this limit. Since $|f(x) - f_n(x)|^p$ converges to zero almost everywhere and $|f(x) - f_n(x)|^p \leq (2G(x))^p \in L^1$, dominated convergence shows $\|f - f_n\|_p \to 0$.

In the case $p = \infty$, note that the Cauchy sequence property $|f_n(x) - f_m(x)| < \varepsilon$ for $n, m > N$ holds except for sets $A_{m,n}$ of measure zero. Since $A = \bigcup_{n,m} A_{n,m}$ is again of measure zero, we see that $f_n(x)$ is a Cauchy sequence for $x \in X \backslash A$. The pointwise limit $f(x) = \lim_{n \to \infty} f_n(x)$, $x \in X \backslash A$, is the required limit in $L^\infty(X, d\mu)$ (show this). □

In particular, in the proof of the last theorem we have seen:

Corollary 0.35. *If $\|f_n - f\|_p \to 0$, then there is a subsequence (of representatives) which converges pointwise almost everywhere.*

Note that the statement is not true in general without passing to a subsequence (Problem 0.38).

Using Hölder's inequality, we can also identify a class of bounded operators in L^p.

Lemma 0.36 (Schur criterion). *Consider $L^p(X, d\mu)$ and $L^p(Y, d\nu)$ and let $\frac{1}{p} + \frac{1}{q} = 1$. Suppose that $K(x, y)$ is measurable and there are measurable functions $K_1(x, y)$, $K_2(x, y)$ such that $|K(x, y)| \leq K_1(x, y) K_2(x, y)$ and*

$$(0.74) \qquad \|K_1(x, .)\|_{L^q(Y, d\nu)} \leq C_1, \qquad \|K_2(., y)\|_{L^p(X, d\mu)} \leq C_2$$

for μ-almost every x, respectively, for ν-almost every y. Then the operator $K : L^p(Y, d\nu) \to L^p(X, d\mu)$ defined by

$$(0.75) \qquad (Kf)(x) = \int_Y K(x, y) f(y) d\nu(y)$$

for μ-almost every x is bounded with $\|K\| \leq C_1 C_2$.

Proof. We assume $1 < p < \infty$ for simplicity and leave the cases $p = 1, \infty$ to the reader. Choose $f \in L^p(Y, d\nu)$. By Fubini's theorem, $\int_Y |K(x, y) f(y)| d\nu(y)$ is measurable and by Hölder's inequality, we have

$$\int_Y |K(x, y) f(y)| d\nu(y) \leq \int_Y K_1(x, y) K_2(x, y) |f(y)| d\nu(y)$$

$$\leq \left(\int_Y K_1(x, y)^q d\nu(y) \right)^{1/q} \left(\int_Y |K_2(x, y) f(y)|^p d\nu(y) \right)^{1/p}$$

$$\leq C_1 \left(\int_Y |K_2(x, y) f(y)|^p d\nu(y) \right)^{1/p}$$

(if $K_2(x, .) f(.) \notin L^p(X, d\nu)$, the inequality is trivially true). Now take this inequality to the p'th power and integrate with respect to x using Fubini:

$$\int_X \left(\int_Y |K(x, y) f(y)| d\nu(y) \right)^p d\mu(x) \leq C_1^p \int_X \int_Y |K_2(x, y) f(y)|^p d\nu(y) d\mu(x)$$

$$= C_1^p \int_Y \int_X |K_2(x, y) f(y)|^p d\mu(x) d\nu(y) \leq C_1^p C_2^p \|f\|_p^p.$$

Hence $\int_Y |K(x, y) f(y)| d\nu(y) \in L^p(X, d\mu)$ and in particular it is finite for μ-almost every x. Thus $K(x, .) f(.)$ is ν integrable for μ-almost every x and $\int_Y K(x, y) f(y) d\nu(y)$ is measurable. $\square$

It even turns out that L^p is separable.

Lemma 0.37. *Suppose X is a second countable topological space (i.e., it has a countable basis) and μ is an outer regular Borel measure. Then $L^p(X, d\mu)$, $1 \leq p < \infty$, is separable. In particular, for every countable basis which is*

0.6. Lebesgue L^p spaces

closed under finite unions, the set of characteristic functions $\chi_O(x)$ with O in this basis is total.

Proof. The set of all characteristic functions $\chi_A(x)$ with $A \in \Sigma$ and $\mu(A) < \infty$ is total by construction of the integral. Now our strategy is as follows: Using outer regularity, we can restrict A to open sets, and using the existence of a countable base, we can restrict A to open sets from this base.

Fix A. By outer regularity, there is a decreasing sequence of open sets O_n such that $\mu(O_n) \to \mu(A)$. Since $\mu(A) < \infty$, it is no restriction to assume $\mu(O_n) < \infty$, and thus $\mu(O_n \backslash A) = \mu(O_n) - \mu(A) \to 0$. Now dominated convergence implies $\|\chi_A - \chi_{O_n}\|_p \to 0$. Thus the set of all characteristic functions $\chi_O(x)$ with O open and $\mu(O) < \infty$ is total. Finally, let $\mathcal{B}$ be a countable basis for the topology. Then, every open set O can be written as $O = \bigcup_{j=1}^\infty \tilde{O}_j$ with $\tilde{O}_j \in \mathcal{B}$. Moreover, by considering the set of all finite unions of elements from $\mathcal{B}$, it is no restriction to assume $\bigcup_{j=1}^n \tilde{O}_j \in \mathcal{B}$. Hence there is an increasing sequence $\tilde{O}_n \nearrow O$ with $\tilde{O}_n \in \mathcal{B}$. By monotone convergence, $\|\chi_O - \chi_{\tilde{O}_n}\|_p \to 0$ and hence the set of all characteristic functions $\chi_{\tilde{O}}$ with $\tilde{O} \in \mathcal{B}$ is total. $\square$

To finish this chapter, let us show that continuous functions are dense in L^p.

Theorem 0.38. *Let X be a locally compact metric space and let μ be a σ-finite regular Borel measure. Then the set $C_c(X)$ of continuous functions with compact support is dense in $L^p(X, d\mu)$, $1 \leq p < \infty$.*

Proof. As in the previous proof, the set of all characteristic functions $\chi_K(x)$ with K compact is total (using inner regularity). Hence it suffices to show that $\chi_K(x)$ can be approximated by continuous functions. By outer regularity there is an open set $O \supset K$ such that $\mu(O \backslash K) \leq \varepsilon$. By Urysohn's lemma (Lemma 0.16) there is a continuous function f_ε which is 1 on K and 0 outside O. Since

$$\int_X |\chi_K - f_\varepsilon|^p d\mu = \int_{O \backslash K} |f_\varepsilon|^p d\mu \leq \mu(O \backslash K) \leq \varepsilon,$$

we have $\|f_\varepsilon - \chi_K\| \to 0$ and we are done. $\square$

If X is some subset of $\mathbb{R}^n$, we can do even better. A nonnegative function $u \in C_c^\infty(\mathbb{R}^n)$ is called a **mollifier** if

(0.76) $$\int_{\mathbb{R}^n} u(x) dx = 1.$$

The standard mollifier is $u(x) = \exp(\frac{1}{|x|^2-1})$ for $|x| < 1$ and $u(x) = 0$ otherwise.

If we scale a mollifier according to $u_k(x) = k^n u(k\,x)$ such that its mass is preserved ($\|u_k\|_1 = 1$) and it concentrates more and more around the origin,

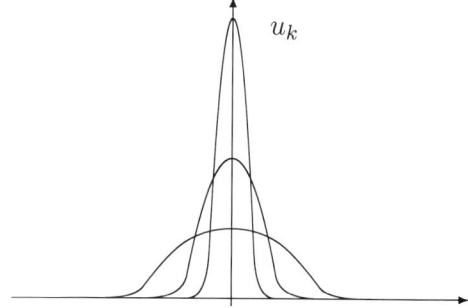

we have the following result (Problem 0.39):

Lemma 0.39. *Let u be a mollifier in $\mathbb{R}^n$ and set $u_k(x) = k^n u(k\,x)$. Then for every (uniformly) continuous function $f : \mathbb{R}^n \to \mathbb{C}$ we have that*

$$(0.77) \qquad f_k(x) = \int_{\mathbb{R}^n} u_k(x-y) f(y) dy$$

is in $C^\infty(\mathbb{R}^n)$ and converges to f (uniformly).

Now we are ready to prove

Theorem 0.40. *If $X \subseteq \mathbb{R}^n$ is open and μ is a regular Borel measure, then the set $C_c^\infty(X)$ of all smooth functions with compact support is dense in $L^p(X, d\mu)$, $1 \leq p < \infty$.*

Proof. By our previous result it suffices to show that every continuous function $f(x)$ with compact support can be approximated by smooth ones. By setting $f(x) = 0$ for $x \notin X$, it is no restriction to assume $X = \mathbb{R}^n$. Now choose a mollifier u and observe that f_k has compact support (since f has). Moreover, since f has compact support, it is uniformly continuous and $f_k \to f$ uniformly. But this implies $f_k \to f$ in L^p. $\square$

We say that $f \in L^p_{loc}(X)$ if $f \in L^p(K)$ for every compact subset $K \subset X$.

Lemma 0.41. *Suppose $f \in L^1_{loc}(\mathbb{R}^n)$. Then*

$$(0.78) \qquad \int_{\mathbb{R}^n} \varphi(x) f(x) dx = 0, \qquad \forall \varphi \in C_c^\infty(\mathbb{R}^n),$$

if and only if $f(x) = 0$ (a.e.).

Proof. First of all, we claim that for every bounded function g with compact support K, there is a sequence of functions $\varphi_n \in C_c^\infty(\mathbb{R}^n)$ with support in K which converges pointwise to g such that $\|\varphi_n\|_\infty \leq \|g\|_\infty$.

0.6. Lebesgue L^p spaces

To see this, take a sequence of continuous functions φ_n with support in K which converges to g in L^1. To make sure that $\|\varphi_n\|_\infty \leq \|g\|_\infty$, just set it equal to $\text{sign}(\varphi_n)\|g\|_\infty$ whenever $|\varphi_n| > \|g\|_\infty$ (show that the resulting sequence still converges). Finally, use (0.77) to make φ_n smooth (note that this operation does not change the sup) and extract a pointwise convergent subsequence.

Now let K be some compact set and choose $g = \text{sign}(f)^* \chi_K$. Then

$$\int_K |f| dx = \int_K f \, \text{sign}(f)^* dx = \lim_{n \to \infty} \int f \varphi_n dx = 0,$$

which shows $f = 0$ for a.e. $x \in K$. Since K is arbitrary, we are done. $\square$

Problem 0.33. *Suppose $\mu(X) < \infty$. Show that $L^\infty(X, d\mu) \subseteq L^p(X, d\mu)$ and*

$$\lim_{p \to \infty} \|f\|_p = \|f\|_\infty, \qquad f \in L^\infty(X, d\mu).$$

Problem 0.34. *Show the following generalization of Hölder's inequality:*

$$(0.79) \qquad \|fg\|_r \leq \|f\|_p \|g\|_q, \qquad \frac{1}{p} + \frac{1}{q} = \frac{1}{r}.$$

Problem 0.35. *Show the iterated Hölder's inequality:*

$$(0.80) \qquad \|f_1 \cdots f_m\|_r \leq \prod_{j=1}^m \|f_j\|_{p_j}, \qquad \frac{1}{p_1} + \cdots + \frac{1}{p_m} = \frac{1}{r}.$$

Problem 0.36. *Show that*

$$\|u\|_{p_0} \leq \mu(X)^{\frac{1}{p_0} - \frac{1}{p}} \|u\|_p, \qquad 1 \leq p_0 \leq p.$$

(Hint: Hölder's inequality.)

Problem 0.37 (Lyapunov inequality). *Let $0 < \theta < 1$. Show that if $f \in L^{p_1} \cap L^{p_2}$, then $f \in L^p$ and*

$$(0.81) \qquad \|f\|_p \leq \|f\|_{p_1}^\theta \|f\|_{p_2}^{1-\theta},$$

where $\frac{1}{p} = \frac{\theta}{p_1} + \frac{1-\theta}{p_2}$.

Problem 0.38. *Find a sequence f_n which converges to 0 in $L^p([0,1], dx)$, $1 \leq p < \infty$, but for which $f_n(x) \to 0$ for a.e. $x \in [0,1]$ does not hold. (Hint: Every $n \in \mathbb{N}$ can be uniquely written as $n = 2^m + k$ with $0 \leq m$ and $0 \leq k < 2^m$. Now consider the characteristic functions of the intervals $I_{m,k} = [k2^{-m}, (k+1)2^{-m}]$.)*

Problem 0.39. *Prove Lemma 0.39. (Hint: To show that f_k is smooth, use Problems A.19 and A.20.)*

Problem 0.40. *Construct a function $f \in L^p(0,1)$ which has a singularity at every rational number in $[0,1]$ (such that the essential supremum is infinite on every open subinterval). (Hint: Start with the function $f_0(x) = |x|^{-\alpha}$ which has a single singularity at 0, then $f_j(x) = f_0(x - x_j)$ has a singularity at x_j.)*

Problem 0.41. *Let μ_j be σ-finite regular Borel measures on some second countable topological spaces X_j, $j = 1, 2$. Show that the set of characteristic functions $\chi_{A_1 \times A_2}$ with A_j Borel sets is total in $L^p(X_1 \times X_2, d(\mu_1 \otimes \mu_2))$ for $1 \le p < \infty$. (Hint: Problem A.21 and Lemma 0.37.)*

0.7. Appendix: The uniform boundedness principle

Recall that the interior of a set is the largest open subset (that is, the union of all open subsets). A set is called **nowhere dense** if its closure has empty interior. The key to several important theorems about Banach spaces is the observation that a Banach space cannot be the countable union of nowhere dense sets.

Theorem 0.42 (Baire category theorem). *Let X be a complete metric space. Then X cannot be the countable union of nowhere dense sets.*

Proof. Suppose $X = \bigcup_{n=1}^{\infty} X_n$. We can assume that the sets X_n are closed and none of them contains a ball; that is, $X \backslash X_n$ is open and nonempty for every n. We will construct a Cauchy sequence x_n which stays away from all X_n.

Since $X \backslash X_1$ is open and nonempty, there is a closed ball $B_{r_1}(x_1) \subseteq X \backslash X_1$. Reducing r_1 a little, we can even assume $\overline{B_{r_1}(x_1)} \subseteq X \backslash X_1$. Moreover, since X_2 cannot contain $B_{r_1}(x_1)$, there is some $x_2 \in B_{r_1}(x_1)$ that is not in X_2. Since $B_{r_1}(x_1) \cap (X \backslash X_2)$ is open, there is a closed ball $\overline{B_{r_2}(x_2)} \subseteq B_{r_1}(x_1) \cap (X \backslash X_2)$. Proceeding by induction, we obtain a sequence of balls such that
$$\overline{B_{r_n}(x_n)} \subseteq B_{r_{n-1}}(x_{n-1}) \cap (X \backslash X_n).$$
Now observe that in every step we can choose r_n as small as we please; hence without loss of generality $r_n \to 0$. Since by construction $x_n \in \overline{B_{r_N}(x_N)}$ for $n \ge N$, we conclude that x_n is Cauchy and converges to some point $x \in X$. But $x \in \overline{B_{r_n}(x_n)} \subseteq X \backslash X_n$ for every n, contradicting our assumption that the X_n cover X. $\square$

(Sets which can be written as the countable union of nowhere dense sets are said to be of first category. All other sets are second category. Hence we have the name category theorem.)

In other words, if $X_n \subseteq X$ is a sequence of closed subsets which cover X, at least one X_n contains a ball of radius $\varepsilon > 0$.

0.7. Appendix: The uniform boundedness principle

Since a closed set is nowhere dense if and only if its complement is open and dense (cf. Problem 0.4), there is a reformulation which is also worthwhile noting:

Corollary 0.43. *Let X be a complete metric space. Then any countable intersection of open dense sets is again dense.*

Proof. Let O_n be open dense sets whose intersection is not dense. Then this intersection must be missing some ball B_ε. The closure of this ball will lie in $\bigcup_n X_n$, where $X_n = X \backslash O_n$ are closed and nowhere dense. But $\overline{B_\varepsilon}$ is a complete metric space, a contradiction. $\square$

Now we come to the following important consequence, the **uniform boundedness principle**.

Theorem 0.44 (Banach–Steinhaus). *Let X be a Banach space and Y some normed vector space. Let $\{A_\alpha\} \subseteq \mathfrak{L}(X,Y)$ be a family of bounded operators. Suppose $\|A_\alpha x\| \leq C(x)$ is bounded for fixed $x \in X$. Then $\{A_\alpha\}$ is uniformly bounded, $\|A_\alpha\| \leq C$.*

Proof. Let
$$X_n = \{x | \, \|A_\alpha x\| \leq n \text{ for all } \alpha\} = \bigcap_\alpha \{x | \, \|A_\alpha x\| \leq n\}.$$

Then $\bigcup_n X_n = X$ by assumption. Moreover, by continuity of A_α and the norm, each X_n is an intersection of closed sets and hence closed. By Baire's theorem, at least one contains a ball of positive radius: $\overline{B_\varepsilon(x_0)} \subset X_n$. Now observe
$$\|A_\alpha y\| \leq \|A_\alpha (y + x_0)\| + \|A_\alpha x_0\| \leq n + C(x_0)$$
for $\|y\| \leq \varepsilon$. Setting $y = \varepsilon \frac{x}{\|x\|}$, we obtain
$$\|A_\alpha x\| \leq \frac{n + C(x_0)}{\varepsilon} \|x\|$$
for every x. $\square$

Part 1

Mathematical Foundations of Quantum Mechanics

Chapter 1

Hilbert spaces

The phase space in classical mechanics is the Euclidean space $\mathbb{R}^{2n}$ (for the n position and n momentum coordinates). In quantum mechanics the phase space is always a Hilbert space $\mathfrak{H}$. Hence the geometry of Hilbert spaces stands at the outset of our investigations.

1.1. Hilbert spaces

Suppose $\mathfrak{H}$ is a vector space. A map $\langle .,..\rangle : \mathfrak{H} \times \mathfrak{H} \to \mathbb{C}$ is called a sesquilinear form if it is conjugate linear in the first argument and linear in the second. A positive definite sesquilinear form is called an **inner product** or **scalar product**. Associated with every scalar product is a norm

(1.1) $$\|\psi\| = \sqrt{\langle \psi, \psi \rangle}.$$

The triangle inequality follows from the **Cauchy–Schwarz–Bunjakowski** inequality:

(1.2) $$|\langle \psi, \varphi \rangle| \leq \|\psi\| \, \|\varphi\|$$

with equality if and only if ψ and φ are parallel.

If $\mathfrak{H}$ is complete with respect to the above norm, it is called a **Hilbert space**. It is no restriction to assume that $\mathfrak{H}$ is complete since one can easily replace it by its completion.

Example. The space $L^2(M, d\mu)$ is a Hilbert space with scalar product given by

(1.3) $$\langle f, g \rangle = \int_M f(x)^* g(x) d\mu(x).$$

Similarly, the set of all square summable sequences $\ell^2(\mathbb{N})$ is a Hilbert space with scalar product

$$(1.4) \qquad \langle f, g \rangle = \sum_{j \in \mathbb{N}} f_j^* g_j.$$

(Note that the second example is a special case of the first one; take $M = \mathbb{R}$ and μ a sum of Dirac measures.) ⋄

A vector $\psi \in \mathfrak{H}$ is called **normalized** or a **unit vector** if $\|\psi\| = 1$. Two vectors $\psi, \varphi \in \mathfrak{H}$ are called **orthogonal** or **perpendicular** ($\psi \perp \varphi$) if $\langle \psi, \varphi \rangle = 0$ and **parallel** if one is a multiple of the other.

If ψ and φ are orthogonal, we have the **Pythagorean theorem**:

$$(1.5) \qquad \|\psi + \varphi\|^2 = \|\psi\|^2 + \|\varphi\|^2, \qquad \psi \perp \varphi,$$

which is one line of computation.

Suppose φ is a unit vector. Then the projection of ψ in the direction of φ is given by

$$(1.6) \qquad \psi_\| = \langle \varphi, \psi \rangle \varphi,$$

and $\psi_\perp$ defined via

$$(1.7) \qquad \psi_\perp = \psi - \langle \varphi, \psi \rangle \varphi$$

is perpendicular to φ.

These results can also be generalized to more than one vector. A set of vectors $\{\varphi_j\}$ is called an **orthonormal set** (ONS) if $\langle \varphi_j, \varphi_k \rangle = 0$ for $j \neq k$ and $\langle \varphi_j, \varphi_j \rangle = 1$. Note that every orthonormal set is linearly independent (show this).

Lemma 1.1. *Suppose $\{\varphi_j\}_{j=0}^n$ is an orthonormal set. Then every $\psi \in \mathfrak{H}$ can be written as*

$$(1.8) \qquad \psi = \psi_\| + \psi_\perp, \qquad \psi_\| = \sum_{j=0}^n \langle \varphi_j, \psi \rangle \varphi_j,$$

where $\psi_\|$ and $\psi_\perp$ are orthogonal. Moreover, $\langle \varphi_j, \psi_\perp \rangle = 0$ for all $1 \leq j \leq n$. In particular,

$$(1.9) \qquad \|\psi\|^2 = \sum_{j=0}^n |\langle \varphi_j, \psi \rangle|^2 + \|\psi_\perp\|^2.$$

Moreover, every $\hat\psi$ in the span of $\{\varphi_j\}_{j=0}^n$ satisfies

$$(1.10) \qquad \|\psi - \hat\psi\| \geq \|\psi_\perp\|$$

with equality holding if and only if $\hat\psi = \psi_\|$. In other words, $\psi_\|$ is uniquely characterized as the vector in the span of $\{\varphi_j\}_{j=0}^n$ closest to ψ.

Proof. A straightforward calculation shows $\langle \varphi_j, \psi - \psi_\| \rangle = 0$ and hence $\psi_\|$ and $\psi_\perp = \psi - \psi_\|$ are orthogonal. The formula for the norm follows by applying (1.5) iteratively.

Now, fix a vector
$$\hat{\psi} = \sum_{j=0}^{n} \alpha_j \varphi_j$$
in the span of $\{\varphi_j\}_{j=0}^n$. Then one computes
$$\|\psi - \hat{\psi}\|^2 = \|\psi_\| + \psi_\perp - \hat{\psi}\|^2 = \|\psi_\perp\|^2 + \|\psi_\| - \hat{\psi}\|^2$$
$$= \|\psi_\perp\|^2 + \sum_{j=0}^{n} |\alpha_j - \langle \varphi_j, \psi \rangle|^2$$
from which the last claim follows. $\square$

From (1.9) we obtain **Bessel's inequality**

(1.11) $$\sum_{j=0}^{n} |\langle \varphi_j, \psi \rangle|^2 \le \|\psi\|^2$$

with equality holding if and only if ψ lies in the span of $\{\varphi_j\}_{j=0}^n$.

Recall that a scalar product can be recovered from its norm by virtue of the **polarization identity**

(1.12) $$\langle \varphi, \psi \rangle = \frac{1}{4} \left(\|\varphi + \psi\|^2 - \|\varphi - \psi\|^2 + i\|\varphi - i\psi\|^2 - i\|\varphi + i\psi\|^2 \right).$$

A bijective linear operator $U \in \mathfrak{L}(\mathfrak{H}_1, \mathfrak{H}_2)$ is called **unitary** if U preserves scalar products:

(1.13) $$\langle U\varphi, U\psi \rangle_2 = \langle \varphi, \psi \rangle_1, \qquad \varphi, \psi \in \mathfrak{H}_1.$$

By the polarization identity, this is the case if and only if U preserves norms: $\|U\psi\|_2 = \|\psi\|_1$ for all $\psi \in \mathfrak{H}_1$. The two Hilbert spaces $\mathfrak{H}_1$ and $\mathfrak{H}_2$ are called **unitarily equivalent** in this case.

Problem 1.1. *The operator*
$$S: \ell^2(\mathbb{N}) \to \ell^2(\mathbb{N}), \qquad (a_1, a_2, a_3, \dots) \mapsto (0, a_1, a_2, \dots)$$
satisfies $\|Sa\| = \|a\|$. *Is it unitary?*

1.2. Orthonormal bases

Of course, since we cannot assume $\mathfrak{H}$ to be a finite dimensional vector space, we need to generalize Lemma 1.1 to arbitrary orthonormal sets $\{\varphi_j\}_{j \in J}$.

We start by assuming that J is countable. Then Bessel's inequality (1.11) shows that

$$\sum_{j \in J} |\langle \varphi_j, \psi \rangle|^2 \tag{1.14}$$

converges absolutely. Moreover, for every finite subset $K \subset J$ we have

$$\|\sum_{j \in K} \langle \varphi_j, \psi \rangle \varphi_j\|^2 = \sum_{j \in K} |\langle \varphi_j, \psi \rangle|^2 \tag{1.15}$$

by the Pythagorean theorem and thus $\sum_{j \in J} \langle \varphi_j, \psi \rangle \varphi_j$ is Cauchy if and only if $\sum_{j \in J} |\langle \varphi_j, \psi \rangle|^2$ is. Now let J be arbitrary. Again, Bessel's inequality shows that for every given $\varepsilon > 0$ there are at most finitely many j for which $|\langle \varphi_j, \psi \rangle| \geq \varepsilon$. Hence there are at most countably many j for which $|\langle \varphi_j, \psi \rangle| > 0$. Thus it follows that

$$\sum_{j \in J} |\langle \varphi_j, \psi \rangle|^2 \tag{1.16}$$

is well-defined and so is

$$\sum_{j \in J} \langle \varphi_j, \psi \rangle \varphi_j. \tag{1.17}$$

In particular, by continuity of the scalar product we see that Lemma 1.1 can be generalized to arbitrary orthonormal sets.

Theorem 1.2. *Suppose $\{\varphi_j\}_{j \in J}$ is an orthonormal set. Then every $\psi \in \mathfrak{H}$ can be written as*

$$\psi = \psi_\| + \psi_\perp, \qquad \psi_\| = \sum_{j \in J} \langle \varphi_j, \psi \rangle \varphi_j, \tag{1.18}$$

where $\psi_\|$ and $\psi_\perp$ are orthogonal. Moreover, $\langle \varphi_j, \psi_\perp \rangle = 0$ for all $j \in J$. In particular,

$$\|\psi\|^2 = \sum_{j \in J} |\langle \varphi_j, \psi \rangle|^2 + \|\psi_\perp\|^2. \tag{1.19}$$

Furthermore, every $\hat{\psi}$ in $\overline{\text{span}\{\varphi_j\}_{j \in J}}$ satisfies

$$\|\psi - \hat{\psi}\| \geq \|\psi_\perp\| \tag{1.20}$$

with equality holding if and only if $\hat{\psi} = \psi_\|$. In other words, $\psi_\|$ is uniquely characterized as the vector in $\overline{\text{span}\{\varphi_j\}_{j \in J}}$ closest to ψ.

Proof. The first part follows as in Lemma 1.1 using continuity of the scalar product. The same is true for the last part except for the fact that every $\psi \in \overline{\text{span}\{\varphi_j\}_{j \in J}}$ can be written as $\psi = \sum_{j \in J} \alpha_j \varphi_j$ (i.e., $\psi = \psi_\|$). To see this, let $\psi_n \in \text{span}\{\varphi_j\}_{j \in J}$ converge to ψ. Then $\|\psi - \psi_n\|^2 = \|\psi_\| - \psi_n\|^2 + \|\psi_\perp\|^2 \to 0$ implies $\psi_n \to \psi_\|$ and $\psi_\perp = 0$. $\square$

1.2. Orthonormal bases

Note that from Bessel's inequality (which of course still holds), it follows that the map $\psi \to \psi_\|$ is continuous.

An orthonormal set which is not a proper subset of any other orthonormal set is called an **orthonormal basis** (ONB) due to the following result:

Theorem 1.3. *For an orthonormal set $\{\varphi_j\}_{j \in J}$, the following conditions are equivalent:*

(i) $\{\varphi_j\}_{j \in J}$ *is a maximal orthonormal set.*

(ii) *For every vector $\psi \in \mathfrak{H}$ we have*

$$\psi = \sum_{j \in J} \langle \varphi_j, \psi \rangle \varphi_j. \tag{1.21}$$

(iii) *For every vector $\psi \in \mathfrak{H}$ we have* **Parseval's relation**

$$\|\psi\|^2 = \sum_{j \in J} |\langle \varphi_j, \psi \rangle|^2. \tag{1.22}$$

(iv) $\langle \varphi_j, \psi \rangle = 0$ *for all $j \in J$ implies $\psi = 0$.*

Proof. We will use the notation from Theorem 1.2.
(i) $\Rightarrow$ (ii): If $\psi_\perp \neq 0$, then we can normalize $\psi_\perp$ to obtain a unit vector $\tilde{\psi}_\perp$ which is orthogonal to all vectors φ_j. But then $\{\varphi_j\}_{j \in J} \cup \{\tilde{\psi}_\perp\}$ would be a larger orthonormal set, contradicting the maximality of $\{\varphi_j\}_{j \in J}$.
(ii) $\Rightarrow$ (iii): This follows since (ii) implies $\psi_\perp = 0$.
(iii) $\Rightarrow$ (iv): If $\langle \psi, \varphi_j \rangle = 0$ for all $j \in J$, we conclude $\|\psi\|^2 = 0$ and hence $\psi = 0$.
(iv) $\Rightarrow$ (i): If $\{\varphi_j\}_{j \in J}$ were not maximal, there would be a unit vector φ such that $\{\varphi_j\}_{j \in J} \cup \{\varphi\}$ is a larger orthonormal set. But $\langle \varphi_j, \varphi \rangle = 0$ for all $j \in J$ implies $\varphi = 0$ by (iv), a contradiction. $\square$

Since $\psi \to \psi_\|$ is continuous, it suffices to check conditions (ii) and (iii) on a dense set.

Example. The set of functions

$$\varphi_n(x) = \frac{1}{\sqrt{2\pi}} e^{i n x}, \qquad n \in \mathbb{Z}, \tag{1.23}$$

forms an orthonormal basis for $\mathfrak{H} = L^2(0, 2\pi)$. The corresponding orthogonal expansion is just the ordinary Fourier series (Problem 1.21). $\diamond$

A Hilbert space is **separable** if and only if there is a countable orthonormal basis. In fact, if $\mathfrak{H}$ is separable, then there exists a countable total set $\{\psi_j\}_{j=0}^N$. Here $N \in \mathbb{N}$ if $\mathfrak{H}$ is finite dimensional and $N = \infty$ otherwise. After throwing away some vectors, we can assume that ψ_{n+1} cannot be expressed as a linear combination of the vectors $\psi_0, \ldots, \psi_n$. Now we can construct an orthonormal basis as follows: We begin by normalizing ψ_0,

$$\varphi_0 = \frac{\psi_0}{\|\psi_0\|}. \tag{1.24}$$

Next we take ψ_1 and remove the component parallel to φ_0 and normalize again:

$$\varphi_1 = \frac{\psi_1 - \langle \varphi_0, \psi_1 \rangle \varphi_0}{\|\psi_1 - \langle \varphi_0, \psi_1 \rangle \varphi_0\|}. \tag{1.25}$$

Proceeding like this, we define recursively

$$\varphi_n = \frac{\psi_n - \sum_{j=0}^{n-1} \langle \varphi_j, \psi_n \rangle \varphi_j}{\|\psi_n - \sum_{j=0}^{n-1} \langle \varphi_j, \psi_n \rangle \varphi_j\|}. \tag{1.26}$$

This procedure is known as **Gram–Schmidt orthogonalization**. Hence we obtain an orthonormal set $\{\varphi_j\}_{j=0}^N$ such that $\mathrm{span}\{\varphi_j\}_{j=0}^n = \mathrm{span}\{\psi_j\}_{j=0}^n$ for every finite n and thus also for N (if $N = \infty$). Since $\{\psi_j\}_{j=0}^N$ is total, so is $\{\varphi_j\}_{j=0}^N$. Now suppose there is some $\psi = \psi_\| + \psi_\perp \in \mathfrak{H}$ for which $\psi_\perp \neq 0$. Since $\{\varphi_j\}_{j=1}^N$ is total, we can find a $\hat\psi$ in its span such that $\|\psi - \hat\psi\| < \|\psi_\perp\|$, contradicting (1.20). Hence we infer that $\{\varphi_j\}_{j=1}^N$ is an orthonormal basis.

Theorem 1.4. *Every separable Hilbert space has a countable orthonormal basis.*

Example. In $L^2(-1, 1)$, we can orthogonalize the polynomial $f_n(x) = x^n$. The resulting polynomials are up to a normalization equal to the Legendre polynomials

$$P_0(x) = 1, \quad P_1(x) = x, \quad P_2(x) = \frac{3x^2 - 1}{2}, \quad \ldots \tag{1.27}$$

(which are normalized such that $P_n(1) = 1$). ◇

In fact, if there is one countable basis, then it follows that every other basis is countable as well.

Theorem 1.5. *If $\mathfrak{H}$ is separable, then every orthonormal basis is countable.*

Proof. We know that there is at least one countable orthonormal basis $\{\varphi_j\}_{j \in J}$. Now let $\{\phi_k\}_{k \in K}$ be a second basis and consider the set $K_j = \{k \in K | \langle \phi_k, \varphi_j \rangle \neq 0\}$. Since these are the expansion coefficients of φ_j with respect to $\{\phi_k\}_{k \in K}$, this set is countable. Hence the set $\tilde K = \bigcup_{j \in J} K_j$ is countable as well. But $k \in K \setminus \tilde K$ implies $\phi_k = 0$ and hence $\tilde K = K$. □

We will assume all Hilbert spaces to be separable.

In particular, it can be shown that $L^2(M, d\mu)$ is separable. Moreover, it turns out that, up to unitary equivalence, there is only one (separable) infinite dimensional Hilbert space:

Let $\mathfrak{H}$ be an infinite dimensional Hilbert space and let $\{\varphi_j\}_{j \in \mathbb{N}}$ be any orthogonal basis. Then the map $U : \mathfrak{H} \to \ell^2(\mathbb{N})$, $\psi \mapsto (\langle \varphi_j, \psi \rangle)_{j \in \mathbb{N}}$ is unitary (by Theorem 1.3 (iii)). In particular,

Theorem 1.6. *Any separable infinite dimensional Hilbert space is unitarily equivalent to $\ell^2(\mathbb{N})$.*

Let me remark that if $\mathfrak{H}$ is not separable, there still exists an orthonormal basis. However, the proof requires Zorn's lemma: The collection of all orthonormal sets in $\mathfrak{H}$ can be partially ordered by inclusion. Moreover, every linearly ordered chain has an upper bound (the union of all sets in the chain). Hence Zorn's lemma implies the existence of a maximal element, that is, an orthonormal basis.

Problem 1.2. *Let $\{\varphi_j\}$ be some orthonormal basis. Show that a bounded linear operator A is uniquely determined by its matrix elements $A_{jk} = \langle \varphi_j, A\varphi_k \rangle$ with respect to this basis.*

Problem 1.3. *Show that $\mathfrak{L}(\mathfrak{H})$ is not separable if $\mathfrak{H}$ is infinite dimensional.*

1.3. The projection theorem and the Riesz lemma

Let $M \subseteq \mathfrak{H}$ be a subset. Then $M^\perp = \{\psi | \langle \varphi, \psi \rangle = 0, \forall \varphi \in M\}$ is called the **orthogonal complement** of M. By continuity of the scalar product it follows that $M^\perp$ is a closed linear subspace and by linearity that $\overline{(\mathrm{span}(M))}^\perp = M^\perp$. For example, we have $\mathfrak{H}^\perp = \{0\}$ since every vector in $\mathfrak{H}^\perp$ must be in particular orthogonal to all vectors in some orthonormal basis.

Theorem 1.7 (Projection theorem). *Let M be a closed linear subspace of a Hilbert space $\mathfrak{H}$. Then every $\psi \in \mathfrak{H}$ can be uniquely written as $\psi = \psi_\| + \psi_\perp$ with $\psi_\| \in M$ and $\psi_\perp \in M^\perp$. One writes*

(1.28) $$M \oplus M^\perp = \mathfrak{H}$$

in this situation.

Proof. Since M is closed, it is a Hilbert space and has an orthonormal basis $\{\varphi_j\}_{j \in J}$. Hence the existence part follows from Theorem 1.2. To see uniqueness, suppose there is another decomposition $\psi = \tilde{\psi}_\| + \tilde{\psi}_\perp$. Then $\psi_\| - \tilde{\psi}_\| = \tilde{\psi}_\perp - \psi_\perp \in M \cap M^\perp = \{0\}$. $\square$

This also shows that every orthogonal set $\{\varphi_j\}_{j\in J}$ can be extended to an orthogonal basis, since we can just add an orthogonal basis for $(\{\varphi_j\}_{j\in J})^\perp$.

Moreover, Theorem 1.7 implies that to every $\psi \in \mathfrak{H}$ we can assign a unique vector $\psi_\|$ which is the vector in M closest to ψ. The rest, $\psi - \psi_\|$, lies in $M^\perp$. The operator $P_M\psi = \psi_\|$ is called the **orthogonal projection** corresponding to M. Note that we have

$$(1.29) \qquad P_M^2 = P_M \quad \text{and} \quad \langle P_M\psi, \varphi\rangle = \langle \psi, P_M\varphi\rangle$$

since $\langle P_M\psi, \varphi\rangle = \langle \psi_\|, \varphi_\|\rangle = \langle \psi, P_M\varphi\rangle$. Clearly we have $P_{M^\perp}\psi = \psi - P_M\psi = \psi_\perp$. Furthermore, (1.29) uniquely characterizes orthogonal projections (Problem 1.6).

Moreover, if M is a closed subspace, we have $P_{M^{\perp\perp}} = \mathbb{I} - P_{M^\perp} = \mathbb{I} - (\mathbb{I} - P_M) = P_M$; that is, $M^{\perp\perp} = M$. If M is an arbitrary subset, we have at least

$$(1.30) \qquad M^{\perp\perp} = \overline{\operatorname{span}(M)}.$$

Note that by $\mathfrak{H}^\perp = \{0\}$ we see that $M^\perp = \{0\}$ if and only if M is total.

Finally we turn to **linear functionals**, that is, to operators $\ell : \mathfrak{H} \to \mathbb{C}$. By the Cauchy–Schwarz inequality we know that $\ell_\varphi : \psi \mapsto \langle \varphi, \psi\rangle$ is a bounded linear functional (with norm $\|\varphi\|$). It turns out that, in a Hilbert space, every bounded linear functional can be written in this way.

Theorem 1.8 (Riesz lemma). *Suppose ℓ is a bounded linear functional on a Hilbert space $\mathfrak{H}$. Then there is a unique vector $\varphi \in \mathfrak{H}$ such that $\ell(\psi) = \langle \varphi, \psi\rangle$ for all $\psi \in \mathfrak{H}$.*

In other words, a Hilbert space is equivalent to its own dual space $\mathfrak{H}^* \cong \mathfrak{H}$ via the map $\varphi \mapsto \langle \varphi, .\rangle$ which is a conjugate linear isometric bijection between $\mathfrak{H}$ and $\mathfrak{H}^*$.

Proof. If $\ell \equiv 0$, we can choose $\varphi = 0$. Otherwise $\operatorname{Ker}(\ell) = \{\psi | \ell(\psi) = 0\}$ is a proper subspace and we can find a unit vector $\tilde{\varphi} \in \operatorname{Ker}(\ell)^\perp$. For every $\psi \in \mathfrak{H}$ we have $\ell(\psi)\tilde{\varphi} - \ell(\tilde{\varphi})\psi \in \operatorname{Ker}(\ell)$ and hence

$$0 = \langle \tilde{\varphi}, \ell(\psi)\tilde{\varphi} - \ell(\tilde{\varphi})\psi\rangle = \ell(\psi) - \ell(\tilde{\varphi})\langle \tilde{\varphi}, \psi\rangle.$$

In other words, we can choose $\varphi = \ell(\tilde{\varphi})^* \tilde{\varphi}$. To see uniqueness, let φ_1, φ_2 be two such vectors. Then $\langle \varphi_1 - \varphi_2, \psi\rangle = \langle \varphi_1, \psi\rangle - \langle \varphi_2, \psi\rangle = \ell(\psi) - \ell(\psi) = 0$ for every $\psi \in \mathfrak{H}$, which shows $\varphi_1 - \varphi_2 \in \mathfrak{H}^\perp = \{0\}$. $\square$

The following easy consequence is left as an exercise.

Corollary 1.9. *Suppose s is a bounded sesquilinear form; that is,*

$$(1.31) \qquad |s(\psi, \varphi)| \leq C\|\psi\|\|\varphi\|.$$

1.3. The projection theorem and the Riesz lemma

Then there is a unique bounded operator A such that

(1.32) $$s(\psi, \varphi) = \langle A\psi, \varphi \rangle.$$

Moreover, the norm of A is given by

(1.33) $$\|A\| = \sup_{\|\psi\|=\|\varphi\|=1} |\langle A\psi, \varphi \rangle| \le C.$$

Note that by the polarization identity (Problem 0.22), A is already uniquely determined by its quadratic form $q_A(\psi) = \langle \psi, A\psi \rangle$.

Problem 1.4. *Suppose $U : \mathfrak{H} \to \mathfrak{H}$ is unitary and $M \subseteq \mathfrak{H}$. Show that $UM^\perp = (UM)^\perp$.*

Problem 1.5. *Show that an orthogonal projection $P_M \ne 0$ has norm one.*

Problem 1.6. *Suppose $P \in \mathfrak{L}(\mathfrak{H})$ satisfies*

$$P^2 = P \quad \text{and} \quad \langle P\psi, \varphi \rangle = \langle \psi, P\varphi \rangle$$

and set $M = \operatorname{Ran}(P)$. Show

- *$P\psi = \psi$ for $\psi \in M$ and M is closed,*
- *$\varphi \in M^\perp$ implies $P\varphi \in M^\perp$ and thus $P\varphi = 0$,*

and conclude $P = P_M$.

Problem 1.7. *Let P_1, P_2 be two orthogonal projections. Show that $P_1 \le P_2$ (that is, $\langle \psi, P_1 \psi \rangle \le \langle \psi, P_2 \psi \rangle$) if and only if $\operatorname{Ran}(P_1) \subseteq \operatorname{Ran}(P_2)$. Show in this case that the two projections commute (that is, $P_1 P_2 = P_2 P_1$) and that $P_2 - P_1$ is also a projection. (Hints: $\|P_j \psi\| = \|\psi\|$ if and only if $P_j \psi = \psi$ and $\operatorname{Ran}(P_1) \subseteq \operatorname{Ran}(P_2)$ if and only if $P_2 P_1 = P_1$.)*

Problem 1.8. *Show $P : L^2(\mathbb{R}) \to L^2(\mathbb{R})$, $f(x) \mapsto \frac{1}{2}(f(x) + f(-x))$ is a projection. Compute its range and kernel.*

Problem 1.9. *Prove Corollary 1.9.*

Problem 1.10. *Consider the sesquilinear form*

$$B(f, g) = \int_0^1 \left(\int_0^x f(t)^* dt \right) \left(\int_0^x g(t) dt \right) dx$$

in $L^2(0,1)$. Show that it is bounded and find the corresponding operator A. (Hint: Integration by parts.)

1.4. Orthogonal sums and tensor products

Given two Hilbert spaces $\mathfrak{H}_1$ and $\mathfrak{H}_2$, we define their **orthogonal sum** $\mathfrak{H}_1 \oplus \mathfrak{H}_2$ to be the set of all pairs $(\psi_1, \psi_2) \in \mathfrak{H}_1 \times \mathfrak{H}_2$ together with the scalar product

$$\langle (\varphi_1, \varphi_2), (\psi_1, \psi_2) \rangle = \langle \varphi_1, \psi_1 \rangle_1 + \langle \varphi_2, \psi_2 \rangle_2. \tag{1.34}$$

It is left as an exercise to verify that $\mathfrak{H}_1 \oplus \mathfrak{H}_2$ is again a Hilbert space. Moreover, $\mathfrak{H}_1$ can be identified with $\{(\psi_1, 0) | \psi_1 \in \mathfrak{H}_1\}$, and we can regard $\mathfrak{H}_1$ as a subspace of $\mathfrak{H}_1 \oplus \mathfrak{H}_2$, and similarly for $\mathfrak{H}_2$. It is also customary to write $\psi_1 + \psi_2$ instead of (ψ_1, ψ_2).

More generally, let $\mathfrak{H}_j$, $j \in \mathbb{N}$, be a countable collection of Hilbert spaces and define

$$\bigoplus_{j=1}^{\infty} \mathfrak{H}_j = \{ \sum_{j=1}^{\infty} \psi_j \, | \, \psi_j \in \mathfrak{H}_j, \, \sum_{j=1}^{\infty} \|\psi_j\|_j^2 < \infty \}, \tag{1.35}$$

which becomes a Hilbert space with the scalar product

$$\langle \sum_{j=1}^{\infty} \varphi_j, \sum_{j=1}^{\infty} \psi_j \rangle = \sum_{j=1}^{\infty} \langle \varphi_j, \psi_j \rangle_j. \tag{1.36}$$

Example. $\bigoplus_{j=1}^{\infty} \mathbb{C} = \ell^2(\mathbb{N})$. ◇

Similarly, if $\mathfrak{H}$ and $\tilde{\mathfrak{H}}$ are two Hilbert spaces, we define their tensor product as follows: The elements should be products $\psi \otimes \tilde{\psi}$ of elements $\psi \in \mathfrak{H}$ and $\tilde{\psi} \in \tilde{\mathfrak{H}}$. Hence we start with the set of all finite linear combinations of elements of $\mathfrak{H} \times \tilde{\mathfrak{H}}$:

$$\mathcal{F}(\mathfrak{H}, \tilde{\mathfrak{H}}) = \{ \sum_{j=1}^{n} \alpha_j (\psi_j, \tilde{\psi}_j) | (\psi_j, \tilde{\psi}_j) \in \mathfrak{H} \times \tilde{\mathfrak{H}}, \, \alpha_j \in \mathbb{C} \}. \tag{1.37}$$

Since we want $(\psi_1 + \psi_2) \otimes \tilde{\psi} = \psi_1 \otimes \tilde{\psi} + \psi_2 \otimes \tilde{\psi}$, $\psi \otimes (\tilde{\psi}_1 + \tilde{\psi}_2) = \psi \otimes \tilde{\psi}_1 + \psi \otimes \tilde{\psi}_2$, and $(\alpha \psi) \otimes \tilde{\psi} = \psi \otimes (\alpha \tilde{\psi})$, we consider $\mathcal{F}(\mathfrak{H}, \tilde{\mathfrak{H}}) / \mathcal{N}(\mathfrak{H}, \tilde{\mathfrak{H}})$, where

$$\mathcal{N}(\mathfrak{H}, \tilde{\mathfrak{H}}) = \mathrm{span}\{ \sum_{j,k=1}^{n} \alpha_j \beta_k (\psi_j, \tilde{\psi}_k) - (\sum_{j=1}^{n} \alpha_j \psi_j, \sum_{k=1}^{n} \beta_k \tilde{\psi}_k) \} \tag{1.38}$$

and write $\psi \otimes \tilde{\psi}$ for the equivalence class of $(\psi, \tilde{\psi})$.

Next, we define

$$\langle \psi \otimes \tilde{\psi}, \phi \otimes \tilde{\phi} \rangle = \langle \psi, \phi \rangle \langle \tilde{\psi}, \tilde{\phi} \rangle \tag{1.39}$$

which extends to a sesquilinear form on $\mathcal{F}(\mathfrak{H}, \tilde{\mathfrak{H}}) / \mathcal{N}(\mathfrak{H}, \tilde{\mathfrak{H}})$. To show that we obtain a scalar product, we need to ensure positivity. Let $\psi = \sum_i \alpha_i \psi_i \otimes \tilde{\psi}_i \neq$

1.4. Orthogonal sums and tensor products

0 and pick orthonormal bases φ_j, $\tilde{\varphi}_k$ for span$\{\psi_i\}$, span$\{\tilde{\psi}_i\}$, respectively. Then

$$\psi = \sum_{j,k} \alpha_{jk} \varphi_j \otimes \tilde{\varphi}_k, \quad \alpha_{jk} = \sum_i \alpha_i \langle \varphi_j, \psi_i \rangle \langle \tilde{\varphi}_k, \tilde{\psi}_i \rangle \tag{1.40}$$

and we compute

$$\langle \psi, \psi \rangle = \sum_{j,k} |\alpha_{jk}|^2 > 0. \tag{1.41}$$

The completion of $\mathcal{F}(\mathfrak{H}, \tilde{\mathfrak{H}})/\mathcal{N}(\mathfrak{H}, \tilde{\mathfrak{H}})$ with respect to the induced norm is called the **tensor product** $\mathfrak{H} \otimes \tilde{\mathfrak{H}}$ of $\mathfrak{H}$ and $\tilde{\mathfrak{H}}$.

Lemma 1.10. *If φ_j, $\tilde{\varphi}_k$ are orthonormal bases for $\mathfrak{H}$, $\tilde{\mathfrak{H}}$, respectively, then $\varphi_j \otimes \tilde{\varphi}_k$ is an orthonormal basis for $\mathfrak{H} \otimes \tilde{\mathfrak{H}}$.*

Proof. That $\varphi_j \otimes \tilde{\varphi}_k$ is an orthonormal set is immediate from (1.39). Moreover, since span$\{\varphi_j\}$, span$\{\tilde{\varphi}_k\}$ are dense in $\mathfrak{H}$, $\tilde{\mathfrak{H}}$, respectively, it is easy to see that $\varphi_j \otimes \tilde{\varphi}_k$ is dense in $\mathcal{F}(\mathfrak{H}, \tilde{\mathfrak{H}})/\mathcal{N}(\mathfrak{H}, \tilde{\mathfrak{H}})$. But the latter is dense in $\mathfrak{H} \otimes \tilde{\mathfrak{H}}$. □

Example. We have $\mathfrak{H} \otimes \mathbb{C}^n = \mathfrak{H}^n$. ◊

Example. Let $(M, d\mu)$ and $(\tilde{M}, d\tilde{\mu})$ be two measure spaces. Then we have $L^2(M, d\mu) \otimes L^2(\tilde{M}, d\tilde{\mu}) = L^2(M \times \tilde{M}, d\mu \times d\tilde{\mu})$.

Clearly we have $L^2(M, d\mu) \otimes L^2(\tilde{M}, d\tilde{\mu}) \subseteq L^2(M \times \tilde{M}, d\mu \times d\tilde{\mu})$. Now take an orthonormal basis $\varphi_j \otimes \tilde{\varphi}_k$ for $L^2(M, d\mu) \otimes L^2(\tilde{M}, d\tilde{\mu})$ as in our previous lemma. Then

$$\int_M \int_{\tilde{M}} (\varphi_j(x) \tilde{\varphi}_k(y))^* f(x,y) d\mu(x) d\tilde{\mu}(y) = 0 \tag{1.42}$$

implies

$$\int_M \varphi_j(x)^* f_k(x) d\mu(x) = 0, \quad f_k(x) = \int_{\tilde{M}} \tilde{\varphi}_k(y)^* f(x,y) d\tilde{\mu}(y) \tag{1.43}$$

and hence $f_k(x) = 0$ μ-a.e. x. But this implies $f(x,y) = 0$ for μ-a.e. x and $\tilde{\mu}$-a.e. y and thus $f = 0$. Hence $\varphi_j \otimes \tilde{\varphi}_k$ is a basis for $L^2(M \times \tilde{M}, d\mu \times d\tilde{\mu})$ and equality follows. ◊

It is straightforward to extend the tensor product to any finite number of Hilbert spaces. We even note

$$\left(\bigoplus_{j=1}^\infty \mathfrak{H}_j \right) \otimes \mathfrak{H} = \bigoplus_{j=1}^\infty (\mathfrak{H}_j \otimes \mathfrak{H}), \tag{1.44}$$

where equality has to be understood in the sense that both spaces are unitarily equivalent by virtue of the identification

$$(\sum_{j=1}^{\infty} \psi_j) \otimes \psi = \sum_{j=1}^{\infty} \psi_j \otimes \psi. \tag{1.45}$$

Problem 1.11. *Show that $\psi \otimes \tilde{\psi} = 0$ if and only if $\psi = 0$ or $\tilde{\psi} = 0$.*

Problem 1.12. *We have $\psi \otimes \tilde{\psi} = \phi \otimes \tilde{\phi} \neq 0$ if and only if there is some $\alpha \in \mathbb{C}\backslash\{0\}$ such that $\psi = \alpha\phi$ and $\tilde{\psi} = \alpha^{-1}\tilde{\phi}$.*

Problem 1.13. *Show* (1.44).

1.5. The C^* algebra of bounded linear operators

We start by introducing a conjugation for operators on a Hilbert space $\mathfrak{H}$. Let $A \in \mathfrak{L}(\mathfrak{H})$. Then the **adjoint operator** is defined via

$$\langle \varphi, A^*\psi \rangle = \langle A\varphi, \psi \rangle \tag{1.46}$$

(compare Corollary 1.9).

Example. If $\mathfrak{H} = \mathbb{C}^n$ and $A = (a_{jk})_{1 \leq j,k \leq n}$, then $A^* = (a_{kj}^*)_{1 \leq j,k \leq n}$. $\diamond$

Lemma 1.11. *Let $A, B \in \mathfrak{L}(\mathfrak{H})$ and $\alpha \in \mathbb{C}$. Then*

(i) $(A + B)^* = A^* + B^*$, $(\alpha A)^* = \alpha^* A^*$,

(ii) $A^{**} = A$,

(iii) $(AB)^* = B^* A^*$,

(iv) $\|A^*\| = \|A\|$ and $\|A\|^2 = \|A^*A\| = \|AA^*\|$.

Proof. (i) is obvious. (ii) follows from $\langle \varphi, A^{**}\psi \rangle = \langle A^*\varphi, \psi \rangle = \langle f\varphi, A\psi \rangle$. (iii) follows from $\langle \varphi, (AB)\psi \rangle = \langle A^*\varphi, B\psi \rangle = \langle B^*A^*\varphi, \psi \rangle$. (iv) follows using (1.33) from

$$\|A^*\| = \sup_{\|\varphi\|=\|\psi\|=1} |\langle \psi, A^*\varphi \rangle| = \sup_{\|\varphi\|=\|\psi\|=1} |\langle A\psi, \varphi \rangle|$$
$$= \sup_{\|\varphi\|=\|\psi\|=1} |\langle \varphi, A\psi \rangle| = \|A\|$$

and

$$\|A^*A\| = \sup_{\|\varphi\|=\|\psi\|=1} |\langle \varphi, A^*A\psi \rangle| = \sup_{\|\varphi\|=\|\psi\|=1} |\langle A\varphi, A\psi \rangle|$$
$$= \sup_{\|\varphi\|=1} \|A\varphi\|^2 = \|A\|^2,$$

where we have used $\|\varphi\| = \sup_{\|\psi\|=1} |\langle \psi, \varphi \rangle|$. $\square$

As a consequence of $\|A^*\| = \|A\|$, observe that taking the adjoint is continuous.

In general, a Banach algebra $\mathcal{A}$ together with an **involution**

(1.47) $\quad (a+b)^* = a^* + b^*, \quad (\alpha a)^* = \alpha^* a^*, \quad a^{**} = a, \quad (ab)^* = b^* a^*$

satisfying

(1.48) $\quad\quad\quad\quad\quad\quad\quad\quad \|a\|^2 = \|a^*a\|$

is called a C^* **algebra**. The element a^* is called the adjoint of a. Note that $\|a^*\| = \|a\|$ follows from (1.48) and $\|aa^*\| \le \|a\|\|a^*\|$.

Any subalgebra which is also closed under involution is called a *-subalgebra. An **ideal** is a subspace $\mathcal{I} \subseteq \mathcal{A}$ such that $a \in \mathcal{I}$, $b \in \mathcal{A}$ imply $ab \in \mathcal{I}$ and $ba \in \mathcal{I}$. If it is closed under the adjoint map, it is called a *-ideal. Note that if there is an identity e, we have $e^* = e$ and hence $(a^{-1})^* = (a^*)^{-1}$ (show this).

Example. The continuous functions $C(I)$ together with complex conjugation form a commutative C^* algebra. $\diamond$

An element $a \in \mathcal{A}$ is called **normal** if $aa^* = a^*a$, **self-adjoint** if $a = a^*$, **unitary** if $aa^* = a^*a = \mathbb{I}$, an (orthogonal) **projection** if $a = a^* = a^2$, and **positive** if $a = bb^*$ for some $b \in \mathcal{A}$. Clearly both self-adjoint and unitary elements are normal.

Problem 1.14. *Let $A \in \mathfrak{L}(\mathfrak{H})$. Show that A is normal if and only if*

$$\|A\psi\| = \|A^*\psi\|, \quad \forall \psi \in \mathfrak{H}.$$

(Hint: Problem 0.22.)

Problem 1.15. *Show that $U : \mathfrak{H} \to \mathfrak{H}$ is unitary if and only if $U^{-1} = U^*$.*

Problem 1.16. *Compute the adjoint of*

$$S : \ell^2(\mathbb{N}) \to \ell^2(\mathbb{N}), \quad (a_1, a_2, a_3, \dots) \mapsto (0, a_1, a_2, \dots).$$

1.6. Weak and strong convergence

Sometimes a weaker notion of convergence is useful: We say that ψ_n **converges weakly** to ψ and write

(1.49) $\quad\quad\quad\quad\quad \text{w-}\lim_{n\to\infty} \psi_n = \psi \quad \text{or} \quad \psi_n \rightharpoonup \psi$

if $\langle \varphi, \psi_n \rangle \to \langle \varphi, \psi \rangle$ for every $\varphi \in \mathfrak{H}$ (show that a weak limit is unique).

Example. Let φ_n be an (infinite) orthonormal set. Then $\langle \psi, \varphi_n \rangle \to 0$ for every ψ since these are just the expansion coefficients of ψ. (φ_n does not converge to 0, since $\|\varphi_n\| = 1$.) $\diamond$

Clearly $\psi_n \to \psi$ implies $\psi_n \rightharpoonup \psi$ and hence this notion of convergence is indeed weaker. Moreover, the weak limit is unique, since $\langle \varphi, \psi_n \rangle \to \langle \varphi, \psi \rangle$ and $\langle \varphi, \psi_n \rangle \to \langle \varphi, \tilde{\psi} \rangle$ imply $\langle \varphi, (\psi - \tilde{\psi}) \rangle = 0$. A sequence ψ_n is called a **weak Cauchy sequence** if $\langle \varphi, \psi_n \rangle$ is Cauchy for every $\varphi \in \mathfrak{H}$.

Lemma 1.12. *Let $\mathfrak{H}$ be a Hilbert space.*

  (i) $\psi_n \rightharpoonup \psi$ *implies* $\|\psi\| \leq \liminf \|\psi_n\|$.
  (ii) *Every weak Cauchy sequence ψ_n is bounded:* $\|\psi_n\| \leq C$.
  (iii) *Every weak Cauchy sequence converges weakly.*
  (iv) *For a weakly convergent sequence $\psi_n \rightharpoonup \psi$ we have $\psi_n \to \psi$ if and only if $\limsup \|\psi_n\| \leq \|\psi\|$.*

Proof. (i) Observe
$$\|\psi\|^2 = \langle \psi, \psi \rangle = \liminf \langle \psi, \psi_n \rangle \leq \|\psi\| \liminf \|\psi_n\|.$$
(ii) For every φ we have that $|\langle \varphi, \psi_n \rangle| \leq C(\varphi)$ is bounded. Hence by the uniform boundedness principle we have $\|\psi_n\| = \|\langle \psi_n, . \rangle\| \leq C$.
(iii) Let φ_m be an orthonormal basis and define $c_m = \lim_{n \to \infty} \langle \varphi_m, \psi_n \rangle$. Then $\psi = \sum_m c_m \varphi_m$ is the desired limit.
(iv) By (i) we have $\lim \|\psi_n\| = \|\psi\|$ and hence
$$\|\psi - \psi_n\|^2 = \|\psi\|^2 - 2\operatorname{Re}(\langle \psi, \psi_n \rangle) + \|\psi_n\|^2 \to 0.$$
The converse is straightforward. $\square$

Clearly, an orthonormal basis does not have a norm convergent subsequence. Hence the unit ball in an infinite dimensional Hilbert space is never compact. However, we can at least extract weakly convergent subsequences:

Lemma 1.13. *Let $\mathfrak{H}$ be a Hilbert space. Every bounded sequence ψ_n has a weakly convergent subsequence.*

Proof. Let φ_k be an orthonormal basis. Then by the usual diagonal sequence argument we can find a subsequence ψ_{n_m} such that $\langle \varphi_k, \psi_{n_m} \rangle$ converges for all k. Since ψ_n is bounded, $\langle \varphi, \psi_{n_m} \rangle$ converges for every $\varphi \in \mathfrak{H}$ and hence ψ_{n_m} is a weak Cauchy sequence. $\square$

Finally, let me remark that similar concepts can be introduced for operators. This is of particular importance for the case of unbounded operators, where convergence in the operator norm makes no sense at all.

A sequence of operators A_n is said to **converge strongly** to A,

(1.50) $\quad \operatorname*{s-lim}_{n \to \infty} A_n = A \quad :\Leftrightarrow \quad A_n \psi \to A \psi \quad \forall \psi \in \mathfrak{D}(A) \subseteq \mathfrak{D}(A_n).$

1.6. Weak and strong convergence

It is said to **converge weakly** to A,

(1.51) $\quad \underset{n\to\infty}{\text{w-lim}} A_n = A \quad :\Leftrightarrow \quad A_n \psi \rightharpoonup A\psi \quad \forall \psi \in \mathfrak{D}(A) \subseteq \mathfrak{D}(A_n).$

Clearly, norm convergence implies strong convergence and strong convergence implies weak convergence.

Example. Consider the operator $S_n \in \mathfrak{L}(\ell^2(\mathbb{N}))$ which shifts a sequence n places to the left, that is,

(1.52) $\quad S_n(x_1, x_2, \dots) = (x_{n+1}, x_{n+2}, \dots),$

and the operator $S_n^* \in \mathfrak{L}(\ell^2(\mathbb{N}))$ which shifts a sequence n places to the right and fills up the first n places with zeros, that is,

(1.53) $\quad S_n^*(x_1, x_2, \dots) = (\underbrace{0, \dots, 0}_{n \text{ places}}, x_1, x_2, \dots).$

Then S_n converges to zero strongly but not in norm (since $\|S_n\| = 1$) and S_n^* converges weakly to zero (since $\langle \varphi, S_n^* \psi \rangle = \langle S_n \varphi, \psi \rangle$) but not strongly (since $\|S_n^* \psi\| = \|\psi\|$). ⋄

Note that this example also shows that taking adjoints is not continuous with respect to strong convergence! If $A_n \xrightarrow{s} A$, we only have

(1.54) $\quad \langle \varphi, A_n^* \psi \rangle = \langle A_n \varphi, \psi \rangle \to \langle A\varphi, \psi \rangle = \langle \varphi, A^* \psi \rangle$

and hence $A_n^* \rightharpoonup A^*$ in general. However, if A_n and A are normal, we have (Problem 1.14)

(1.55) $\quad \|A_n^* \psi\| = \|A_n \psi\| \to \|A\psi\| = \|A^* \psi\|$

and hence $A_n^* \xrightarrow{s} A^*$ by Lemma 1.12 (iv) in this case. Thus, at least for normal operators, taking adjoints is continuous with respect to strong convergence.

Lemma 1.14. *Suppose $A_n \in \mathfrak{L}(\mathfrak{H})$ is a sequence of bounded operators.*

(i) *$\underset{n\to\infty}{\text{s-lim}} A_n = A$ implies $\|A\| \leq \underset{n\to\infty}{\liminf} \|A_n\|$.*

(ii) *Every strong Cauchy sequence A_n is bounded: $\|A_n\| \leq C$.*

(iii) *If $A_n \psi \to A\psi$ for ψ in some dense set and $\|A_n\| \leq C$, then $\underset{n\to\infty}{\text{s-lim}} A_n = A$.*

The same result holds if strong convergence is replaced by weak convergence.

Proof. (i) follows from

$$\|A\psi\| = \lim_{n\to\infty} \|A_n \psi\| \leq \liminf_{n\to\infty} \|A_n\|$$

for every ψ with $\|\psi\| = 1$.

(ii) follows as in Lemma 1.12 (ii).

(iii) Just use

$$\|A_n\psi - A\psi\| \leq \|A_n\psi - A_n\varphi\| + \|A_n\varphi - A\varphi\| + \|A\varphi - A\psi\|$$
$$\leq 2C\|\psi - \varphi\| + \|A_n\varphi - A\varphi\|$$

and choose φ in the dense subspace such that $\|\psi - \varphi\| \leq \frac{\varepsilon}{4C}$ and n large such that $\|A_n\varphi - A\varphi\| \leq \frac{\varepsilon}{2}$.

The case of weak convergence is left as an exercise. (Hint: (2.14).) □

Lemma 1.15. *Suppose $A_n, B_n \in \mathfrak{L}(\mathfrak{H})$ are two sequences of bounded operators.*

(i) $\operatorname*{s-lim}_{n\to\infty} A_n = A$ *and* $\operatorname*{s-lim}_{n\to\infty} B_n = B$ *implies* $\operatorname*{s-lim}_{n\to\infty} A_n B_n = AB$.

(ii) $\operatorname*{w-lim}_{n\to\infty} A_n = A$ *and* $\operatorname*{s-lim}_{n\to\infty} B_n = B$ *implies* $\operatorname*{w-lim}_{n\to\infty} A_n B_n = AB$.

(iii) $\lim_{n\to\infty} A_n = A$ *and* $\operatorname*{w-lim}_{n\to\infty} B_n = B$ *implies* $\operatorname*{w-lim}_{n\to\infty} A_n B_n = AB$.

Proof. For the first case, just observe

$$\|(A_n B_n - AB)\psi\| \leq \|(A_n - A)B\psi\| + \|A_n\|\|(B_n - B)\psi\| \to 0.$$

The remaining cases are similar and again left as an exercise. □

Example. Consider again the last example. Then

$$S_n^* S_n (x_1, x_2, \dots) = (\underbrace{0, \dots, 0}_{n \text{ places}}, x_{n+1}, x_{n+2}, \dots)$$

converges to 0 weakly (in fact even strongly) but

$$S_n S_n^*(x_1, x_2, \dots) = (x_1, x_2, \dots)$$

does not! Hence the order in the second claim is important. ⋄

Problem 1.17. *Suppose $\psi_n \to \psi$ and $\varphi_n \rightharpoonup \varphi$. Then $\langle \psi_n, \varphi_n \rangle \to \langle \psi, \varphi \rangle$.*

Problem 1.18. *Show that $\psi_n \rightharpoonup \psi$ implies $A\psi_n \rightharpoonup A\psi$ for $A \in \mathfrak{L}(\mathfrak{H})$.*

Problem 1.19. *Let $\{\varphi_j\}_{j=1}^\infty$ be some orthonormal basis. Show that $\psi_n \rightharpoonup \psi$ if and only if ψ_n is bounded and $\langle \varphi_j, \psi_n \rangle \to \langle \varphi_j, \psi \rangle$ for every j. Show that this is wrong without the boundedness assumption.*

Problem 1.20. *A subspace $M \subseteq \mathfrak{H}$ is closed if and only if every weak Cauchy sequence in M has a limit in M. (Hint: $\overline{M} = M^{\perp\perp}$.)*

1.7. Appendix: The Stone–Weierstraß theorem

In case of a self-adjoint operator, the spectral theorem will show that the closed $*$-subalgebra generated by this operator is isomorphic to the C^* algebra of continuous functions $C(K)$ over some compact set $K \subseteq \mathbb{C}$. Hence it is important to be able to identify dense sets. We will be slightly more general and assume that K is some compact metric space. Then it is straightforward to check that the same proof as in the case $K = [a, b]$ (Section 0.2) shows that $C(K, \mathbb{R})$ and $C(K) = C(K, \mathbb{C})$ are Banach spaces when equipped with the maximum norm $\|f\|_\infty = \max_{x \in K} |f(x)|$.

Theorem 1.16 (Stone–Weierstraß, real version). *Suppose K is a compact metric space and let $C(K, \mathbb{R})$ be the Banach algebra of continuous functions (with the maximum norm).*

If $F \subset C(K, \mathbb{R})$ contains the identity 1 and separates points (i.e., for every $x_1 \neq x_2$ there is some function $f \in F$ such that $f(x_1) \neq f(x_2)$), then the algebra generated by F is dense.

Proof. Denote by A the algebra generated by F. Note that if $f \in \overline{A}$, we have $|f| \in \overline{A}$: By the Weierstraß approximation theorem (Theorem 0.20) there is a polynomial $p_n(t)$ such that $\bigl||t| - p_n(t)\bigr| < \frac{1}{n}$ for $t \in f(K)$ and hence $p_n(f) \to |f|$.

In particular, if f, g are in $\overline{A}$, we also have

$$\max\{f, g\} = \frac{(f+g) + |f-g|}{2}, \quad \min\{f, g\} = \frac{(f+g) - |f-g|}{2}$$

in $\overline{A}$.

Now fix $f \in C(K, \mathbb{R})$. We need to find some $f_\varepsilon \in \overline{A}$ with $\|f - f_\varepsilon\|_\infty < \varepsilon$.

First of all, since A separates points, observe that for given $y, z \in K$ there is a function $f_{y,z} \in A$ such that $f_{y,z}(y) = f(y)$ and $f_{y,z}(z) = f(z)$ (show this). Next, for every $y \in K$ there is a neighborhood $U(y)$ such that

$$f_{y,z}(x) > f(x) - \varepsilon, \quad x \in U(y),$$

and since K is compact, finitely many, say $U(y_1), \ldots, U(y_j)$, cover K. Then

$$f_z = \max\{f_{y_1,z}, \ldots, f_{y_j,z}\} \in \overline{A}$$

and satisfies $f_z > f - \varepsilon$ by construction. Since $f_z(z) = f(z)$ for every $z \in K$, there is a neighborhood $V(z)$ such that

$$f_z(x) < f(x) + \varepsilon, \quad x \in V(z),$$

and a corresponding finite cover $V(z_1), \ldots, V(z_k)$. Now

$$f_\varepsilon = \min\{f_{z_1}, \ldots, f_{z_k}\} \in \overline{A}$$

satisfies $f_\varepsilon < f + \varepsilon$. Since $f - \varepsilon < f_{z_l} < f_\varepsilon$, we have found a required function. □

Theorem 1.17 (Stone–Weierstraß)**.** *Suppose K is a compact metric space and let $C(K)$ be the C^* algebra of continuous functions (with the maximum norm).*

If $F \subset C(K)$ contains the identity 1 and separates points, then the $*$-subalgebra generated by F is dense.

Proof. Just observe that $\tilde{F} = \{\text{Re}(f), \text{Im}(f) | f \in F\}$ satisfies the assumption of the real version. Hence every real-valued continuous function can be approximated by elements from $\tilde{F}$; in particular, this holds for the real and imaginary parts for every given complex-valued function. □

Note that the additional requirement of being closed under complex conjugation is crucial: The functions holomorphic on the unit ball and continuous on the boundary separate points, but they are not dense (since the uniform limit of holomorphic functions is again holomorphic).

Corollary 1.18. *Suppose K is a compact set and let $C(K)$ be the C^* algebra of continuous functions (with the maximum norm).*

If $F \subset C(K)$ separates points, then the closure of the $*$-subalgebra generated by F is either $C(K)$ or $\{f \in C(K) | f(t_0) = 0\}$ for some $t_0 \in K$.

Proof. There are two possibilities: either all $f \in F$ vanish at one point $t_0 \in K$ (there can be at most one such point since F separates points) or there is no such point. If there is no such point, we can proceed as in the proof of the Stone–Weierstraß theorem to show that the identity can be approximated by elements in $\overline{A}$ (note that to show $|f| \in \overline{A}$ if $f \in \overline{A}$, we do not need the identity, since p_n can be chosen to contain no constant term). If there is such a t_0, the identity is clearly missing from $\overline{A}$. However, adding the identity to $\overline{A}$, we get $\overline{A} + \mathbb{C} = C(K)$, and it is easy to see that $\overline{A} = \{f \in C(K) | f(t_0) = 0\}$. □

Problem 1.21. Show that the functions $\varphi_n(x) = \frac{1}{\sqrt{2\pi}} e^{inx}$, $n \in \mathbb{Z}$, form an orthonormal basis for $\mathfrak{H} = L^2(0, 2\pi)$.

Problem 1.22. Let $k \in \mathbb{N}$ and $I \subseteq \mathbb{R}$. Show that the $*$-subalgebra generated by $f_{z_0}(t) = \frac{1}{(t-z_0)^k}$ for one $z_0 \in \mathbb{C}$ is dense in the C^* algebra $C_\infty(I)$ of continuous functions vanishing at infinity:

- for $I = \mathbb{R}$ if $z_0 \in \mathbb{C} \setminus \mathbb{R}$ and $k = 1$ or $k = 2$,
- for $I = [a, \infty)$ if $z_0 \in (-\infty, a)$ and k arbitrary,
- for $I = (-\infty, a] \cup [b, \infty)$ if $z_0 \in (a, b)$ and k odd.

1.7. Appendix: The Stone–Weierstraß theorem

(Hint: Add ∞ to $\mathbb{R}$ to make it compact.)

Problem 1.23. Let $U \subseteq \mathbb{C}\setminus\mathbb{R}$ be a set which has a limit point and is symmetric under complex conjugation. Show that the span of $\{(t-z)^{-1}|z \in U\}$ is dense in $C^\infty(\mathbb{R})$. (Hint: The product of two such functions is in the span provided they are different.)

Problem 1.24. Let $K \subseteq \mathbb{C}$ be a compact set. Show that the set of all functions $f(z) = p(x,y)$, where $p: \mathbb{R}^2 \to \mathbb{C}$ is polynomial and $z = x + iy$, is dense in $C(K)$.

Chapter 2

Self-adjointness and spectrum

2.1. Some quantum mechanics

In quantum mechanics, a single particle living in $\mathbb{R}^3$ is described by a complex-valued function (the **wave function**)

$$\psi(x,t), \qquad (x,t) \in \mathbb{R}^3 \times \mathbb{R}, \tag{2.1}$$

where x corresponds to a point in space and t corresponds to time. The quantity $\rho_t(x) = |\psi(x,t)|^2$ is interpreted as the **probability density** of the particle at the time t. In particular, ψ must be normalized according to

$$\int_{\mathbb{R}^3} |\psi(x,t)|^2 d^3x = 1, \qquad t \in \mathbb{R}. \tag{2.2}$$

The location x of the particle is a quantity which can be observed (i.e., measured) and is hence called **observable**. Due to our probabilistic interpretation, it is also a random variable whose **expectation** is given by

$$\mathbb{E}_\psi(x) = \int_{\mathbb{R}^3} x |\psi(x,t)|^2 d^3x. \tag{2.3}$$

In a real-life setting, it will not be possible to measure x directly, and one will only be able to measure certain functions of x. For example, it is possible to check whether the particle is inside a certain area Ω of space (e.g., inside a detector). The corresponding observable is the characteristic function $\chi_\Omega(x)$ of this set. In particular, the number

$$\mathbb{E}_\psi(\chi_\Omega) = \int_{\mathbb{R}^3} \chi_\Omega(x) |\psi(x,t)|^2 d^3x = \int_\Omega |\psi(x,t)|^2 d^3x \tag{2.4}$$

corresponds to the probability of finding the particle inside $\Omega \subseteq \mathbb{R}^3$ at time $t \in \mathbb{R}$. An important point to observe is that, in contradistinction to classical mechanics, the particle is no longer localized at a certain point. In particular, the **mean-square deviation** (or **variance**) $\Delta_\psi(x)^2 = \mathbb{E}_\psi(x^2) - \mathbb{E}_\psi(x)^2$ is always nonzero.

In general, the **configuration space** (or **phase space**) of a quantum system is a (complex) Hilbert space $\mathfrak{H}$, and the possible states of this system are represented by the elements ψ having norm one, $\|\psi\| = 1$.

An observable a corresponds to a linear operator A in this Hilbert space, and its expectation, if the system is in the state ψ, is given by the real number

$$(2.5) \qquad \mathbb{E}_\psi(A) = \langle \psi, A\psi \rangle = \langle A\psi, \psi \rangle,$$

where $\langle ., .. \rangle$ denotes the scalar product of $\mathfrak{H}$. Similarly, the mean-square deviation is given by

$$(2.6) \qquad \Delta_\psi(A)^2 = \mathbb{E}_\psi(A^2) - \mathbb{E}_\psi(A)^2 = \|(A - \mathbb{E}_\psi(A))\psi\|^2.$$

Note that $\Delta_\psi(A)$ vanishes if and only if ψ is an eigenstate corresponding to the eigenvalue $\mathbb{E}_\psi(A)$; that is, $A\psi = \mathbb{E}_\psi(A)\psi$.

From a physical point of view, (2.5) should make sense for every $\psi \in \mathfrak{H}$. However, this is not in the cards as our simple example of one particle already shows. In fact, the reader is invited to find a square integrable function $\psi(x)$ for which $x\psi(x)$ is no longer square integrable. The deeper reason behind this nuisance is that $\mathbb{E}_\psi(x)$ can attain arbitrarily large values if the particle is not confined to a finite domain, which renders the corresponding operator unbounded. But unbounded operators cannot be defined on the entire Hilbert space in a natural way by the closed graph theorem (Theorem 2.9 below).

Hence, A will only be defined on a subset $\mathfrak{D}(A) \subseteq \mathfrak{H}$, called the **domain** of A. Since we want A to be defined for at least *most* states, we require $\mathfrak{D}(A)$ to be dense.

However, it should be noted that there is no general prescription for how to find the operator corresponding to a given observable.

Now let us turn to the time evolution of such a quantum mechanical system. Given an initial state $\psi(0)$ of the system, there should be a unique $\psi(t)$ representing the state of the system at time $t \in \mathbb{R}$. We will write

$$(2.7) \qquad \psi(t) = U(t)\psi(0).$$

Moreover, it follows from physical experiments that **superposition of states** holds; that is, $U(t)(\alpha_1\psi_1(0) + \alpha_2\psi_2(0)) = \alpha_1\psi_1(t) + \alpha_2\psi_2(t)$. In other words, $U(t)$ should be a linear operator. Moreover, since $\psi(t)$ is a

2.1. Some quantum mechanics

state (i.e., $\|\psi(t)\| = 1$), we have

(2.8) $$\|U(t)\psi\| = \|\psi\|.$$

Such operators are called **unitary**. Next, since we have assumed uniqueness of solutions to the initial value problem, we must have

(2.9) $$U(0) = \mathbb{I}, \qquad U(t+s) = U(t)U(s).$$

A family of unitary operators $U(t)$ having this property is called a **one-parameter unitary group**. In addition, it is natural to assume that this group is **strongly continuous**; that is,

(2.10) $$\lim_{t \to t_0} U(t)\psi = U(t_0)\psi, \qquad \psi \in \mathfrak{H}.$$

Each such group has an **infinitesimal generator**, defined by

(2.11) $$H\psi = \lim_{t \to 0} \frac{\mathrm{i}}{t}(U(t)\psi - \psi), \quad \mathfrak{D}(H) = \{\psi \in \mathfrak{H} | \lim_{t \to 0} \frac{1}{t}(U(t)\psi - \psi) \text{ exists}\}.$$

This operator is called the **Hamiltonian** and corresponds to the energy of the system. If $\psi(0) \in \mathfrak{D}(H)$, then $\psi(t)$ is a solution of the **Schrödinger equation** (in suitable units)

(2.12) $$\mathrm{i}\frac{d}{dt}\psi(t) = H\psi(t).$$

This equation will be the main subject of our course.

In summary, we have the following **axioms of quantum mechanics**.

Axiom 1. The configuration space of a quantum system is a complex separable Hilbert space $\mathfrak{H}$, and the possible states of this system are represented by the elements of $\mathfrak{H}$ which have norm one.

Axiom 2. Each observable a corresponds to a linear operator A defined maximally on a dense subset $\mathfrak{D}(A)$. Moreover, the operator corresponding to a polynomial $P_n(a) = \sum_{j=0}^n \alpha_j a^j$, $\alpha_j \in \mathbb{R}$, is $P_n(A) = \sum_{j=0}^n \alpha_j A^j$, $\mathfrak{D}(P_n(A)) = \mathfrak{D}(A^n) = \{\psi \in \mathfrak{D}(A)|A\psi \in \mathfrak{D}(A^{n-1})\}$ ($A^0 = \mathbb{I}$).

Axiom 3. The expectation value for a measurement of a, when the system is in the state $\psi \in \mathfrak{D}(A)$, is given by (2.5), which must be real for all $\psi \in \mathfrak{D}(A)$.

Axiom 4. The time evolution is given by a strongly continuous one-parameter unitary group $U(t)$. The generator of this group corresponds to the energy of the system.

In the following sections we will try to draw some mathematical consequences from these assumptions:

First, we will see that Axioms 2 and 3 imply that observables correspond to self-adjoint operators. Hence these operators play a central role in quantum mechanics and we will derive some of their basic properties. Another crucial role is played by the set of all possible expectation values for the measurement of a, which is connected with the spectrum $\sigma(A)$ of the corresponding operator A.

The problem of defining functions of an observable will lead us to the spectral theorem (in the next chapter), which generalizes the diagonalization of symmetric matrices.

Axiom 4 will be the topic of Chapter 5.

2.2. Self-adjoint operators

Let $\mathfrak{H}$ be a (complex separable) Hilbert space. A **linear operator** is a linear mapping

$$(2.13) \qquad A : \mathfrak{D}(A) \to \mathfrak{H},$$

where $\mathfrak{D}(A)$ is a linear subspace of $\mathfrak{H}$, called the **domain** of A. It is called **bounded** if the operator norm

$$(2.14) \qquad \|A\| = \sup_{\|\psi\|=1} \|A\psi\| = \sup_{\|\varphi\|=\|\psi\|=1} |\langle \psi, A\varphi \rangle|$$

is finite. The second equality follows since equality in $|\langle \psi, A\varphi \rangle| \leq \|\psi\| \, \|A\varphi\|$ is attained when $A\varphi = z\psi$ for some $z \in \mathbb{C}$. If A is bounded, it is no restriction to assume $\mathfrak{D}(A) = \mathfrak{H}$ and we will usually do so. The Banach space of all bounded linear operators is denoted by $\mathfrak{L}(\mathfrak{H})$. Sums and products of (unbounded) operators are defined naturally; that is, $(A+B)\psi = A\psi + B\psi$ for $\psi \in \mathfrak{D}(A+B) = \mathfrak{D}(A) \cap \mathfrak{D}(B)$ and $AB\psi = A(B\psi)$ for $\psi \in \mathfrak{D}(AB) = \{\psi \in \mathfrak{D}(B) | B\psi \in \mathfrak{D}(A)\}$.

Example. (Multiplication operator). Consider the multiplication operator

$$(2.15) \quad (Af)(x) = A(x)f(x), \quad \mathfrak{D}(A) = \{f \in L^2(\mathbb{R}^d, d\mu) \,|\, Af \in L^2(\mathbb{R}^d, d\mu)\}$$

given by multiplication with the measurable function $A : \mathbb{R}^n \to \mathbb{C}$. First of all note that $\mathfrak{D}(A)$ is dense. In fact, consider $\Omega_n = \{x \in \mathbb{R}^d \,|\, |A(x)| \leq n\} \nearrow \mathbb{R}^d$. Then, for every $f \in L^2(\mathbb{R}^d, d\mu)$ the function $f_n = \chi_{\Omega_n} f \in \mathfrak{D}(A)$ converges to f as $n \to \infty$ by dominated convergence.

Moreover, A is a bounded operator if and only if $A(x)$ is an (essentially) bounded function and $\|A\| = \|A\|_\infty$ in this case.

If $\|A\|_\infty < \infty$ we of course have $|A(x)| \leq \|A\|_\infty$ (a.e.) and $|(Af)(x)|^2 = |A(x)|^2 |f(x)|^2 \leq \|A\|_\infty |f(x)|^2$ shows $\mathfrak{D}(A) = \mathfrak{H}$. Furthermore,

$$\|Af\|^2 = \int_{\mathbb{R}^d} |A(x)f(x)|^2 d\mu(x) \leq \|A\|_\infty^2 \int_{\mathbb{R}^d} |f(x)|^2 d\mu(x) = \|A\|_\infty^2 \|f\|^2$$

shows $\|A\| \leq \|A\|_\infty$. To see the converse inequality, consider the set $\Omega_\varepsilon = \{x \,|\, |A(x)| \geq \|A\|_\infty - \varepsilon\}$ which satisfies $\mu(\Omega_\varepsilon) > 0$ for $\varepsilon > 0$. Thus we can

2.2. Self-adjoint operators

choose $\tilde{\Omega}_\varepsilon \subseteq \Omega_\varepsilon$ with $0 < \mu(\tilde{\Omega}_\varepsilon) < \infty$, implying

$$\|A\chi_{\tilde{\Omega}_\varepsilon}\|^2 = \int_{\tilde{\Omega}_\varepsilon} |A(x)|^2 d\mu(x) \geq (\|A\|_\infty - \varepsilon)^2 \|\chi_{\tilde{\Omega}_\varepsilon}\|^2$$

which shows $\|A\| \geq \|A\|_\infty$. $\diamond$

The expression $\langle \psi, A\psi \rangle$ encountered in the previous section is called the **quadratic form**,

(2.16) $\qquad q_A(\psi) = \langle \psi, A\psi \rangle, \qquad \psi \in \mathfrak{D}(A),$

associated to A. An operator can be reconstructed from its quadratic form via the **polarization identity**

(2.17) $\langle \varphi, A\psi \rangle = \dfrac{1}{4} \left(q_A(\varphi + \psi) - q_A(\varphi - \psi) + iq_A(\varphi - i\psi) - iq_A(\varphi + i\psi) \right).$

A densely defined linear operator A is called **symmetric** (or **hermitian**) if

(2.18) $\qquad \langle \varphi, A\psi \rangle = \langle A\varphi, \psi \rangle, \qquad \psi, \varphi \in \mathfrak{D}(A).$

The justification for this definition is provided by the following

Lemma 2.1. *A densely defined operator A is symmetric if and only if the corresponding quadratic form is real-valued.*

Proof. Clearly (2.18) implies that $\text{Im}(q_A(\psi)) = 0$. Conversely, taking the imaginary part of the identity

$$q_A(\psi + i\varphi) = q_A(\psi) + q_A(\varphi) + i(\langle \psi, A\varphi \rangle - \langle \varphi, A\psi \rangle)$$

shows $\text{Re}\langle A\varphi, \psi \rangle = \text{Re}\langle \varphi, A\psi \rangle$. Replacing φ by $i\varphi$ in this last equation shows $\text{Im}\langle A\varphi, \psi \rangle = \text{Im}\langle \varphi, A\psi \rangle$ and finishes the proof. $\square$

In other words, a densely defined operator A is symmetric if and only if

(2.19) $\qquad \langle \psi, A\psi \rangle = \langle A\psi, \psi \rangle, \qquad \psi \in \mathfrak{D}(A).$

This already narrows the class of admissible operators to the class of symmetric operators by Axiom 3. Next, let us tackle the issue of the correct domain.

By Axiom 2, A should be defined maximally; that is, if $\tilde{A}$ is another symmetric operator such that $A \subseteq \tilde{A}$, then $A = \tilde{A}$. Here we write $A \subseteq \tilde{A}$ if $\mathfrak{D}(A) \subseteq \mathfrak{D}(\tilde{A})$ and $A\psi = \tilde{A}\psi$ for all $\psi \in \mathfrak{D}(A)$. The operator $\tilde{A}$ is called an **extension** of A in this case. In addition, we write $A = \tilde{A}$ if both $\tilde{A} \subseteq A$ and $A \subseteq \tilde{A}$ hold.

The **adjoint operator** A^* of a densely defined linear operator A is defined by

(2.20) $\begin{aligned} \mathfrak{D}(A^*) &= \{\psi \in \mathfrak{H} | \exists \tilde{\psi} \in \mathfrak{H} : \langle \psi, A\varphi \rangle = \langle \tilde{\psi}, \varphi \rangle, \forall \varphi \in \mathfrak{D}(A)\}, \\ A^*\psi &= \tilde{\psi}. \end{aligned}$

The requirement that $\mathfrak{D}(A)$ be dense implies that A^* is well-defined. However, note that $\mathfrak{D}(A^*)$ might not be dense in general. In fact, it might contain no vectors other than 0.

Clearly we have $(\alpha A)^* = \alpha^* A^*$ for $\alpha \in \mathbb{C}$ and $(A + B)^* \supseteq A^* + B^*$ provided $\mathfrak{D}(A + B) = \mathfrak{D}(A) \cap \mathfrak{D}(B)$ is dense. However, equality will not hold in general unless one operator is bounded (Problem 2.2).

For later use, note that (Problem 2.4)

$$\text{Ker}(A^*) = \text{Ran}(A)^\perp. \tag{2.21}$$

For symmetric operators, we clearly have $A \subseteq A^*$. If, in addition, $A = A^*$ holds, then A is called **self-adjoint**. Our goal is to show that observables correspond to self-adjoint operators. This is for example true in the case of the position operator x, which is a special case of a multiplication operator.

Example. (Multiplication operator). Consider again the multiplication operator

$$(Af)(x) = A(x)f(x), \quad \mathfrak{D}(A) = \{f \in L^2(\mathbb{R}^d, d\mu) \,|\, Af \in L^2(\mathbb{R}^d, d\mu)\} \tag{2.22}$$

given by multiplication with the measurable function $A : \mathbb{R}^d \to \mathbb{C}$ and let us compute its adjoint.

Performing a formal computation, we have for $h, f \in \mathfrak{D}(A)$ that

$$\langle h, Af \rangle = \int h(x)^* A(x) f(x) d\mu(x) = \int (A(x)^* h(x))^* f(x) d\mu(x) = \langle \tilde{A}h, f \rangle, \tag{2.23}$$

where $\tilde{A}$ is multiplication by $A(x)^*$,

$$(\tilde{A}f)(x) = A(x)^* f(x), \quad \mathfrak{D}(\tilde{A}) = \{f \in L^2(\mathbb{R}^d, d\mu) \,|\, \tilde{A}f \in L^2(\mathbb{R}^d, d\mu)\}. \tag{2.24}$$

Note $\mathfrak{D}(\tilde{A}) = \mathfrak{D}(A)$. At first sight this seems to show that the adjoint of A is $\tilde{A}$. But for our calculation we had to assume $h \in \mathfrak{D}(A)$, and there might be some functions in $\mathfrak{D}(A^*)$ which do not satisfy this requirement! In particular, our calculation only shows $\tilde{A} \subseteq A^*$. To show that equality holds, we need to work a little harder:

If $h \in \mathfrak{D}(A^*)$, there is some $g \in L^2(\mathbb{R}^d, d\mu)$ such that

$$\int h(x)^* A(x) f(x) d\mu(x) = \int g(x)^* f(x) d\mu(x), \quad f \in \mathfrak{D}(A), \tag{2.25}$$

and thus

$$\int (h(x) A(x)^* - g(x))^* f(x) d\mu(x) = 0, \quad f \in \mathfrak{D}(A). \tag{2.26}$$

In particular, using $\Omega_n = \{x \in \mathbb{R}^d \,|\, |A(x)| \le n\} \nearrow \mathbb{R}^d$,

$$\int \chi_{\Omega_n}(x)(h(x) A(x)^* - g(x))^* f(x) d\mu(x) = 0, \quad f \in L^2(\mathbb{R}^d, d\mu), \tag{2.27}$$

2.2. Self-adjoint operators

which shows that $\chi_{\Omega_n}(h(x)A(x)^* - g(x))^* \in L^2(\mathbb{R}^d, d\mu)$ vanishes. Since n is arbitrary, we even have $h(x)A(x)^* = g(x) \in L^2(\mathbb{R}^d, d\mu)$ and thus A^* is multiplication by $A(x)^*$ and $\mathfrak{D}(A^*) = \mathfrak{D}(A)$.

In particular, A is self-adjoint if A is real-valued. In the general case we have at least $\|Af\| = \|A^*f\|$ for all $f \in \mathfrak{D}(A) = \mathfrak{D}(A^*)$. Such operators are called **normal**. ◇

Now note that

$$(2.28) \qquad A \subseteq B \quad \Rightarrow \quad B^* \subseteq A^*;$$

that is, increasing the domain of A implies decreasing the domain of A^*. Thus there is no point in trying to extend the domain of a self-adjoint operator further. In fact, if A is self-adjoint and B is a symmetric extension, we infer $A \subseteq B \subseteq B^* \subseteq A^* = A$ implying $A = B$.

Corollary 2.2. *Self-adjoint operators are maximal; that is, they do not have any symmetric extensions.*

Furthermore, if A^* is densely defined (which is the case if A is symmetric), we can consider A^{**}. From the definition (2.20) it is clear that $A \subseteq A^{**}$ and thus A^{**} is an extension of A. This extension is closely related to extending a linear subspace M via $M^{\perp\perp} = \overline{M}$ (as we will see a bit later) and thus is called the **closure** $\overline{A} = A^{**}$ of A.

If A is symmetric, we have $A \subseteq A^*$ and hence $\overline{A} = A^{**} \subseteq A^*$; that is, $\overline{A}$ lies between A and A^*. Moreover, $\langle \psi, A^*\varphi \rangle = \langle \overline{A}\psi, \varphi \rangle$ for all $\psi \in \mathfrak{D}(\overline{A})$, $\varphi \in \mathfrak{D}(A^*)$ implies that $\overline{A}$ is symmetric since $A^*\varphi = \overline{A}\varphi$ for $\varphi \in \mathfrak{D}(\overline{A})$.

Example. (Differential operator). Take $\mathfrak{H} = L^2(0, 2\pi)$.

(i) Consider the operator

$$(2.29) \qquad A_0 f = -\mathrm{i}\frac{d}{dx}f, \quad \mathfrak{D}(A_0) = \{f \in C^1[0, 2\pi] \mid f(0) = f(2\pi) = 0\}.$$

That A_0 is symmetric can be shown by a simple integration by parts (do this). Note that the *boundary conditions* $f(0) = f(2\pi) = 0$ are chosen such that the boundary terms occurring from integration by parts vanish. However, this will also follow once we have computed A_0^*. If $g \in \mathfrak{D}(A_0^*)$, we must have

$$(2.30) \qquad \int_0^{2\pi} g(x)^*(-\mathrm{i}f'(x))dx = \int_0^{2\pi} \tilde{g}(x)^* f(x) dx$$

for some $\tilde{g} \in L^2(0, 2\pi)$. Integration by parts (cf. (2.119)) shows

$$(2.31) \qquad \int_0^{2\pi} f'(x)\left(g(x) - \mathrm{i}\int_0^x \tilde{g}(t)dt\right)^* dx = 0.$$

In fact, this formula holds for $\tilde{g} \in C[0, 2\pi]$. Since the set of continuous functions is dense, the general case $\tilde{g} \in L^2(0, 2\pi)$ follows by approximating $\tilde{g}$ with continuous functions and taking limits on both sides using dominated convergence.

Hence $g(x) - \mathrm{i}\int_0^x \tilde{g}(t)dt \in \{f'|f \in \mathfrak{D}(A_0)\}^\perp$. But $\{f'|f \in \mathfrak{D}(A_0)\} = \{h \in C[0, 2\pi]| \int_0^{2\pi} h(t)dt = 0\}$ (show this) implying $g(x) = g(0) + \mathrm{i}\int_0^x \tilde{g}(t)dt$ since $\overline{\{f'|f \in \mathfrak{D}(A_0)\}} = \{h \in \mathfrak{H}|\langle 1, h\rangle = 0\} = \{1\}^\perp$ and $\{1\}^{\perp\perp} = \mathrm{span}\{1\}$. Thus $g \in AC[0, 2\pi]$, where

$$(2.32) \qquad AC[a, b] = \{f \in C[a, b]| f(x) = f(a) + \int_a^x g(t)dt,\ g \in L^1(a, b)\}$$

denotes the set of all absolutely continuous functions (see Section 2.7). In summary, $g \in \mathfrak{D}(A_0^*)$ implies $g \in AC[0, 2\pi]$ and $A_0^* g = \tilde{g} = -\mathrm{i}g'$. Conversely, for every $g \in H^1(0, 2\pi) = \{f \in AC[0, 2\pi]| f' \in L^2(0, 2\pi)\}$, (2.30) holds with $\tilde{g} = -\mathrm{i}g'$ and we conclude

$$(2.33) \qquad A_0^* f = -\mathrm{i}\frac{d}{dx}f, \quad \mathfrak{D}(A_0^*) = H^1(0, 2\pi).$$

In particular, A_0 is symmetric but not self-adjoint. Since $\overline{A_0} = A_0^{**} \subseteq A_0^*$, we can use integration by parts to compute

$$(2.34) \qquad 0 = \langle g, \overline{A_0}f\rangle - \langle A_0^* g, f\rangle = \mathrm{i}(f(0)g(0)^* - f(2\pi)g(2\pi)^*)$$

and since the boundary values of $g \in \mathfrak{D}(A_0^*)$ can be prescribed arbitrarily, we must have $f(0) = f(2\pi) = 0$. Thus

$$(2.35) \qquad \overline{A_0}f = -\mathrm{i}\frac{d}{dx}f, \quad \mathfrak{D}(\overline{A_0}) = \{f \in \mathfrak{D}(A_0^*)\,|\,f(0) = f(2\pi) = 0\}.$$

(ii) Now let us take

$$(2.36) \qquad Af = -\mathrm{i}\frac{d}{dx}f, \quad \mathfrak{D}(A) = \{f \in C^1[0, 2\pi]\,|\,f(0) = f(2\pi)\},$$

which is clearly an extension of A_0. Thus $A^* \subseteq A_0^*$ and we compute

$$(2.37) \qquad 0 = \langle g, Af\rangle - \langle A^*g, f\rangle = \mathrm{i}f(0)(g(0)^* - g(2\pi)^*).$$

Since this must hold for all $f \in \mathfrak{D}(A)$, we conclude $g(0) = g(2\pi)$ and

$$(2.38) \qquad A^*f = -\mathrm{i}\frac{d}{dx}f, \quad \mathfrak{D}(A^*) = \{f \in H^1(0, 2\pi)\,|\,f(0) = f(2\pi)\}.$$

Similarly, as before, $\overline{A} = A^*$ and thus $\overline{A}$ is self-adjoint. ◇

One might suspect that there is no big difference between the two symmetric operators A_0 and A from the previous example, since they coincide on a dense set of vectors. However, the converse is true: For example, the first operator A_0 has no eigenvectors at all (i.e., solutions of the equation

2.2. Self-adjoint operators

$A_0\psi = z\psi$, $z \in \mathbb{C}$) whereas the second one has an orthonormal basis of eigenvectors!

Example. Compute the eigenvectors of A_0 and A from the previous example.

(i) By definition, an eigenvector is a (nonzero) solution of $A_0 u = zu$, $z \in \mathbb{C}$, that is, a solution of the ordinary differential equation

$$-i\,u'(x) = zu(x) \tag{2.39}$$

satisfying the boundary conditions $u(0) = u(2\pi) = 0$ (since we must have $u \in \mathfrak{D}(A_0)$). The general solution of the differential equation is $u(x) = u(0)e^{izx}$ and the boundary conditions imply $u(x) = 0$. Hence there are no eigenvectors.

(ii) Now we look for solutions of $Au = zu$, that is, the same differential equation as before, but now subject to the boundary condition $u(0) = u(2\pi)$. Again the general solution is $u(x) = u(0)e^{izx}$, and the boundary condition requires $u(0) = u(0)e^{2\pi i z}$. Thus there are two possibilities. Either $u(0) = 0$ (which is of no use for us) or $z \in \mathbb{Z}$. In particular, we see that all eigenvectors are given by

$$u_n(x) = \frac{1}{\sqrt{2\pi}} e^{inx}, \qquad n \in \mathbb{Z}, \tag{2.40}$$

which are well known to form an orthonormal basis. ◇

We will see a bit later that this is a consequence of self-adjointness of $\overline{A}$. Hence it will be important to know whether a given operator is self-adjoint. Our example shows that symmetry is easy to check (in case of differential operators it usually boils down to integration by parts), but computing the adjoint of an operator is a nontrivial job even in simple situations. However, we will learn soon that self-adjointness is a much stronger property than symmetry, justifying the additional effort needed to prove it.

On the other hand, if a given symmetric operator A turns out not to be self-adjoint, this raises the question of self-adjoint extensions. Two cases need to be distinguished. If $\overline{A}$ is self-adjoint, then there is only one self-adjoint extension (if B is another one, we have $\overline{A} \subseteq B$ and hence $\overline{A} = B$ by Corollary 2.2). In this case A is called **essentially self-adjoint** and $\mathfrak{D}(A)$ is called a **core** for $\overline{A}$. Otherwise there might be more than one self-adjoint extension or none at all. This situation is more delicate and will be investigated in Section 2.6.

Since we have seen that computing A^* is not always easy, a criterion for self-adjointness not involving A^* will be useful.

Lemma 2.3. *Let A be symmetric such that* $\mathrm{Ran}(A+z) = \mathrm{Ran}(A+z^*) = \mathfrak{H}$ *for one $z \in \mathbb{C}$. Then A is self-adjoint.*

Proof. Let $\psi \in \mathfrak{D}(A^*)$ and $A^*\psi = \tilde{\psi}$. Since $\operatorname{Ran}(A + z^*) = \mathfrak{H}$, there is a $\vartheta \in \mathfrak{D}(A)$ such that $(A + z^*)\vartheta = \tilde{\psi} + z^*\psi$. Now we compute

$$\langle \psi, (A+z)\varphi \rangle = \langle \tilde{\psi} + z^*\psi, \varphi \rangle = \langle (A+z^*)\vartheta, \varphi \rangle = \langle \vartheta, (A+z)\varphi \rangle, \quad \varphi \in \mathfrak{D}(A),$$

and hence $\psi = \vartheta \in \mathfrak{D}(A)$ since $\operatorname{Ran}(A + z) = \mathfrak{H}$. □

To proceed further, we will need more information on the closure of an operator. We will use a different approach which avoids the use of the adjoint operator. We will establish equivalence with our original definition in Lemma 2.4.

The simplest way of extending an operator A is to take the closure of its **graph** $\Gamma(A) = \{(\psi, A\psi) | \psi \in \mathfrak{D}(A)\} \subset \mathfrak{H}^2$. That is, if $(\psi_n, A\psi_n) \to (\psi, \tilde{\psi})$, we might try to define $A\psi = \tilde{\psi}$. For $A\psi$ to be well-defined, we need that $(\psi_n, A\psi_n) \to (0, \tilde{\psi})$ implies $\tilde{\psi} = 0$. In this case A is called **closable**, and the unique operator $\overline{A}$ which satisfies $\Gamma(\overline{A}) = \overline{\Gamma(A)}$ is called the **closure** of A. Clearly, A is called **closed** if $\overline{A} = A$, which is the case if and only if the graph of A is closed. Equivalently, A is closed if and only if $\Gamma(A)$ equipped with the **graph norm** $\|(\psi, A\psi)\|^2_{\Gamma(A)} = \|\psi\|^2 + \|A\psi\|^2$ is a Hilbert space (i.e., closed). By construction, $\overline{A}$ is the smallest closed extension of A.

Clearly we have $\overline{\alpha A} = \alpha \overline{A}$ for $\alpha \in \mathbb{C}$ and $\overline{A + B} \supseteq \overline{A} + \overline{B}$ provided A, B, and $A + B$ are closable. However, equality will not hold in general unless one operator is bounded (Problem 2.8).

Example. Suppose A is bounded. Then the closure was already computed in Theorem 0.29. In particular, $\mathfrak{D}(\overline{A}) = \overline{\mathfrak{D}(A)}$ and a bounded operator is closed if and only if its domain is closed. ◇

Example. Consider again the differential operator A_0 from (2.29) and let us compute the closure without the use of the adjoint operator.

Let $f \in \mathfrak{D}(\overline{A_0})$ and let $f_n \in \mathfrak{D}(A_0)$ be a sequence such that $f_n \to f$, $A_0 f_n \to -ig$. Then $f'_n \to g$ and hence $f(x) = \int_0^x g(t)dt$. Thus $f \in AC[0, 2\pi]$ and $f(0) = 0$. Moreover, $f(2\pi) = \lim_{n\to\infty} \int_0^{2\pi} f'_n(t)dt = 0$. Conversely, any such f can be approximated by functions in $\mathfrak{D}(A_0)$ (show this). ◇

Example. Consider again the multiplication operator by $A(x)$ in $L^2(\mathbb{R}^d, d\mu)$ but now defined on functions with compact support, that is,

(2.41) $\qquad \mathfrak{D}(A_0) = \{f \in \mathfrak{D}(A) \mid \operatorname{supp}(f) \text{ is compact}\}.$

Then its closure is given by $\overline{A_0} = A$. In particular, if $A(x)$ is real-valued, then A_0 is essentially self-adjoint and $\mathfrak{D}(A_0)$ is a core for A.

To prove $\overline{A_0} = A$, let some $f \in \mathfrak{D}(A)$ be given and consider $f_n = \chi_{\{x \mid |x| \leq n\}} f$. Then $f_n \in \mathfrak{D}(A_0)$ and $f_n(x) \to f(x)$ as well as $A(x)f_n(x) \to$

2.2. Self-adjoint operators

$A(x)f(x)$ in $L^2(\mathbb{R}^d, d\mu)$ by dominated convergence. Thus $\mathfrak{D}(A) \subseteq \mathfrak{D}(\overline{A_0})$ and since A is closed, we even get equality. $\diamond$

Example. Consider the multiplication operator by $A(x) = x$ in $L^2(\mathbb{R})$ defined on

$$\mathfrak{D}(A_0) = \{f \in \mathfrak{D}(A) \mid \int_{\mathbb{R}} f(x)dx = 0\}. \tag{2.42}$$

Then A_0 is closed. Hence $\mathfrak{D}(A_0)$ is not a core for A.

Note that $\mathfrak{D}(A_0)$ is well-defined since $f \in \mathfrak{D}(A)$ implies $f \in L^1(\mathbb{R})$ as we can write $f(x) = (1+|x|)^{-1}((1+|x|)f(x))$, where both $(1+|x|)^{-1}$, $(1+|x|)f(x) \in L^2(\mathbb{R})$. Moreover, $\mathfrak{D}(A_0)$ is dense. To see this, it suffices to show that every element of $\mathfrak{D}(A)$ can be approximated by elements from $\mathfrak{D}(A_0)$ (why?). So choose $f \in \mathfrak{D}(A)$ and consider

$$f_n(x) = \chi_{[-n,n]}(x)\left(f(x) - \frac{1}{2n}\int_{-n}^n f(t)dt\right).$$

Clearly $f_n \in \mathfrak{D}(A_0)$. Moreover, $f_n(x) \to f(x)$ pointwise for every x and

$$|f_n(x)| \leq |f(x)| + \frac{\|f\|_1}{2n}\chi_{[-n,n]}(x) \leq |f(x)| + \frac{\|f\|_1}{2}\min(1, \frac{1}{|x|}) \in L^2(\mathbb{R}).$$

Thus dominated convergence shows $f_n \to f$ in $L^2(\mathbb{R})$.

To show that A_0 is closed, suppose there is a sequence $f_n(x) \to f(x)$ such that $xf_n(x) \to g(x)$. Since A is closed, we necessarily have $f \in \mathfrak{D}(A)$ and $g(x) = xf(x)$. But then

$$0 = \lim_{n\to\infty} \int_{\mathbb{R}} f_n(x)dx = \lim_{n\to\infty} \int_{\mathbb{R}} \frac{1}{1+|x|}(f_n(x) + \text{sign}(x)xf_n(x))dx$$

$$= \int_{\mathbb{R}} \frac{1}{1+|x|}(f(x) + \text{sign}(x)g(x))dx = \int_{\mathbb{R}} f(x)dx \tag{2.43}$$

which shows $f \in \mathfrak{D}(A_0)$. $\diamond$

Next, let us collect a few important results.

Lemma 2.4. *Suppose A is a densely defined operator.*

(i) *A^* is closed.*

(ii) *A is closable if and only if $\mathfrak{D}(A^*)$ is dense and $\overline{A} = A^{**}$, respectively, $(\overline{A})^* = A^*$, in this case.*

(iii) *If A is injective and $\text{Ran}(A)$ is dense, then $(A^*)^{-1} = (A^{-1})^*$. If A is closable and $\overline{A}$ is injective, then $\overline{A}^{-1} = \overline{A^{-1}}$.*

Proof. Let us consider the following two unitary operators from $\mathfrak{H}^2$ to itself:

$$U(\varphi, \psi) = (\psi, -\varphi), \qquad V(\varphi, \psi) = (\psi, \varphi).$$

Note $U^{-1} = -U$ and $V^{-1} = V$.

(i) From
$$\Gamma(A^*) = \{(\varphi, \tilde{\varphi}) \in \mathfrak{H}^2 | \langle \varphi, A\psi \rangle = \langle \tilde{\varphi}, \psi \rangle, \forall \psi \in \mathfrak{D}(A)\}$$
$$= \{(\varphi, \tilde{\varphi}) \in \mathfrak{H}^2 | \langle (\varphi, \tilde{\varphi}), (\tilde{\psi}, -\psi) \rangle_{\mathfrak{H}^2} = 0, \forall (\psi, \tilde{\psi}) \in \Gamma(A)\}$$
$$\tag{2.44} = (U\Gamma(A))^\perp,$$
we conclude that A^* is closed.

(ii) Similarly, using $U\Gamma^\perp = (U\Gamma)^\perp$ (Problem 1.4) and (i), by
$$\overline{\Gamma(A)} = \Gamma(A)^{\perp\perp} = (U^{-1}\Gamma(A^*))^\perp$$
$$\tag{2.45} = \{(\psi, \tilde{\psi}) | \langle \psi, A^*\varphi \rangle - \langle \tilde{\psi}, \varphi \rangle = 0, \forall \varphi \in \mathfrak{D}(A^*)\},$$
we see that $(0, \tilde{\psi}) \in \overline{\Gamma(A)}$ if and only if $\tilde{\psi} \in \mathfrak{D}(A^*)^\perp$. Hence A is closable if and only if $\mathfrak{D}(A^*)$ is dense. In this case, equation (2.44) also shows $\overline{A}^* = A^*$. Moreover, replacing A by A^* in (2.44) and comparing with (2.45) shows $A^{**} = \overline{A}$.

(iii) Next note that (provided A is injective)
$$\Gamma(A^{-1}) = V\Gamma(A).$$
Hence if $\mathrm{Ran}(A)$ is dense, then $\mathrm{Ker}(A^*) = \mathrm{Ran}(A)^\perp = \{0\}$ and
$$\Gamma((A^*)^{-1}) = V\Gamma(A^*) = VU\Gamma(A)^\perp = UV\Gamma(A)^\perp = U(V\Gamma(A))^\perp$$
shows that $(A^*)^{-1} = (A^{-1})^*$. Similarly, if A is closable and $\overline{A}$ is injective, then $\overline{A}^{-1} = \overline{A^{-1}}$ by
$$\Gamma(\overline{A}^{-1}) = V\Gamma(\overline{A}) = \overline{V\Gamma(A)} = \overline{\Gamma(A^{-1})}.$$

$\square$

Corollary 2.5. *If A is self-adjoint and injective, then A^{-1} is also self-adjoint.*

Proof. Equation (2.21) in the case $A = A^*$ implies $\mathrm{Ran}(A)^\perp = \mathrm{Ker}(A) = \{0\}$ and hence (iii) is applicable. $\square$

If A is densely defined and bounded, we clearly have $\mathfrak{D}(A^*) = \mathfrak{H}$ and by Corollary 1.9, $A^* \in \mathcal{L}(\mathfrak{H})$. In particular, since $\overline{A} = A^{**}$, we obtain

Theorem 2.6. *We have $\overline{A} \in \mathcal{L}(\mathfrak{H})$ if and only if $A^* \in \mathcal{L}(\mathfrak{H})$.*

Next note the following estimates for the inverse of symmetric operators.

Lemma 2.7. *Let A be symmetric. Then $A - z$ is injective for $z \in \mathbb{C}\backslash\mathbb{R}$ with its inverse being bounded $\|(A-z)^{-1}\| \leq |\mathrm{Im}(z)|^{-1}$. Moreover, $\mathrm{Ran}(\overline{A} - z) = \overline{\mathrm{Ran}(A - z)}$ and A is closed if and only if $\mathrm{Ran}(A - z)$ is closed.*

2.2. Self-adjoint operators

If A is nonnegative, that is, $\langle \psi, A\psi \rangle \geq 0$ for all $\psi \in \mathfrak{D}(A)$, we can also admit $z \in (-\infty, 0)$ and the estimate for the inverse now reads $\|(A+\lambda)^{-1}\| \leq \lambda^{-1}$, $\lambda > 0$.

Proof. Let $z = x + \mathrm{i}y$. From

$$\|(A - z)\psi\|^2 = \|(A + x)\psi - \mathrm{i}y\psi\|^2$$
(2.46)
$$= \|(A + x)\psi\|^2 + y^2\|\psi\|^2 \geq y^2\|\psi\|^2,$$

we infer that $\mathrm{Ker}(A - z) = \{0\}$ and hence $(A - z)^{-1}$ exists. Moreover, setting $\psi = (A-z)^{-1}\varphi$ ($y \neq 0$) shows $\|(A-z)^{-1}\| \leq |y|^{-1}$. Hence $(A-z)^{-1}$ is bounded and hence the domain of its closure equals the closure of its domain. Moreover, by Lemma 2.4 (iii), $A - z$ (and hence A) is closed if and only if $(A - z)^{-1}$ is closed.

The argument for the nonnegative case with $z < 0$ is similar using $\lambda\|\psi\|^2 \leq \langle \psi, (A + \lambda)\psi \rangle \leq \|\psi\|\|(A + \lambda)\psi\|$ which shows $\|(A + \lambda)^{-1}\| \leq \lambda^{-1}$, $\lambda > 0$. □

Now we can also generalize Lemma 2.3 to the case of essentially self-adjoint operators.

Corollary 2.8. *A symmetric operator A is essentially self-adjoint if and only if one of the following conditions holds for one $z \in \mathbb{C} \backslash \mathbb{R}$:*

- $\overline{\mathrm{Ran}(A - z)} = \overline{\mathrm{Ran}(A - z^*)} = \mathfrak{H}$,
- $\mathrm{Ker}(A^* - z) = \mathrm{Ker}(A^* - z^*) = \{0\}$.

If A is nonnegative we can also admit $z \in (-\infty, 0)$.

Proof. First of all, note that by (2.21) the two conditions are equivalent. By the previous lemma our assumption implies $\mathrm{Ran}(\overline{A} - z) = \mathrm{Ran}(\overline{A} - z^*) = \mathfrak{H}$ which shows that $\overline{A}$ is self-adjoint by Lemma 2.3.

Conversely, Lemma 2.7 shows that for symmetric A we have $\mathrm{Ker}(\overline{A}-z) = \{0\}$. If $\overline{A} = A^*$ this implies $\mathrm{Ker}(A^* - z) = \{0\}$ which finishes the proof. □

In addition, we can also prove the closed graph theorem which shows that an unbounded closed operator cannot be defined on the entire Hilbert space.

Theorem 2.9 (Closed graph). *Let $\mathfrak{H}_1$ and $\mathfrak{H}_2$ be two Hilbert spaces and $A : \mathfrak{H}_1 \to \mathfrak{H}_2$ an operator defined on all of $\mathfrak{H}_1$. Then A is bounded if and only if $\Gamma(A)$ is closed.*

Proof. If A is bounded, then it is easy to see that $\Gamma(A)$ is closed. So let us assume that $\Gamma(A)$ is closed. Then A^* is well-defined and for all unit vectors

$\varphi \in \mathfrak{D}(A^*)$ we have that the linear functional $\ell_\varphi(\psi) = \langle A^*\varphi, \psi \rangle$ is pointwise bounded, that is,
$$|\ell_\varphi(\psi)| = |\langle \varphi, A\psi \rangle| \leq \|A\psi\|.$$
Hence, by the uniform boundedness principle, there is a constant C such that $\|\ell_\varphi\| = \|A^*\varphi\| \leq C$. That is, A^* is bounded and so is $A = A^{**}$. $\square$

Note that since symmetric operators are closable, they are automatically closed if they are defined on the entire Hilbert space.

Theorem 2.10 (Hellinger–Toeplitz). *A symmetric operator defined on the entire Hilbert space is bounded.*

Problem 2.1 (Jacobi operator). *Let a and b be some real-valued sequences in $\ell^\infty(\mathbb{Z})$. Consider the operator*
$$Jf_n = a_n f_{n+1} + a_{n-1} f_{n-1} + b_n f_n, \qquad f \in \ell^2(\mathbb{Z}).$$
Show that J is a bounded self-adjoint operator.

Problem 2.2. *Show that $(\alpha A)^* = \alpha^* A^*$ and $(A+B)^* \supseteq A^* + B^*$ (where $\mathfrak{D}(A^* + B^*) = \mathfrak{D}(A^*) \cap \mathfrak{D}(B^*)$) with equality if one operator is bounded. Give an example where equality does not hold.*

Problem 2.3. *Suppose AB is densely defined. Show that $(AB)^* \supseteq B^* A^*$. Moreover, if B is bounded, then $(BA)^* = A^* B^*$.*

Problem 2.4. *Show (2.21).*

Problem 2.5. *An operator is called **normal** if $\|A\psi\| = \|A^*\psi\|$ for all $\psi \in \mathfrak{D}(A) = \mathfrak{D}(A^*)$.*

Show that if A is normal, so is $A + z$ for every $z \in \mathbb{C}$.

Problem 2.6. *Show that normal operators are closed. (Hint: A^* is closed.)*

Problem 2.7. *Show that the kernel of a closed operator is closed.*

Problem 2.8. *Suppose A, B, and $A + B$ (defined on $\mathfrak{D}(A + B) = \mathfrak{D}(A) \cap \mathfrak{D}(B)$) are closable. Show that $\overline{\alpha A} = \alpha \overline{A}$ and $\overline{A + B} \supseteq \overline{A} + \overline{B}$ with equality if one operator is bounded. Give an example where equality does not hold.*

Problem 2.9. *Show that if A is closed and B is bounded, then AB is closed. Moreover, if B is injective and B^{-1} is bounded, then BA is closed.*

2.3. Quadratic forms and the Friedrichs extension

Finally we want to draw some further consequences of Axiom 2 and show that observables correspond to self-adjoint operators. Since self-adjoint operators are already maximal, the difficult part remaining is to show that an observable has at least one self-adjoint extension. There is a good way of

2.3. Quadratic forms and the Friedrichs extension

doing this for nonnegative operators and hence we will consider this case first.

A densely defined operator is called **nonnegative** (resp. **positive**) if $\langle \psi, A\psi \rangle \geq 0$ (resp. > 0 for $\psi \neq 0$) for all $\psi \in \mathfrak{D}(A)$. Recall that by Lemma 2.1 any nonnegative operator is automatically symmetric. If A is positive, the map $(\varphi, \psi) \mapsto \langle \varphi, A\psi \rangle$ is a scalar product. However, there might be sequences which are Cauchy with respect to this scalar product but not with respect to our original one. To avoid this, we introduce the scalar product

$$(2.47) \quad \langle \varphi, \psi \rangle_A = \langle \varphi, (A+1)\psi \rangle, \qquad A \geq 0,$$

defined on $\mathfrak{D}(A)$, which satisfies $\|\psi\| \leq \|\psi\|_A$. Let $\mathfrak{H}_A$ be the completion of $\mathfrak{D}(A)$ with respect to the above scalar product. We claim that $\mathfrak{H}_A$ can be regarded as a subspace of $\mathfrak{H}$; that is, $\mathfrak{D}(A) \subseteq \mathfrak{H}_A \subseteq \mathfrak{H}$.

If (ψ_n) is a Cauchy sequence in $\mathfrak{D}(A)$, then it is also Cauchy in $\mathfrak{H}$ (since $\|\psi\| \leq \|\psi\|_A$ by assumption), and hence we can identify the limit in $\mathfrak{H}_A$ with the limit of (ψ_n) regarded as a sequence in $\mathfrak{H}$. For this identification to be unique, we need to show that if $(\psi_n) \subset \mathfrak{D}(A)$ is a Cauchy sequence in $\mathfrak{H}_A$ such that $\|\psi_n\| \to 0$, then $\|\psi_n\|_A \to 0$. This follows from

$$(2.48) \quad \begin{aligned} \|\psi_n\|_A^2 &= \langle \psi_n, \psi_n - \psi_m \rangle_A + \langle \psi_n, \psi_m \rangle_A \\ &\leq \|\psi_n\|_A \|\psi_n - \psi_m\|_A + \|(A+1)\psi_n\| \|\psi_m\| \end{aligned}$$

since the right-hand side can be made arbitrarily small by first choosing n, m large such that the first term gets small (note that $\|\psi_n\|_A$ is bounded) and then further increasing m such that the second term gets small.

Clearly, the quadratic form q_A can be extended to every $\psi \in \mathfrak{H}_A$ by setting

$$(2.49) \quad q_A(\psi) = \langle \psi, \psi \rangle_A - \|\psi\|^2, \qquad \psi \in \mathfrak{Q}(A) = \mathfrak{H}_A.$$

The set $\mathfrak{Q}(A)$ is also called the **form domain** of A.

Example. (Multiplication operator). Let A be multiplication by $A(x) \geq 0$ in $L^2(\mathbb{R}^d, d\mu)$. Then

$$(2.50) \quad \mathfrak{Q}(A) = \mathfrak{D}(A^{1/2}) = \{ f \in L^2(\mathbb{R}^d, d\mu) \mid A^{1/2} f \in L^2(\mathbb{R}^d, d\mu) \}$$

and

$$(2.51) \quad q_A(f) = \int_{\mathbb{R}^d} A(x)|f(x)|^2 d\mu(x)$$

(show this). $\diamond$

Now we come to our extension result. Note that $A + 1$ is injective and the best we can hope for is that for a nonnegative extension $\tilde{A}$, the operator $\tilde{A} + 1$ is a bijection from $\mathfrak{D}(\tilde{A})$ onto $\mathfrak{H}$.

Lemma 2.11. *Suppose A is a nonnegative operator. Then there is a nonnegative extension $\tilde{A}$, given by restricting A^* to $\mathfrak{H}_A$, such that $\text{Ran}(\tilde{A}+1) = \mathfrak{H}$.*

Proof. Let us define an operator $\tilde{A}$ by
$$\mathfrak{D}(\tilde{A}) = \{\psi \in \mathfrak{H}_A | \exists \tilde{\psi} \in \mathfrak{H} : \langle \varphi, \psi \rangle_A = \langle \varphi, \tilde{\psi} \rangle, \forall \varphi \in \mathfrak{H}_A\},$$
$$\tilde{A}\psi = A^*\psi = \tilde{\psi} - \psi.$$

Since $\mathfrak{H}_A$ is dense, $\tilde{\psi}$ is well-defined. Moreover, it is straightforward to see that $\tilde{A}$ is a nonnegative extension of A and that
$$\mathfrak{D}(\tilde{A}) = \{\psi \in \mathfrak{H}_A | \exists \tilde{\psi} \in \mathfrak{H} : \langle \varphi, \psi \rangle_A = \langle \varphi, \tilde{\psi} \rangle, \forall \varphi \in \mathfrak{H}_A\} = \mathfrak{H}_A \cap \mathfrak{D}(A^*)$$
as $\mathfrak{D}(A) \subset \mathfrak{H}_A$ is dense.

It is also not hard to see that $\text{Ran}(\tilde{A}+1) = \mathfrak{H}$. Indeed, for every $\tilde{\psi} \in \mathfrak{H}$, $\varphi \mapsto \langle \tilde{\psi}, \varphi \rangle$ is a bounded linear functional on $\mathfrak{H}_A$. Hence there is an element $\psi \in \mathfrak{H}_A$ such that $\langle \tilde{\psi}, \varphi \rangle = \langle \psi, \varphi \rangle_A$ for all $\varphi \in \mathfrak{H}_A$. By the definition of $\tilde{A}$, $(\tilde{A}+1)\psi = \tilde{\psi}$ and hence $\tilde{A} + 1$ is onto. $\square$

Example. Let us take $\mathfrak{H} = L^2(0,\pi)$ and consider the operator

(2.52) $$Af = -\frac{d^2}{dx^2}f, \quad \mathfrak{D}(A) = \{f \in C^2[0,\pi] \,|\, f(0) = f(\pi) = 0\},$$

which corresponds to the one-dimensional model of a particle confined to a box.

(i) First of all, using integration by parts twice, it is straightforward to check that A is symmetric:

(2.53) $$\int_0^\pi g(x)^*(-f'')(x)dx = \int_0^\pi g'(x)^* f'(x)dx = \int_0^\pi (-g'')(x)^* f(x)dx.$$

Note that the compact support assumption ensures that the boundary terms occurring from integration by parts vanish. Moreover, the same calculation also shows that A is positive:

(2.54) $$\int_0^\pi f(x)^*(-f'')(x)dx = \int_0^\pi |f'(x)|^2 dx > 0, \quad f \neq 0.$$

(ii) Next, let us show $\mathfrak{H}_A = \{f \in H^1(0,\pi) \,|\, f(0) = f(\pi) = 0\}$. In fact, since

(2.55) $$\langle g, f \rangle_A = \int_0^\pi \left(g'(x)^* f'(x) + g(x)^* f(x) \right) dx,$$

we see that f_n is Cauchy in $\mathfrak{H}_A$ if and only if both f_n and f_n' are Cauchy in $L^2(0,\pi)$. Thus $f_n \to f$ and $f_n' \to g$ in $L^2(0,\pi)$ and $f_n(x) = \int_0^x f_n'(t)dt$ implies $f(x) = \int_0^x g(t)dt$. Thus $f \in AC[0,\pi]$. Moreover, $f(0) = 0$ is obvious and from $0 = f_n(\pi) = \int_0^\pi f_n'(t)dt$ we have $f(\pi) = \lim_{n\to\infty} \int_0^\pi f_n'(t)dt = 0$.

2.3. Quadratic forms and the Friedrichs extension

So we have $\mathfrak{H}_A \subseteq \{f \in H^1(0,\pi) \mid f(0) = f(\pi) = 0\}$. To see the converse, approximate f' by smooth functions g_n. Using $g_n - \frac{1}{\pi}\int_0^\pi g_n(t)dt$ instead of g_n, it is no restriction to assume $\int_0^\pi g_n(t)dt = 0$. Now define $f_n(x) = \int_0^x g_n(t)dt$ and note that $f_n \in \mathfrak{D}(A) \to f$.

(iii) Finally, let us compute the extension $\tilde{A}$. We have $f \in \mathfrak{D}(\tilde{A})$ if for all $g \in \mathfrak{H}_A$ there is an $\tilde{f}$ such that $\langle g, f \rangle_A = \langle g, \tilde{f} \rangle$. That is,

$$(2.56) \qquad \int_0^\pi g'(x)^* f'(x) dx = \int_0^\pi g(x)^* (\tilde{f}(x) - f(x)) dx.$$

Integration by parts on the right-hand side shows

$$(2.57) \qquad \int_0^\pi g'(x)^* f'(x) dx = -\int_0^\pi g'(x)^* \int_0^x (\tilde{f}(t) - f(t)) dt\, dx$$

or, equivalently,

$$(2.58) \qquad \int_0^\pi g'(x)^* \left(f'(x) + \int_0^x (\tilde{f}(t) - f(t)) dt \right) dx = 0.$$

Now observe $\{g' \in \mathfrak{H} \mid g \in \mathfrak{H}_A\} = \{h \in \mathfrak{H} \mid \int_0^\pi h(t)dt = 0\} = \{1\}^\perp$ and thus $f'(x) + \int_0^x (\tilde{f}(t) - f(t)) dt \in \{1\}^{\perp\perp} = \text{span}\{1\}$. So we see $f \in H^2(0,\pi) = \{f \in AC[0,\pi] \mid f' \in H^1(0,\pi)\}$ and $\tilde{A}f = -f''$. The converse is easy and hence

$$(2.59) \qquad \tilde{A}f = -\frac{d^2}{dx^2}f, \quad \mathfrak{D}(\tilde{A}) = \{f \in H^2[0,\pi] \mid f(0) = f(\pi) = 0\}.$$

◇

Now let us apply this result to operators A corresponding to observables. Since A will, in general, not satisfy the assumptions of our lemma, we will consider A^2 instead, which has a symmetric extension $\tilde{A}^2$ with $\text{Ran}(\tilde{A}^2 + 1) = \mathfrak{H}$. By our requirement for observables, A^2 is maximally defined and hence is equal to this extension. In other words, $\text{Ran}(A^2 + 1) = \mathfrak{H}$. Moreover, for every $\varphi \in \mathfrak{H}$ there is a $\psi \in \mathfrak{D}(A^2)$ such that

$$(2.60) \qquad (A-i)(A+i)\psi = (A+i)(A-i)\psi = \varphi$$

and since $(A \pm i)\psi \in \mathfrak{D}(A)$, we infer $\text{Ran}(A \pm i) = \mathfrak{H}$. As an immediate consequence we obtain

Corollary 2.12. *Observables correspond to self-adjoint operators.*

But there is another important consequence of the results which is worthwhile mentioning. A symmetric operator is called **semi-bounded**, respectively, **bounded from below**, if there exists a $\gamma \in \mathbb{R}$ such that

$$(2.61) \qquad q_A(\psi) = \langle \psi, A\psi \rangle \geq \gamma \|\psi\|^2, \qquad \psi \in \mathfrak{D}(A).$$

We will write $A \geq \gamma$ for short.

Theorem 2.13 (Friedrichs extension). *Let A be a symmetric operator which is bounded from below by γ. Then there is a self-adjoint extension*

$$(2.62) \qquad \tilde{A}\psi = A^*\psi, \qquad \mathfrak{D}(\tilde{A}) = \mathfrak{D}(A^*) \cap \mathfrak{H}_{A-\gamma}$$

which is also bounded from below by γ. Moreover, $\tilde{A}$ is the only self-adjoint extension with $\mathfrak{D}(\tilde{A}) \subseteq \mathfrak{H}_{A-\gamma}$.

Proof. If we replace A by $A - \gamma$ we can assume $\gamma = 0$ without loss of generality. Existence follows from Lemma 2.11. To see uniqueness, let $\hat{A}$ be another self-adjoint extension with $\mathfrak{D}(\hat{A}) \subseteq \mathfrak{H}_A$. Choose $\varphi \in \mathfrak{D}(A)$ and $\psi \in \mathfrak{D}(\hat{A})$. Then

$$\langle \varphi, (\hat{A}+1)\psi \rangle = \langle (A+1)\varphi, \psi \rangle = \langle \psi, (A+1)\varphi \rangle^* = \langle \psi, \varphi \rangle_A^* = \langle \varphi, \psi \rangle_A,$$

and by continuity we even get $\langle \varphi, (\hat{A}+1)\psi \rangle = \langle \varphi, \psi \rangle_A$ for every $\varphi \in \mathfrak{H}_A$. Hence by the definition of $\tilde{A}$ we have $\psi \in \mathfrak{D}(\tilde{A})$ and $\tilde{A}\psi = \hat{A}\psi$; that is, $\hat{A} \subseteq \tilde{A}$. But self-adjoint operators are maximal by Corollary 2.2 and thus $\hat{A} = \tilde{A}$. $\square$

Clearly $\mathfrak{Q}(A) = \mathfrak{H}_A$ and q_A can be defined for semi-bounded operators as before by using $\|\psi\|_A^2 = \langle \psi, (A-\gamma)\psi \rangle + \|\psi\|^2$.

In many physical applications, the converse of this result is also of importance: given a quadratic form q, when is there a corresponding operator A such that $q = q_A$?

So let $q : \mathfrak{Q} \to \mathbb{C}$ be a densely defined **quadratic form** corresponding to a sesquilinear form $s : \mathfrak{Q} \times \mathfrak{Q} \to \mathbb{C}$; that is, $q(\psi) = s(\psi, \psi)$. As with a scalar product, s can be recovered from q via the polarization identity (cf. Problem 0.22). Furthermore, as in Lemma 2.1, one can show that s is symmetric; that is, $s(\varphi, \psi) = s(\psi, \varphi)^*$, if and only if q is real-valued. In this case q will be called **hermitian**.

A hermitian form q is called **nonnegative** if $q(\psi) \geq 0$ and **semi-bounded** if $q(\psi) \geq \gamma \|\psi\|^2$ for some $\gamma \in \mathbb{R}$. As before, we can associate a scalar product $\langle \psi, \varphi \rangle_q = s(\psi, \varphi) + (1-\gamma)\langle \psi, \varphi \rangle$ and norm $\|\psi\|_q^2 = q(\psi) + (1-\gamma)\|\psi\|^2$ with every semi-bounded q and look at the completion $\mathfrak{H}_q$ of $\mathfrak{Q}$ with respect to this norm. However, since we are not assuming that q is steaming from a semi-bounded operator, we do not know whether $\mathfrak{H}_q$ can be regarded as a subspace of $\mathfrak{H}$! Hence we will call q **closable** if for every Cauchy sequence $\psi_n \in \mathfrak{Q}$ with respect to $\|.\|_q$, $\|\psi_n\| \to 0$ implies $\|\psi_n\|_q \to 0$. In this case we have $\mathfrak{H}_q \subseteq \mathfrak{H}$ and we call the extension of q to $\mathfrak{H}_q$ the closure of q. In particular, we will call q **closed** if $\mathfrak{Q} = \mathfrak{H}_q$.

Example. Let $\mathfrak{H} = L^2(0,1)$. Then

$$q(f) = |f(c)|^2, \qquad f \in C[0,1], \quad c \in [0,1],$$

2.3. Quadratic forms and the Friedrichs extension

is a well-defined nonnegative form. However, let $f_n(x) = \max(0, 1-n|x-c|)$. Then f_n is a Cauchy sequence with respect to $\|.\|_q$ such that $\|f_n\| \to 0$ but $\|f_n\|_q \to 1$. Hence q is not closable and hence also not associated with a nonnegative operator. Formally, one can interpret q as the quadratic form of the multiplication *operator* with the delta distribution at $x = c$. Exercise: Show $\mathfrak{H}_q = \mathfrak{H} \oplus \mathbb{C}$. ◇

From our previous considerations, we already know that the quadratic form q_A of a semi-bounded operator A is closable and its closure is associated with a self-adjoint operator. It turns out that the converse is also true (compare also Corollary 1.9 for the case of bounded operators):

Theorem 2.14. *To every closed semi-bounded quadratic form q there corresponds a unique self-adjoint operator A such that $\mathfrak{Q} = \mathfrak{Q}(A)$ and $q = q_A$. If s is the sesquilinear form corresponding to q, then A is given by*

$$(2.63) \quad \begin{aligned} \mathfrak{D}(A) &= \{\psi \in \mathfrak{H}_q | \exists \tilde{\psi} \in \mathfrak{H} : s(\varphi, \psi) = \langle \varphi, \tilde{\psi} \rangle, \forall \varphi \in \mathfrak{H}_q\}, \\ A\psi &= \tilde{\psi}. \end{aligned}$$

Proof. Since $\mathfrak{H}_q$ is dense, $\tilde{\psi}$ and hence A is well-defined. Moreover, replacing q by $q(.) - \gamma \|.\|^2$ and A by $A - \gamma$, it is no restriction to assume $\gamma = 0$. By construction, $q_A(\psi) = q(\psi)$ for $\psi \in \mathfrak{D}(A)$, which shows that A is nonnegative (and hence symmetric). Moreover, as in the proof of Lemma 2.11, it follows that $\text{Ran}(A+1) = \mathfrak{H}$. In particular, $(A+1)^{-1}$ exists and is bounded. Furthermore, for every $\varphi_j \in \mathfrak{H}$ we can find $\psi_j \in \mathfrak{D}(A)$ such that $\varphi_j = (A+1)\psi_j$. Finally,

$$\begin{aligned} \langle (A+1)^{-1}\varphi_1, \varphi_2 \rangle &= \langle \psi_1, (A+1)\psi_2 \rangle = s(\psi_1, \psi_2) = s(\psi_2, \psi_1)^* \\ &= \langle \psi_2, (A+1)\psi_1 \rangle^* = \langle (A+1)\psi_1, \psi_2 \rangle \\ &= \langle \varphi_1, (A+1)^{-1}\varphi_2 \rangle \end{aligned}$$

shows that $(A+1)^{-1}$ is self-adjoint and so is $A+1$ by Corollary 2.5. The rest is straightforward. □

Any subspace $\tilde{\mathfrak{Q}} \subseteq \mathfrak{Q}(A)$ which is dense with respect to $\|.\|_A$ is called a **form core** of A and uniquely determines A.

Example. We have already seen that the operator

$$(2.64) \quad Af = -\frac{d^2}{dx^2}f, \quad \mathfrak{D}(A) = \{f \in H^2[0,\pi] \,|\, f(0) = f(\pi) = 0\}$$

is associated with the closed form

$$(2.65) \quad q_A(f) = \int_0^\pi |f'(x)|^2 dx, \quad \mathfrak{Q}(A) = \{f \in H^1[0,\pi] \,|\, f(0) = f(\pi) = 0\}.$$

However, this quadratic form even makes sense on the larger form domain $\mathfrak{Q} = H^1[0,\pi]$. What is the corresponding self-adjoint operator? (See Problem 2.14.) ◇

A hermitian form q is called **bounded** if $|q(\psi)| \leq C\|\psi\|^2$ and we call

(2.66) $$\|q\| = \sup_{\|\psi\|=1} |q(\psi)|$$

the norm of q. In this case the norm $\|.\|_q$ is equivalent to $\|.\|$. Hence $\mathfrak{H}_q = \mathfrak{H}$ and the corresponding operator is bounded by the Hellinger–Toeplitz theorem (Theorem 2.10). In fact, the operator norm is equal to the norm of q (see also Problem 0.23):

Lemma 2.15. *A semi-bounded form q is bounded if and only if the associated operator A is. Moreover, in this case*

(2.67) $$\|q\| = \|A\|.$$

Proof. Using the polarization identity and the parallelogram law (Problem 0.22), we infer

$$2\operatorname{Re}\langle \varphi, A\psi\rangle = \frac{1}{2}\big(q(\psi+\varphi) - q(\psi-\varphi)\big) \leq q(\psi) + q(\varphi) \leq \|q\|(\|\psi\|^2 + \|\varphi\|^2)$$

and choosing $\varphi = \|A\psi\|^{-1} A\psi$ as well as $\|\psi\| = 1$ shows that $\|A\| \leq \|q\|$. The converse is easy. □

As a consequence we see that for symmetric operators we have

(2.68) $$\|A\| = \sup_{\|\psi\|=1} |\langle \psi, A\psi\rangle|$$

generalizing (2.14) in this case.

Problem 2.10. *Let A be invertible. Show $A > 0$ if and only if $A^{-1} > 0$.*

Problem 2.11. *Let $A = -\frac{d^2}{dx^2}$, $\mathfrak{D}(A) = \{f \in H^2(0,\pi) \,|\, f(0) = f(\pi) = 0\}$ and let $\psi(x) = \frac{1}{2\sqrt{\pi}} x(\pi - x)$. Find the error in the following argument: Since A is symmetric, we have $1 = \langle A\psi, A\psi\rangle = \langle \psi, A^2\psi\rangle = 0$.*

Problem 2.12. *Suppose A is a densely defined closed operator. Show that A^*A (with $\mathfrak{D}(A^*A) = \{\psi \in \mathfrak{D}(A) | A\psi \in \mathfrak{D}(A^*)\}$) is self-adjoint. Show $\mathfrak{Q}(A^*A) = \mathfrak{D}(A)$. (Hint: $A^*A \geq 0$.)*

Problem 2.13. *Let A and B be two closed operators. Show $\|A\psi\| = \|B\psi\|$ for all $\psi \in \mathfrak{D}(A) = \mathfrak{D}(B)$ if and only if $A^*A = B^*B$.*

Conclude that a closed operator A is normal if and only if $AA^* = A^*A$. (Hint: Problem 2.12 and Theorem 2.14.)

2.4. Resolvents and spectra

Problem 2.14. *Suppose a densely defined operator A_0 can be written as $A_0 = S^*S$, where S is a closable operator with $\mathfrak{D}(S) = \mathfrak{D}(A_0)$. Show that the Friedrichs extension is given by $A = S^*\overline{S}$.*

Use this to compute the Friedrichs extension of $A_0 = -\frac{d^2}{dx^2}$, $\mathfrak{D}(A) = \{f \in C^2(0, \pi) | f(0) = f(\pi) = 0\}$. Compute also the self-adjoint operator $\overline{S}S^*$ and its form domain.

Problem 2.15. *Use the previous problem to compute the Friedrichs extension A of $A_0 = -\frac{d^2}{dx^2}$, $\mathfrak{D}(A_0) = C_c^\infty(\mathbb{R})$. Show that $\mathfrak{Q}(A) = H^1(\mathbb{R})$ and $\mathfrak{D}(A) = H^2(\mathbb{R})$. (Hint: Section 2.7.)*

Problem 2.16. *Let A be self-adjoint. Suppose $\mathfrak{D} \subseteq \mathfrak{D}(A)$ is a core. Then $\mathfrak{D}$ is also a form core.*

Problem 2.17. *Show that (2.68) is wrong if A is not symmetric.*

2.4. Resolvents and spectra

Let A be a (densely defined) closed operator. The **resolvent set** of A is defined by

$$\rho(A) = \{z \in \mathbb{C} | (A - z)^{-1} \in \mathfrak{L}(\mathfrak{H})\}. \tag{2.69}$$

More precisely, $z \in \rho(A)$ if and only if $(A - z) : \mathfrak{D}(A) \to \mathfrak{H}$ is bijective and its inverse is bounded. By the closed graph theorem (Theorem 2.9), it suffices to check that $A - z$ is bijective. The complement of the resolvent set is called the **spectrum**

$$\sigma(A) = \mathbb{C} \setminus \rho(A) \tag{2.70}$$

of A. In particular, $z \in \sigma(A)$ if $A - z$ has a nontrivial kernel. A nonzero vector $\psi \in \text{Ker}(A-z)$ is called an **eigenvector** and z is called an **eigenvalue** in this case.

The function

$$R_A : \rho(A) \to \mathfrak{L}(\mathfrak{H}) \tag{2.71}$$
$$z \mapsto (A - z)^{-1}$$

is called the **resolvent** of A. Note the convenient formula

$$R_A(z)^* = ((A - z)^{-1})^* = ((A - z)^*)^{-1} = (A^* - z^*)^{-1} = R_{A^*}(z^*). \tag{2.72}$$

In particular,

$$\rho(A^*) = \rho(A)^*. \tag{2.73}$$

Example. (Multiplication operator). Consider again the multiplication operator

$$(Af)(x) = A(x)f(x), \quad \mathfrak{D}(A) = \{f \in L^2(\mathbb{R}^d, d\mu) \,|\, Af \in L^2(\mathbb{R}^d, d\mu)\}, \tag{2.74}$$

given by multiplication with the measurable function $A : \mathbb{R}^d \to \mathbb{C}$. Clearly $(A-z)^{-1}$ is given by the multiplication operator

$$(A-z)^{-1}f(x) = \frac{1}{A(x)-z}f(x),$$

(2.75) $$\mathfrak{D}((A-z)^{-1}) = \{f \in L^2(\mathbb{R}^d, d\mu) \mid \frac{1}{A-z}f \in L^2(\mathbb{R}^d, d\mu)\}$$

whenever this operator is bounded. But $\|(A-z)^{-1}\| = \|\frac{1}{A-z}\|_\infty \leq \frac{1}{\varepsilon}$ is equivalent to $\mu(\{x \mid |A(x)-z| < \varepsilon\}) = 0$ and hence

(2.76) $$\rho(A) = \{z \in \mathbb{C} \mid \exists \varepsilon > 0 : \mu(\{x \mid |A(x)-z| < \varepsilon\}) = 0\}.$$

The spectrum

(2.77) $$\sigma(A) = \{z \in \mathbb{C} \mid \forall \varepsilon > 0 : \mu(\{x \mid |A(x)-z| < \varepsilon\}) > 0\}$$

is also known as the **essential range** of $A(x)$. Moreover, z is an eigenvalue of A if $\mu(A^{-1}(\{z\})) > 0$ and $\chi_{A^{-1}(\{z\})}$ is a corresponding eigenfunction in this case. ◇

Example. (Differential operator). Consider again the differential operator

(2.78) $$Af = -i\frac{d}{dx}f, \quad \mathfrak{D}(A) = \{f \in AC[0, 2\pi] \mid f' \in L^2, \ f(0) = f(2\pi)\}$$

in $L^2(0, 2\pi)$. We already know that the eigenvalues of A are the integers and that the corresponding normalized eigenfunctions

(2.79) $$u_n(x) = \frac{1}{\sqrt{2\pi}}e^{inx}$$

form an orthonormal basis.

To compute the resolvent, we must find the solution of the corresponding inhomogeneous equation $-if'(x) - zf(x) = g(x)$. By the variation of constants formula, the solution is given by (this can also be easily verified directly)

(2.80) $$f(x) = f(0)e^{izx} + i\int_0^x e^{iz(x-t)}g(t)dt.$$

Since f must lie in the domain of A, we must have $f(0) = f(2\pi)$ which gives

(2.81) $$f(0) = \frac{i}{e^{-2\pi iz} - 1}\int_0^{2\pi} e^{-izt}g(t)dt, \quad z \in \mathbb{C}\setminus\mathbb{Z}.$$

(Since $z \in \mathbb{Z}$ are the eigenvalues, the inverse cannot exist in this case.) Hence

(2.82) $$(A-z)^{-1}g(x) = \int_0^{2\pi} G(z, x, t)g(t)dt,$$

2.4. Resolvents and spectra

where

(2.83) $$G(z,x,t) = e^{iz(x-t)} \begin{cases} \frac{-i}{1-e^{-2\pi i z}}, & t > x, \\ \frac{i}{1-e^{2\pi i z}}, & t < x, \end{cases} \quad z \in \mathbb{C}\setminus\mathbb{Z}.$$

In particular, $\sigma(A) = \mathbb{Z}$. ◇

If $z, z' \in \rho(A)$, we have the **first resolvent formula**

(2.84) $$R_A(z) - R_A(z') = (z - z')R_A(z)R_A(z') = (z - z')R_A(z')R_A(z).$$

In fact,

(2.85) $$\begin{aligned}(A - z)^{-1} &- (z - z')(A - z)^{-1}(A - z')^{-1} \\ &= (A - z)^{-1}(1 - (z - A + A - z')(A - z')^{-1}) = (A - z')^{-1},\end{aligned}$$

which proves the first equality. The second follows after interchanging z and z'. Now fix $z' = z_0$ and use (2.84) recursively to obtain

(2.86) $$R_A(z) = \sum_{j=0}^{n} (z - z_0)^j R_A(z_0)^{j+1} + (z - z_0)^{n+1} R_A(z_0)^{n+1} R_A(z).$$

The sequence of bounded operators

(2.87) $$R_n = \sum_{j=0}^{n} (z - z_0)^j R_A(z_0)^{j+1}$$

converges to a bounded operator if $|z - z_0| < \|R_A(z_0)\|^{-1}$, and clearly we expect $z \in \rho(A)$ and $R_n \to R_A(z)$ in this case. Let $R_\infty = \lim_{n\to\infty} R_n$ and set $\varphi_n = R_n \psi$, $\varphi = R_\infty \psi$ for some $\psi \in \mathfrak{H}$. Then a quick calculation shows

(2.88) $$AR_n\psi = (A - z_0)R_n\psi + z_0\varphi_n = \psi + (z - z_0)\varphi_{n-1} + z_0\varphi_n.$$

Hence $(\varphi_n, A\varphi_n) \to (\varphi, \psi + z\varphi)$ shows $\varphi \in \mathfrak{D}(A)$ (since A is closed) and $(A - z)R_\infty\psi = \psi$. Similarly, for $\psi \in \mathfrak{D}(A)$,

(2.89) $$R_n A\psi = \psi + (z - z_0)\varphi_{n-1} + z_0\varphi_n$$

and hence $R_\infty(A - z)\psi = \psi$ after taking the limit. Thus $R_\infty = R_A(z)$ as anticipated.

If A is bounded, a similar argument verifies the **Neumann series** for the resolvent

(2.90) $$\begin{aligned}R_A(z) &= -\sum_{j=0}^{n-1} \frac{A^j}{z^{j+1}} + \frac{1}{z^n} A^n R_A(z) \\ &= -\sum_{j=0}^{\infty} \frac{A^j}{z^{j+1}}, \quad |z| > \|A\|.\end{aligned}$$

In summary, we have proved the following:

Theorem 2.16. *The resolvent set $\rho(A)$ is open and $R_A : \rho(A) \to \mathfrak{L}(\mathfrak{H})$ is holomorphic; that is, it has an absolutely convergent power series expansion around every point $z_0 \in \rho(A)$. In addition,*

$$(2.91) \qquad \|R_A(z)\| \geq \operatorname{dist}(z, \sigma(A))^{-1}$$

and if A is bounded, we have $\{z \in \mathbb{C} |\, |z| > \|A\|\} \subseteq \rho(A)$.

As a consequence we obtain the useful

Lemma 2.17. *We have $z \in \sigma(A)$ if there is a sequence $\psi_n \in \mathfrak{D}(A)$ such that $\|\psi_n\| = 1$ and $\|(A - z)\psi_n\| \to 0$. If z is a boundary point of $\rho(A)$, then the converse is also true. Such a sequence is called a **Weyl sequence**.*

Proof. Let ψ_n be a Weyl sequence. Then $z \in \rho(A)$ is impossible by $1 = \|\psi_n\| = \|R_A(z)(A - z)\psi_n\| \leq \|R_A(z)\|\|(A - z)\psi_n\| \to 0$. Conversely, by (2.91), there is a sequence $z_n \to z$ and corresponding vectors $\varphi_n \in \mathfrak{H}$ such that $\|R_A(z_n)\varphi_n\|\|\varphi_n\|^{-1} \to \infty$. Let $\psi_n = R_A(z_n)\varphi_n$ and rescale φ_n such that $\|\psi_n\| = 1$. Then $\|\varphi_n\| \to 0$ and hence

$$\|(A - z)\psi_n\| = \|\varphi_n + (z_n - z)\psi_n\| \leq \|\varphi_n\| + |z - z_n| \to 0$$

shows that ψ_n is a Weyl sequence. $\square$

Let us also note the following spectral mapping result.

Lemma 2.18. *Suppose A is injective and $\operatorname{Ran}(A)$ is dense. Then*

$$(2.92) \qquad \sigma(A^{-1})\backslash\{0\} = (\sigma(A)\backslash\{0\})^{-1}.$$

In addition, we have $A\psi = z\psi$ if and only if $A^{-1}\psi = z^{-1}\psi$.

Proof. Suppose $z \in \rho(A)\backslash\{0\}$. Then we claim

$$R_{A^{-1}}(z^{-1}) = -zAR_A(z) = -z - z^2 R_A(z).$$

In fact, the right-hand side is a bounded operator from $\mathfrak{H} \to \operatorname{Ran}(A) = \mathfrak{D}(A^{-1})$ and

$$(A^{-1} - z^{-1})(-zAR_A(z))\varphi = (-z + A)R_A(z)\varphi = \varphi, \quad \varphi \in \mathfrak{H}.$$

Conversely, if $\psi \in \mathfrak{D}(A^{-1}) = \operatorname{Ran}(A)$, we have $\psi = A\varphi$ and hence

$$(-zAR_A(z))(A^{-1} - z^{-1})\psi = AR_A(z)((A - z)\varphi) = A\varphi = \psi.$$

Thus $z^{-1} \in \rho(A^{-1})$. The rest follows after interchanging the roles of A and A^{-1}. $\square$

Next, let us characterize the spectra of self-adjoint operators.

2.4. Resolvents and spectra

Theorem 2.19. *Let A be symmetric. Then A is self-adjoint if and only if $\sigma(A) \subseteq \mathbb{R}$ and $(A - E) \geq 0$, $E \in \mathbb{R}$, if and only if $\sigma(A) \subseteq [E, \infty)$. Moreover, for self-adjoint A we have $\|R_A(z)\| \leq |\operatorname{Im}(z)|^{-1}$ and, if $(A - E) \geq 0$, $\|R_A(\lambda)\| \leq |\lambda - E|^{-1}$, $\lambda < E$.*

Proof. If $\sigma(A) \subseteq \mathbb{R}$, then $\operatorname{Ran}(A - z) = \mathfrak{H}$, $z \in \mathbb{C}\backslash\mathbb{R}$, and hence A is self-adjoint by Lemma 2.3. Next, if $\sigma(A) \subseteq [E, \infty)$ we show $\langle \psi, (A - E)\psi \rangle \geq 0$ for every $\psi \in \mathfrak{D}(A)$ or, since $E - \lambda \in \rho(A)$ for every $\lambda > 0$, that $f(\lambda) = \langle \varphi, (A - E + \lambda)^{-1}\varphi \rangle > 0$ for every $\lambda > 0$ and $\|\varphi\| = 1$. By the first resolvent identity (2.84) and Cauchy–Schwarz, we see

$$f'(\lambda) = -\|(A + \lambda)^{-1}\varphi\|^2 \leq -f(\lambda)^2$$

and in particular f is a decreasing function. Suppose $f(\lambda_0) < 0$ for some $\lambda_0 > 0$. Then $f(\lambda_0) < 0$ for $\lambda > \lambda_0$ and integrating $f'/f^2 \leq -1$ from λ_0 to λ shows

$$f(\lambda) \leq \frac{f(\lambda_0)}{1 + f(\lambda_0)(\lambda - \lambda_0)}.$$

Hence $f(\lambda) \to -\infty$ as $\lambda \to \lambda_0 - f(\lambda_0)^{-1}$ contradicting the fact that f must be bounded for all λ.

Conversely, if A is self-adjoint (resp. $A \geq E$), then $R_A(z)$ exists for $z \in \mathbb{C}\backslash\mathbb{R}$ (resp. $z \in \mathbb{C}\backslash[E, \infty)$) and satisfies the given estimates as has been shown in Lemma 2.7. □

In particular, we obtain (show this!)

Theorem 2.20. *Let A be self-adjoint. Then*

$$\inf \sigma(A) = \inf_{\psi \in \mathfrak{D}(A), \|\psi\|=1} \langle \psi, A\psi \rangle \tag{2.93}$$

and

$$\sup \sigma(A) = \sup_{\psi \in \mathfrak{D}(A), \|\psi\|=1} \langle \psi, A\psi \rangle. \tag{2.94}$$

For the eigenvalues and corresponding eigenfunctions we have

Lemma 2.21. *Let A be symmetric. Then all eigenvalues are real and eigenvectors corresponding to different eigenvalues are orthogonal.*

Proof. If $A\psi_j = \lambda_j \psi_j$, $j = 1, 2$, we have

$$\lambda_1 \|\psi_1\|^2 = \langle \psi_1, \lambda_1 \psi_1 \rangle = \langle \psi_1, A\psi_1 \rangle = \langle A\psi_1, \psi_1 \rangle = \langle \lambda_1 \psi_1, \psi_1 \rangle = \lambda_1^* \|\psi_1\|^2$$

and

$$(\lambda_1 - \lambda_2)\langle \psi_1, \psi_2 \rangle = \langle A\psi_1, \psi_2 \rangle - \langle A\psi_1, \psi_2 \rangle = 0,$$

finishing the proof. □

The result does not imply that two linearly independent eigenfunctions to the same eigenvalue are orthogonal. However, it is no restriction to assume that they are since we can use Gram–Schmidt to find an orthonormal basis for $\text{Ker}(A - \lambda)$. If $\mathfrak{H}$ is finite dimensional, we can always find an orthonormal basis of eigenvectors. In the infinite dimensional case this is no longer true in general. However, if there is an orthonormal basis of eigenvectors, then A is essentially self-adjoint.

Theorem 2.22. *Suppose A is a symmetric operator which has an orthonormal basis of eigenfunctions $\{\varphi_j\}$. Then A is essentially self-adjoint. In particular, it is essentially self-adjoint on $\text{span}\{\varphi_j\}$.*

Proof. Consider the set of all finite linear combinations $\psi = \sum_{j=0}^{n} c_j \varphi_j$ which is dense in $\mathfrak{H}$. Then $\phi = \sum_{j=0}^{n} \frac{c_j}{\lambda_j \pm \mathrm{i}} \varphi_j \in \mathfrak{D}(A)$ and $(A \pm \mathrm{i})\phi = \psi$ shows that $\text{Ran}(A \pm \mathrm{i})$ is dense. □

Similarly, we can characterize the spectra of unitary operators. Recall that a bijection U is called unitary if $\langle U\psi, U\psi \rangle = \langle \psi, U^*U\psi \rangle = \langle \psi, \psi \rangle$. Thus U is unitary if and only if

$$U^* = U^{-1}. \tag{2.95}$$

Theorem 2.23. *Let U be unitary. Then $\sigma(U) \subseteq \{z \in \mathbb{C} |\, |z| = 1\}$. All eigenvalues have modulus one, and eigenvectors corresponding to different eigenvalues are orthogonal.*

Proof. Since $\|U\| \leq 1$, we have $\sigma(U) \subseteq \{z \in \mathbb{C} |\, |z| \leq 1\}$. Moreover, U^{-1} is also unitary and hence $\sigma(U) \subseteq \{z \in \mathbb{C} |\, |z| \geq 1\}$ by Lemma 2.18. If $U\psi_j = z_j \psi_j$, $j = 1, 2$, we have

$$(z_1 - z_2)\langle \psi_1, \psi_2 \rangle = \langle U^* \psi_1, \psi_2 \rangle - \langle \psi_1, U\psi_2 \rangle = 0$$

since $U\psi = z\psi$ implies $U^*\psi = U^{-1}\psi = z^{-1}\psi = z^*\psi$. □

Problem 2.18. *Suppose A is closed and B bounded:*
- *Show that $\mathbb{I} + B$ has a bounded inverse if $\|B\| < 1$.*
- *Suppose A has a bounded inverse. Then so does $A + B$ if $\|B\| \leq \|A^{-1}\|^{-1}$.*

Problem 2.19. *What is the spectrum of an orthogonal projection?*

Problem 2.20. *Compute the resolvent of*

$$Af = f', \qquad \mathfrak{D}(A) = \{f \in H^1[0,1] \,|\, f(0) = 0\}$$

and show that unbounded operators can have empty spectrum.

Problem 2.21. *Compute the eigenvalues and eigenvectors of $A = -\frac{d^2}{dx^2}$, $\mathfrak{D}(A) = \{f \in H^2(0, \pi) | f(0) = f(\pi) = 0\}$. Compute the resolvent of A.*

2.5. Orthogonal sums of operators

Problem 2.22. *Find a Weyl sequence for the self-adjoint operator $A = -\frac{d^2}{dx^2}$, $\mathfrak{D}(A) = H^2(\mathbb{R})$ for $z \in (0, \infty)$. What is $\sigma(A)$? (Hint: Cut off the solutions of $-u''(x) = z\, u(x)$ outside a finite ball.)*

Problem 2.23. *Suppose $A = \overline{A_0}$. If $\psi_n \in \mathfrak{D}(A)$ is a Weyl sequence for $z \in \sigma(A)$, then there is also one with $\tilde\psi_n \in \mathfrak{D}(A_0)$.*

Problem 2.24. *Suppose A is bounded. Show that the spectra of AA^* and A^*A coincide away from 0 by showing*
(2.96)
$$R_{AA^*}(z) = \frac{1}{z}(A R_{A^*A}(z) A^* - 1), \qquad R_{A^*A}(z) = \frac{1}{z}(A^* R_{AA^*}(z) A - 1).$$

2.5. Orthogonal sums of operators

Let $\mathfrak{H}_j$, $j = 1, 2$, be two given Hilbert spaces and let $A_j : \mathfrak{D}(A_j) \to \mathfrak{H}_j$ be two given operators. Setting $\mathfrak{H} = \mathfrak{H}_1 \oplus \mathfrak{H}_2$, we can define an operator

(2.97) $\qquad A_1 \oplus A_2, \qquad \mathfrak{D}(A_1 \oplus A_2) = \mathfrak{D}(A_1) \oplus \mathfrak{D}(A_2)$

by setting $(A_1 \oplus A_2)(\psi_1 + \psi_2) = A_1 \psi_1 + A_2 \psi_2$ for $\psi_j \in \mathfrak{D}(A_j)$. Clearly $A_1 \oplus A_2$ is closed, (essentially) self-adjoint, etc., if and only if both A_1 and A_2 are. The same considerations apply to countable orthogonal sums. Let $\mathfrak{H} = \bigoplus_j \mathfrak{H}_j$ and define an operator

(2.98) $\qquad \bigoplus_j A_j, \qquad \mathfrak{D}(\bigoplus_j A_j) = \{\sum_j \psi_j \in \bigoplus_j \mathfrak{D}(A_j) \mid \sum_j \|A_j \psi_j\|^2 < \infty\}$

by setting $\bigoplus_j A_j \psi = \sum_j A_j \psi_j$ for $\psi = \sum_j \psi_j \in \mathfrak{D}(\bigoplus_j A_j)$. Then we have

Theorem 2.24. *Suppose A_j are self-adjoint operators on $\mathfrak{H}_j$. Then $A = \bigoplus_j A_j$ is self-adjoint and*

(2.99) $\qquad R_A(z) = \bigoplus_j R_{A_j}(z), \qquad z \in \rho(A) = \mathbb{C} \backslash \sigma(A)$

where

(2.100) $\qquad \sigma(A) = \overline{\bigcup_j \sigma(A_j)}$

(the closure can be omitted if there are only finitely many terms).

Proof. Fix $z \notin \overline{\bigcup_j \sigma(A_j)}$ and let $\varepsilon = \text{Im}(z)$. Then, by Theorem 2.19, $\|R_{A_j}(z)\| \le \varepsilon^{-1}$ and so $R(z) = \bigoplus_j R_{A_j}(z)$ is a bounded operator with $\|R(z)\| \le \varepsilon^{-1}$ (cf. Problem 2.27). It is straightforward to check that $R(z)$ is in fact the resolvent of A and thus $\sigma(A) \subseteq \mathbb{R}$. In particular, A is self-adjoint by Theorem 2.19. To see that $\sigma(A) \subseteq \overline{\bigcup_j \sigma(A_j)}$, note that the above argument can be repeated with $\varepsilon = \text{dist}(z, \bigcup_j \sigma(A_j)) > 0$, which will follow from the spectral theorem (Problem 3.5) to be proven in the next chapter. Conversely, if $z \in \sigma(A_j)$, there is a corresponding Weyl sequence $\psi_n \in \mathfrak{D}(A_j) \subseteq \mathfrak{D}(A)$ and hence $z \in \sigma(A)$. □

Conversely, given an operator A, it might be useful to write A as an orthogonal sum and investigate each part separately.

Let $\mathfrak{H}_1 \subseteq \mathfrak{H}$ be a closed subspace and let P_1 be the corresponding projector. We say that $\mathfrak{H}_1$ **reduces** the operator A if $P_1 A \subseteq A P_1$. Note that this is equivalent to $P_1 \mathfrak{D}(A) \subseteq \mathfrak{D}(A)$ and $P_1 A \psi = A P_1 \psi$ for $\psi \in \mathfrak{D}(A)$. Moreover, if we set $\mathfrak{H}_2 = \mathfrak{H}_1^\perp$, we have $\mathfrak{H} = \mathfrak{H}_1 \oplus \mathfrak{H}_2$ and $P_2 = 1 - P_1$ reduces A as well.

Lemma 2.25. *Suppose $\mathfrak{H} = \bigoplus_j \mathfrak{H}_j$ where each $\mathfrak{H}_j$ reduces A. Then $A \subseteq \bigoplus_j A_j$, where*

(2.101) $$A_j \psi = A\psi, \qquad \mathfrak{D}(A_j) = P_j \mathfrak{D}(A) \subseteq \mathfrak{D}(A).$$

If A is closed, then so are the operators A_j and $A = \bigoplus_j A_j$. If A is closable, then $\mathfrak{H}_j$ also reduces $\overline{A}$ and

(2.102) $$\overline{A} = \bigoplus_j \overline{A_j}.$$

Proof. As already noted, $P_j \mathfrak{D}(A) \subseteq \mathfrak{D}(A)$ and thus A_j is well-defined. Moreover, if $\psi_j \in \mathfrak{D}(A_j)$, we have $A\psi_j = A P_j \psi_j = P_j A \psi_j \in \mathfrak{H}_j$ and thus $A_j : \mathfrak{D}(A_j) \to \mathfrak{H}_j$. Furthermore, every $\psi \in \mathfrak{D}(A)$ can be written as $\psi = \sum_j P_j \psi$ and $A\psi = \sum_j P_j A \psi = \sum_j A P_j \psi$ shows $\|A\psi\|^2 = \sum_j \|A_j P_j \psi\|^2$ implying $\mathfrak{D}(A) \subseteq \mathfrak{D}(\bigoplus_j A_j)$. This proves the first claim.

If A is closed, it is straightforward to see that the same is true for A_j. Now consider $\psi = \sum_j \psi_j \in \mathfrak{D}(\bigoplus_j A_j)$ and abbreviate $\psi^n = \sum_{j=1}^n \psi_j$. Then $\psi^n \to \psi$ and $A\psi^n \to \bigoplus_j A_j \psi$ which implies $\psi \in \mathfrak{D}(A)$ and $\bigoplus_j A_j \psi = A\psi$ since A is closed.

Now let us turn to the last claim. Suppose $\psi \in \mathfrak{D}(\overline{A})$. Then there is a sequence $\psi_n \in \mathfrak{D}(A)$ such that $\psi_n \to \psi$ and $A\psi_n \to \varphi = \overline{A}\psi$. Thus $P_j \psi_n \to P_j \psi$ and $A P_j \psi_n = P_j A \psi_n \to P_j \varphi$ which shows $P_j \psi \in \mathfrak{D}(\overline{A})$ and $P_j \overline{A} \psi = \overline{A} P_j \psi$; that is, $\mathfrak{H}_j$ reduces $\overline{A}$. Moreover, this argument also shows $P_j \mathfrak{D}(\overline{A}) \subseteq \mathfrak{D}(\overline{A_j})$, and the converse follows analogously. □

If A is self-adjoint, then $\mathfrak{H}_1$ reduces A if $P_1 \mathfrak{D}(A) \subseteq \mathfrak{D}(A)$ and $A P_1 \psi \in \mathfrak{H}_1$ for every $\psi \in \mathfrak{D}(A)$. In fact, if $\psi \in \mathfrak{D}(A)$, we can write $\psi = \psi_1 \oplus \psi_2$, with $P_2 = 1 - P_1$ and $\psi_j = P_j \psi \in \mathfrak{D}(A)$. Since $A P_1 \psi = A \psi_1$ and $P_1 A \psi = P_1 A \psi_1 + P_1 A \psi_2 = A\psi_1 + P_1 A \psi_2$, we need to show $P_1 A \psi_2 = 0$. But this follows since

(2.103) $$\langle \varphi, P_1 A \psi_2 \rangle = \langle A P_1 \varphi, \psi_2 \rangle = 0$$

2.6. Self-adjoint extensions

for every $\varphi \in \mathfrak{D}(A)$.

Problem 2.25. *Show* $(\bigoplus_j A_j)^* = \bigoplus_j A_j^*$.

Problem 2.26. *Show that A defined in (2.98) is closed if and only if all A_j are.*

Problem 2.27. *Show that for A defined in (2.98), we have $\|A\| = \sup_j \|A_j\|$.*

2.6. Self-adjoint extensions

It is safe to skip this entire section on first reading.

In many physical applications, a symmetric operator is given. If this operator turns out to be essentially self-adjoint, there is a unique self-adjoint extension and everything is fine. However, if it is not, it is important to find out if there are self-adjoint extensions at all (for physical problems there better be) and to classify them.

In Section 2.2 we saw that A is essentially self-adjoint if $\mathrm{Ker}(A^* - z) = \mathrm{Ker}(A^* - z^*) = \{0\}$ for one $z \in \mathbb{C}\backslash\mathbb{R}$. Hence self-adjointness is related to the dimension of these spaces, and one calls the numbers

$$(2.104) \qquad d_\pm(A) = \dim K_\pm, \qquad K_\pm = \mathrm{Ran}(A \pm \mathrm{i})^\perp = \mathrm{Ker}(A^* \mp \mathrm{i}),$$

defect indices of A (we have chosen $z = \mathrm{i}$ for simplicity; any other $z \in \mathbb{C}\backslash\mathbb{R}$ would be as good). If $d_-(A) = d_+(A) = 0$, there is one self-adjoint extension of A, namely $\overline{A}$. But what happens in the general case? Is there more than one extension, or maybe none at all? These questions can be answered by virtue of the **Cayley transform**

$$(2.105) \qquad V = (A - \mathrm{i})(A + \mathrm{i})^{-1} : \mathrm{Ran}(A + \mathrm{i}) \to \mathrm{Ran}(A - \mathrm{i}).$$

Theorem 2.26. *The Cayley transform is a bijection from the set of all symmetric operators A to the set of all isometric operators V (i.e., $\|V\varphi\| = \|\varphi\|$ for all $\varphi \in \mathfrak{D}(V)$) for which $\mathrm{Ran}(1 - V)$ is dense.*

Proof. Since A is symmetric, we have $\|(A \pm \mathrm{i})\psi\|^2 = \|A\psi\|^2 + \|\psi\|^2$ for all $\psi \in \mathfrak{D}(A)$ by a straightforward computation. Thus for every $\varphi = (A + \mathrm{i})\psi \in \mathfrak{D}(V) = \mathrm{Ran}(A + \mathrm{i})$ we have

$$\|V\varphi\| = \|(A - \mathrm{i})\psi\| = \|(A + \mathrm{i})\psi\| = \|\varphi\|.$$

Next, observe that

$$1 \pm V = ((A + \mathrm{i}) \pm (A - \mathrm{i}))(A + \mathrm{i})^{-1} = \begin{cases} 2A(A + \mathrm{i})^{-1}, \\ 2\mathrm{i}(A + \mathrm{i})^{-1}, \end{cases}$$

which shows that $\mathrm{Ran}(1 - V) = \mathfrak{D}(A)$ is dense and

$$A = \mathrm{i}(1 + V)(1 - V)^{-1}.$$

Conversely, let V be given and use the last equation to define A.

Since V is isometric, we have $\langle (1 \pm V)\varphi, (1 \mp V)\varphi \rangle = \pm 2\mathrm{i}\,\mathrm{Im}\langle V\varphi, \varphi \rangle$ for all $\varphi \in \mathfrak{D}(V)$ by a straightforward computation. Thus for every $\psi = (1-V)\varphi \in \mathfrak{D}(A) = \mathrm{Ran}(1-V)$ we have
$$\langle A\psi, \psi \rangle = -\mathrm{i}\langle (1+V)\varphi, (1-V)\varphi \rangle = \mathrm{i}\langle (1-V)\varphi, (1+V)\varphi \rangle = \langle \psi, A\psi \rangle;$$
that is, A is symmetric. Finally, observe that
$$A \pm \mathrm{i} = ((1+V) \pm (1-V))(1-V)^{-1} = \begin{cases} 2\mathrm{i}(1-V)^{-1}, \\ 2\mathrm{i}V(1-V)^{-1}, \end{cases}$$
which shows that A is the Cayley transform of V and finishes the proof. $\square$

Thus A is self-adjoint if and only if its Cayley transform V is unitary. Moreover, finding a self-adjoint extension of A is equivalent to finding a unitary extension of V and this in turn is equivalent to (taking the closure and) finding a unitary operator from $\mathfrak{D}(V)^\perp$ to $\mathrm{Ran}(V)^\perp$. This is possible if and only if both spaces have the same dimension, that is, if and only if $d_+(A) = d_-(A)$.

Theorem 2.27. *A symmetric operator A has self-adjoint extensions if and only if its defect indices are equal.*

In this case let A_1 be a self-adjoint extension and V_1 its Cayley transform. Then
(2.106)
$$\mathfrak{D}(A_1) = \mathfrak{D}(A) + (1-V_1)K_+ = \{\psi + \varphi_+ - V_1\varphi_+ | \psi \in \mathfrak{D}(A),\ \varphi_+ \in K_+\}$$
and
(2.107)
$$A_1(\psi + \varphi_+ - V_1\varphi_+) = A\psi + \mathrm{i}\varphi_+ + \mathrm{i}V_1\varphi_+.$$
Moreover,
(2.108)
$$(A_1 \pm \mathrm{i})^{-1} = (A \pm \mathrm{i})^{-1} \oplus \frac{\mp \mathrm{i}}{2} \sum_j \langle \varphi_j^\pm, . \rangle (\varphi_j^\pm - \varphi_j^\mp),$$
where $\{\varphi_j^+\}$ is an orthonormal basis for K_+ and $\varphi_j^- = V_1\varphi_j^+$.

Proof. From the proof of the previous theorem we know that $\mathfrak{D}(A_1) = \mathrm{Ran}(1-V_1) = \mathrm{Ran}(1+V) + (1-V_1)K_+ = \mathfrak{D}(A) + (1-V_1)K_+$. Moreover, $A_1(\psi+\varphi_+-V_1\varphi_+) = A\psi+\mathrm{i}(1+V_1)(1-V_1)^{-1}(1-V_1)\varphi_+ = A\psi+\mathrm{i}(1+V_1)\varphi_+$.

Similarly, $\mathrm{Ran}(A_1 \pm \mathrm{i}) = \mathrm{Ran}(A \pm \mathrm{i}) \oplus K_\pm$ and $(A_1+\mathrm{i})^{-1} = -\frac{\mathrm{i}}{2}(1-V_1)$, respectively, $(A_1+\mathrm{i})^{-1} = -\frac{\mathrm{i}}{2}(1-V_1^{-1})$. $\square$

Note that instead of $z = \mathrm{i}$ we could use $V(z) = (A+z^*)(A+z)^{-1}$ for any $z \in \mathbb{C}\backslash\mathbb{R}$. We remark that in this case one can show that the defect indices are independent of $z \in \mathbb{C}_+ = \{z \in \mathbb{C}|\,\mathrm{Im}(z) > 0\}$.

2.6. Self-adjoint extensions

Example. Recall the operator $A = -\mathrm{i}\frac{d}{dx}$, $\mathfrak{D}(A) = \{f \in H^1(0, 2\pi)|f(0) = f(2\pi) = 0\}$ with adjoint $A^* = -\mathrm{i}\frac{d}{dx}$, $\mathfrak{D}(A^*) = H^1(0, 2\pi)$.

Clearly

$$K_\pm = \mathrm{span}\{e^{\mp x}\} \tag{2.109}$$

is one-dimensional and hence all unitary maps are of the form

$$V_\theta e^{2\pi - x} = e^{\mathrm{i}\theta} e^x, \quad \theta \in [0, 2\pi). \tag{2.110}$$

The functions in the domain of the corresponding operator A_θ are given by

$$f_\theta(x) = f(x) + \alpha(e^{2\pi - x} - e^{\mathrm{i}\theta} e^x), \quad f \in \mathfrak{D}(A), \alpha \in \mathbb{C}. \tag{2.111}$$

In particular, f_θ satisfies

$$f_\theta(2\pi) = e^{\mathrm{i}\tilde{\theta}} f_\theta(0), \quad e^{\mathrm{i}\tilde{\theta}} = \frac{1 - e^{\mathrm{i}\theta} e^{2\pi}}{e^{2\pi} - e^{\mathrm{i}\theta}}, \tag{2.112}$$

and thus we have

$$\mathfrak{D}(A_\theta) = \{f \in H^1(0, 2\pi)|f(2\pi) = e^{\mathrm{i}\tilde{\theta}} f(0)\}. \tag{2.113}$$

⋄

Concerning closures, we can combine the fact that a bounded operator is closed if and only if its domain is closed with item (iii) from Lemma 2.4 to obtain

Lemma 2.28. *Suppose A is symmetric. Then the following items are equivalent.*

- *A is closed.*
- *$\mathfrak{D}(V) = \mathrm{Ran}(A + \mathrm{i})$ is closed.*
- *$\mathrm{Ran}(V) = \mathrm{Ran}(A - \mathrm{i})$ is closed.*
- *V is closed.*

Next, we give a useful criterion for the existence of self-adjoint extensions. A conjugate linear map $C : \mathfrak{H} \to \mathfrak{H}$ is called a **conjugation** if it satisfies $C^2 = \mathbb{I}$ and $\langle C\psi, C\varphi \rangle = \langle \psi, \varphi \rangle$. The prototypical example is, of course, complex conjugation $C\psi = \psi^*$. An operator A is called C-**real** if

$$C\mathfrak{D}(A) \subseteq \mathfrak{D}(A), \quad \text{and} \quad AC\psi = CA\psi, \quad \psi \in \mathfrak{D}(A). \tag{2.114}$$

Note that in this case $C\mathfrak{D}(A) = \mathfrak{D}(A)$, since $\mathfrak{D}(A) = C^2\mathfrak{D}(A) \subseteq C\mathfrak{D}(A)$.

Theorem 2.29. *Suppose the symmetric operator A is C-real. Then its defect indices are equal.*

Proof. Let $\{\varphi_j\}$ be an orthonormal set in $\operatorname{Ran}(A+\mathrm{i})^\perp$. Then $\{C\varphi_j\}$ is an orthonormal set in $\operatorname{Ran}(A-\mathrm{i})^\perp$. Hence $\{\varphi_j\}$ is an orthonormal basis for $\operatorname{Ran}(A+\mathrm{i})^\perp$ if and only if $\{C\varphi_j\}$ is an orthonormal basis for $\operatorname{Ran}(A-\mathrm{i})^\perp$. Hence the two spaces have the same dimension. $\square$

Finally, we note the following useful formula for the difference of resolvents of self-adjoint extensions.

Lemma 2.30. *If A_j, $j = 1, 2$, are self-adjoint extensions of A and if $\{\varphi_j(z)\}$ is an orthonormal basis for $\operatorname{Ker}(A^* - z)$, then*

$$(2.115) \quad (A_1 - z)^{-1} - (A_2 - z)^{-1} = \sum_{j,k} (\alpha^1_{jk}(z) - \alpha^2_{jk}(z))\langle \varphi_j(z^*), . \rangle \varphi_k(z),$$

where

$$(2.116) \qquad \alpha^l_{jk}(z) = \langle \varphi_k(z), (A_l - z)^{-1} \varphi_j(z^*) \rangle.$$

Proof. First observe that $((A_1 - z)^{-1} - (A_2 - z)^{-1})\varphi$ is zero for every $\varphi \in \operatorname{Ran}(A - z)$. Hence it suffices to consider vectors of the form $\varphi = \sum_j \langle \varphi_j(z^*), \varphi \rangle \varphi_j(z^*) \in \operatorname{Ran}(A - z)^\perp = \operatorname{Ker}(A^* - z^*)$. Hence we have

$$(A_1 - z)^{-1} - (A_2 - z)^{-1} = \sum_j \langle \varphi_j(z^*), . \rangle \psi_j(z),$$

where

$$\psi_j(z) = ((A_1 - z)^{-1} - (A_2 - z)^{-1})\varphi_j(z^*).$$

Now computing the adjoint once using $((A_l - z)^{-1})^* = (A_l - z^*)^{-1}$ and once using $(\sum_j \langle \varphi_j, . \rangle \psi_j)^* = \sum_j \langle \psi_j, . \rangle \varphi_j$, we obtain

$$\sum_j \langle \varphi_j(z), . \rangle \psi_j(z^*) = \sum_j \langle \psi_j(z), . \rangle \varphi_k(z^*).$$

Evaluating at $\varphi_k(z)$ implies

$$\psi_k(z) = \sum_j \langle \psi_j(z^*), \varphi_k(z^*) \rangle \varphi_j(z) = \sum_j (\alpha^1_{kj}(z) - \alpha^2_{kj}(z))\varphi_j(z)$$

and finishes the proof. $\square$

Problem 2.28. *Compute the defect indices of*

$$A_0 = \mathrm{i} \frac{d}{dx}, \qquad \mathfrak{D}(A_0) = C_c^\infty((0, \infty)).$$

Can you give a self-adjoint extension of A_0?

Problem 2.29. *Let A_1 be a self-adjoint extension of A and suppose $\varphi \in \operatorname{Ker}(A^* - z_0)$. Show that $\varphi(z) = \varphi + (z - z_0)(A_1 - z)^{-1}\varphi \in \operatorname{Ker}(A^* - z)$.*

2.7. Appendix: Absolutely continuous functions

Let $(a,b) \subseteq \mathbb{R}$ be some interval. We denote by
$$AC(a,b) = \{f \in C(a,b) | f(x) = f(c) + \int_c^x g(t)dt,\ c \in (a,b),\ g \in L^1_{loc}(a,b)\} \tag{2.117}$$
the set of all **absolutely continuous functions**. That is, f is absolutely continuous if and only if it can be written as the integral of some locally integrable function. Note that $AC(a,b)$ is a vector space.

By Corollary A.44, $f(x) = f(c) + \int_c^x g(t)dt$ is differentiable a.e. (with respect to Lebesgue measure) and $f'(x) = g(x)$. In particular, g is determined uniquely a.e.

If $[a,b]$ is a compact interval, we set
$$AC[a,b] = \{f \in AC(a,b) | g \in L^1(a,b)\} \subseteq C[a,b]. \tag{2.118}$$

If $f, g \in AC[a,b]$, we have the integration by parts formula (Problem 2.30)
$$\int_a^b f(x) g'(x) dx = f(b)g(b) - f(a)g(a) - \int_a^b f'(x) g(x) dx, \tag{2.119}$$
which also implies that the product rule holds for absolutely continuous functions.

We define the usual **Sobolev spaces** via
$$H^m(a,b) = \{f \in L^2(a,b) | f^{(j)} \in AC(a,b),\ f^{(j+1)} \in L^2(a,b),\ 0 \le j \le m-1\}. \tag{2.120}$$
Note that $H^m(a,b)$ is a Hilbert space when equipped with the norm
$$\|f\|_{2,m}^2 = \sum_{j=0}^m \int_a^b |f^{(j)}(t)|^2 dt \tag{2.121}$$
(cf. Problem 2.32).

Then we have

Lemma 2.31. *Suppose $f \in H^m(a,b)$, $m \ge 1$. Then f is bounded and $\lim_{x \downarrow a} f^{(j)}(x)$, respectively, $\lim_{x \uparrow b} f^{(j)}(x)$, exists for $0 \le j \le m-1$. Moreover, the limit is zero if the endpoint is infinite.*

Proof. If the endpoint is finite, then $f^{(j+1)}$ is integrable near this endpoint and hence the claim follows. If the endpoint is infinite, note that
$$|f^{(j)}(x)|^2 = |f^{(j)}(c)|^2 + 2\int_c^x \operatorname{Re}(f^{(j)}(t)^* f^{(j+1)}(t)) dt$$
shows that the limit exists (dominated convergence). Since $f^{(j)}$ is square integrable, the limit must be zero. $\square$

Hence we can set $f^{(j)}(a) = \lim_{x\downarrow a} f^{(j)}(x)$, $f^{(j)}(b) = \lim_{x\uparrow b} f^{(j)}(x)$ and introduce

(2.122) $\quad H_0^m(a,b) = \{f \in H^m(a,b) | f^{(j)}(a) = f^{(j)}(b) = 0,\ 0 \le j \le m-1\},$

which could equivalently be defined as the closure of $C_0^\infty(a,b)$ in $H^m(a,b)$ (cf. Problem 2.33). Of course if both endpoints are infinite, we have $H_0^m(\mathbb{R}) = H^m(\mathbb{R})$.

Finally, let me remark that it suffices to check that the function plus the highest derivative are in L^2; the lower derivatives are then automatically in L^2. That is,
(2.123)
$H^m(a,b) = \{f \in L^2(a,b) | f^{(j)} \in AC(a,b),\ 0 \le j \le m-1,\ f^{(m)} \in L^2(a,b)\}.$

For a finite endpoint, this is straightforward. For an infinite endpoint, this can also be shown directly, but it is much easier to use the Fourier transform (compare Section 7.1).

Problem 2.30. *Show* (2.119). *(Hint: Fubini.)*

Problem 2.31. *A function $u \in L^1(0,1)$ is called weakly differentiable if for some $v \in L^1(0,1)$ we have*
$$\int_0^1 v(x)\varphi(x)dx = -\int_0^1 u(x)\varphi'(x)dx$$
*for all test functions $\varphi \in C_c^\infty(0,1)$. Show that u is **weakly differentiable** if and only if u is absolutely continuous and $u' = v$ in this case. (Hint: You will need that $\int_0^1 u(t)\varphi'(t)dt = 0$ for all $\varphi \in C_c^\infty(0,1)$ if and only if u is constant. To see this, choose some $\varphi_0 \in C_c^\infty(0,1)$ with $I(\varphi_0) = \int_0^1 \varphi_0(t)dt = 1$. Then invoke Lemma 0.41 and use that every $\varphi \in C_c^\infty(0,1)$ can be written as $\varphi(t) = \Phi'(t) + I(\varphi)\varphi_0(t)$ with $\Phi(t) = \int_0^t \varphi(s)ds - I(\varphi)\int_0^t \varphi_0(s)ds$.)*

Problem 2.32. *Show that $H^m(a,b)$ together with the norm (2.121) is a Hilbert space.*

Problem 2.33. *Show*

(2.124) $\qquad \|f\|_\infty^2 \le \dfrac{\|f\|^2}{b-a} + 2\|f\|\|f'\|, \qquad f \in H^1(a,b),$

(where the first term on the left-hand side is zero if (a,b) is unbounded) and

(2.125) $\qquad \|f\|_\infty^2 \le 2\|f\|\|f'\|, \qquad f \in H_0^1(a,b).$

*Conclude that the **trace operator***

(2.126) $\qquad T : H^1(a,b) \to \mathbb{C}^2, \qquad f \mapsto T(f) = (f(a), f(b)),$

is continuous and that $H_0^1(a,b)$ is a closed subspace.

2.7. Appendix: Absolutely continuous functions

Moreover, show that the closure of $C_0^\infty(a,b)$ in $H^1(a,b)$ is $H_0^1(a,b)$. Similarly, show that the closure of $C_0^\infty(a,b)$ in $H^m(a,b)$ is $H_0^m(a,b)$. (Hint: Start with the case where (a,b) is finite.)

Problem 2.34. Show that if $f \in AC(a,b)$ and $f' \in L^p(a,b)$, then f is Hölder continuous:
$$|f(x) - f(y)| \leq \|f'\|_p |x-y|^{1-\frac{1}{p}}.$$

Chapter 3

The spectral theorem

The time evolution of a quantum mechanical system is governed by the Schrödinger equation

$$\text{(3.1)} \qquad i\frac{d}{dt}\psi(t) = H\psi(t).$$

If $\mathfrak{H} = \mathbb{C}^n$ and H is hence a matrix, this system of ordinary differential equations is solved by the matrix exponential

$$\text{(3.2)} \qquad \psi(t) = \exp(-itH)\psi(0).$$

This matrix exponential can be defined by a convergent power series

$$\text{(3.3)} \qquad \exp(-itH) = \sum_{n=0}^{\infty} \frac{(-it)^n}{n!} H^n.$$

For this approach, the boundedness of H is crucial, which might not be the case for a quantum system. However, the best way to compute the matrix exponential and to understand the underlying dynamics is to diagonalize H. But how do we diagonalize a self-adjoint operator? The answer is known as the spectral theorem.

3.1. The spectral theorem

In this section, we want to address the problem of defining functions of a self-adjoint operator A in a natural way, that is, such that

$$\text{(3.4)} \quad (f+g)(A) = f(A)+g(A), \quad (fg)(A) = f(A)g(A), \quad (f^*)(A) = f(A)^*.$$

As long as f and g are polynomials, no problems arise. If we want to extend this definition to a larger class of functions, we will need to perform some limiting procedure. Hence we could consider convergent power series or equip the space of polynomials on the spectrum with the sup norm. In both cases this only works if the operator A is bounded. To overcome this limitation, we will use characteristic functions $\chi_\Omega(A)$ instead of powers A^j. Since $\chi_\Omega(\lambda)^2 = \chi_\Omega(\lambda)$, the corresponding operators should be orthogonal projections. Moreover, we should also have $\chi_\mathbb{R}(A) = \mathbb{I}$ and $\chi_\Omega(A) = \sum_{j=1}^n \chi_{\Omega_j}(A)$ for any finite union $\Omega = \bigcup_{j=1}^n \Omega_j$ of disjoint sets. The only remaining problem is of course the definition of $\chi_\Omega(A)$. However, we will defer this problem and begin by developing a functional calculus for a family of characteristic functions $\chi_\Omega(A)$.

Denote the Borel sigma algebra of $\mathbb{R}$ by $\mathfrak{B}$. A **projection-valued measure** is a map

$$(3.5) \qquad P: \mathfrak{B} \to \mathfrak{L}(\mathfrak{H}), \qquad \Omega \mapsto P(\Omega),$$

from the Borel sets to the set of orthogonal projections, that is, $P(\Omega)^* = P(\Omega)$ and $P(\Omega)^2 = P(\Omega)$, such that the following two conditions hold:

(i) $P(\mathbb{R}) = \mathbb{I}$.

(ii) If $\Omega = \bigcup_n \Omega_n$ with $\Omega_n \cap \Omega_m = \emptyset$ for $n \neq m$, then $\sum_n P(\Omega_n)\psi = P(\Omega)\psi$ for every $\psi \in \mathfrak{H}$ (strong σ-additivity).

Note that we require strong convergence, $\sum_n P(\Omega_n)\psi = P(\Omega)\psi$, rather than norm convergence, $\sum_n P(\Omega_n) = P(\Omega)$. In fact, norm convergence does not even hold in the simplest case where $\mathfrak{H} = L^2(I)$ and $P(\Omega) = \chi_\Omega$ (multiplication operator), since for a multiplication operator the norm is just the sup norm of the function. Furthermore, it even suffices to require weak convergence, since w-$\lim P_n = P$ for some orthogonal projections implies s-$\lim P_n = P$ by $\langle \psi, P_n\psi \rangle = \langle \psi, P_n^2\psi \rangle = \langle P_n\psi, P_n\psi \rangle = \|P_n\psi\|^2$ together with Lemma 1.12 (iv).

Example. Let $\mathfrak{H} = \mathbb{C}^n$ and let $A \in \mathfrak{L}(\mathbb{C}^n)$ be some symmetric matrix. Let $\lambda_1, \ldots, \lambda_m$ be its (distinct) eigenvalues and let P_j be the projections onto the corresponding eigenspaces. Then

$$(3.6) \qquad P_A(\Omega) = \sum_{\{j \mid \lambda_j \in \Omega\}} P_j$$

is a projection-valued measure. $\diamond$

Example. Let $\mathfrak{H} = L^2(\mathbb{R})$ and let f be a real-valued measurable function. Then

$$(3.7) \qquad P(\Omega) = \chi_{f^{-1}(\Omega)}$$

is a projection-valued measure (Problem 3.3). $\diamond$

3.1. The spectral theorem

It is straightforward to verify that every projection-valued measure satisfies

$$(3.8) \qquad P(\emptyset) = 0, \qquad P(\mathbb{R}\backslash\Omega) = \mathbb{I} - P(\Omega),$$

and

$$(3.9) \qquad P(\Omega_1 \cup \Omega_2) + P(\Omega_1 \cap \Omega_2) = P(\Omega_1) + P(\Omega_2).$$

Moreover, we also have

$$(3.10) \qquad P(\Omega_1)P(\Omega_2) = P(\Omega_1 \cap \Omega_2).$$

Indeed, first suppose $\Omega_1 \cap \Omega_2 = \emptyset$. Then, taking the square of (3.9), we infer

$$(3.11) \qquad P(\Omega_1)P(\Omega_2) + P(\Omega_2)P(\Omega_1) = 0.$$

Multiplying this equation from the right by $P(\Omega_2)$ shows that $P(\Omega_1)P(\Omega_2) = -P(\Omega_2)P(\Omega_1)P(\Omega_2)$ is self-adjoint and thus $P(\Omega_1)P(\Omega_2) = P(\Omega_2)P(\Omega_1) = 0$. For the general case $\Omega_1 \cap \Omega_2 \neq \emptyset$, we now have

$$\begin{aligned}P(\Omega_1)P(\Omega_2) &= (P(\Omega_1 - \Omega_2) + P(\Omega_1 \cap \Omega_2))(P(\Omega_2 - \Omega_1) + P(\Omega_1 \cap \Omega_2)) \\ (3.12) \qquad &= P(\Omega_1 \cap \Omega_2)\end{aligned}$$

as stated.

Moreover, a projection-valued measure is monotone, that is,

$$(3.13) \qquad \Omega_1 \subseteq \Omega_2 \quad \Rightarrow \quad P(\Omega_1) \leq P(\Omega_2),$$

in the sense that $\langle \psi, P(\Omega_1)\psi \rangle \leq \langle \psi, P(\Omega_2)\psi \rangle$ or equivalently $\text{Ran}(P(\Omega_1)) \subseteq \text{Ran}(P(\Omega_2))$ (cf. Problem 1.7). As a useful consequence, note that $P(\Omega_2) = 0$ implies $P(\Omega_1) = 0$ for every subset $\Omega_1 \subseteq \Omega_2$.

To every projection-valued measure there corresponds a **resolution of the identity**

$$(3.14) \qquad P(\lambda) = P((-\infty, \lambda])$$

which has the following properties (Problem 3.4):

(i) $P(\lambda)$ is an orthogonal projection.
(ii) $P(\lambda_1) \leq P(\lambda_2)$ for $\lambda_1 \leq \lambda_2$.
(iii) s-$\lim_{\lambda_n \downarrow \lambda} P(\lambda_n) = P(\lambda)$ (strong right continuity).
(iv) s-$\lim_{\lambda \to -\infty} P(\lambda) = 0$ and s-$\lim_{\lambda \to +\infty} P(\lambda) = \mathbb{I}$.

As before, strong right continuity is equivalent to weak right continuity.

Picking $\psi \in \mathfrak{H}$, we obtain a finite Borel measure $\mu_\psi(\Omega) = \langle \psi, P(\Omega)\psi \rangle = \|P(\Omega)\psi\|^2$ with $\mu_\psi(\mathbb{R}) = \|\psi\|^2 < \infty$. The corresponding distribution function is given by $\mu_\psi(\lambda) = \langle \psi, P(\lambda)\psi \rangle$ and since for every distribution function there is a unique Borel measure (Theorem A.3), for every resolution of the identity there is a unique projection-valued measure.

Using the polarization identity (2.17), we also have the complex Borel measures
(3.15)
$$\mu_{\varphi,\psi}(\Omega) = \langle \varphi, P(\Omega)\psi \rangle = \frac{1}{4}(\mu_{\varphi+\psi}(\Omega) - \mu_{\varphi-\psi}(\Omega) + i\mu_{\varphi-i\psi}(\Omega) - i\mu_{\varphi+i\psi}(\Omega)).$$
Note also that, by Cauchy–Schwarz, $|\mu_{\varphi,\psi}(\Omega)| \leq \|\varphi\|\|\psi\|$.

Now let us turn to integration with respect to our projection-valued measure. For every simple function $f = \sum_{j=1}^n \alpha_j \chi_{\Omega_j}$ (where $\Omega_j = f^{-1}(\alpha_j)$), we set

(3.16) $$P(f) \equiv \int_\mathbb{R} f(\lambda)dP(\lambda) = \sum_{j=1}^n \alpha_j P(\Omega_j).$$

In particular, $P(\chi_\Omega) = P(\Omega)$. Then $\langle \varphi, P(f)\psi \rangle = \sum_j \alpha_j \mu_{\varphi,\psi}(\Omega_j)$ shows

(3.17) $$\langle \varphi, P(f)\psi \rangle = \int_\mathbb{R} f(\lambda) d\mu_{\varphi,\psi}(\lambda)$$

and, by linearity of the integral, the operator P is a linear map from the set of simple functions into the set of bounded linear operators on $\mathfrak{H}$. Moreover, $\|P(f)\psi\|^2 = \sum_j |\alpha_j|^2 \mu_\psi(\Omega_j)$ (the sets Ω_j are disjoint) shows

(3.18) $$\|P(f)\psi\|^2 = \int_\mathbb{R} |f(\lambda)|^2 d\mu_\psi(\lambda).$$

Equipping the set of simple functions with the sup norm, we infer

(3.19) $$\|P(f)\psi\| \leq \|f\|_\infty \|\psi\|,$$

which implies that P has norm one. Since the simple functions are dense in the Banach space of bounded Borel functions $B(\mathbb{R})$, there is a unique extension of P to a bounded linear operator $P: B(\mathbb{R}) \to \mathfrak{L}(\mathfrak{H})$ (whose norm is one) from the bounded Borel functions on $\mathbb{R}$ (with sup norm) to the set of bounded linear operators on $\mathfrak{H}$. In particular, (3.17) and (3.18) remain true.

There is some additional structure behind this extension. Recall that the set $\mathfrak{L}(\mathfrak{H})$ of all bounded linear mappings on $\mathfrak{H}$ forms a C^* algebra. A C^* algebra homomorphism ϕ is a linear map between two C^* algebras which respects both the multiplication and the adjoint; that is, $\phi(ab) = \phi(a)\phi(b)$ and $\phi(a^*) = \phi(a)^*$.

Theorem 3.1. *Let $P(\Omega)$ be a projection-valued measure. Then the operator*

(3.20) $$\begin{aligned} P: B(\mathbb{R}) &\to \mathfrak{L}(\mathfrak{H}) \\ f &\mapsto \int_\mathbb{R} f(\lambda) dP(\lambda) \end{aligned}$$

is a C^* algebra homomorphism with norm one such that

(3.21) $$\langle P(g)\varphi, P(f)\psi \rangle = \int_\mathbb{R} g^*(\lambda) f(\lambda) d\mu_{\varphi,\psi}(\lambda).$$

3.1. The spectral theorem

In addition, if $f_n(\lambda) \to f(\lambda)$ pointwise and if the sequence $\|f_n\|_\infty$ is bounded, then $P(f_n) \xrightarrow{s} P(f)$ strongly.

Proof. The properties $P(1) = \mathbb{I}$, $P(f^*) = P(f)^*$, and $P(fg) = P(f)P(g)$ are straightforward for simple functions f. For general f they follow from continuity. Hence P is a C^* algebra homomorphism.

Equation (3.21) is a consequence of $\langle P(g)\varphi, P(f)\psi \rangle = \langle \varphi, P(g^*f)\psi \rangle$.

The last claim follows from the dominated convergence theorem and (3.18). □

As a consequence of (3.21), observe

$$(3.22) \quad \mu_{P(g)\varphi,P(f)\psi}(\Omega) = \langle P(g)\varphi, P(\Omega)P(f)\psi \rangle = \int_\Omega g^*(\lambda) f(\lambda) d\mu_{\varphi,\psi}(\lambda),$$

which implies

$$(3.23) \quad d\mu_{P(g)\varphi,P(f)\psi} = g^* f d\mu_{\varphi,\psi}.$$

Example. Let $\mathfrak{H} = \mathbb{C}^n$ and $A = A^* \in \mathcal{L}(\mathbb{C}^n)$, respectively, P_A, as in the previous example. Then

$$(3.24) \quad P_A(f) = \sum_{j=1}^m f(\lambda_j) P_j.$$

In particular, $P_A(f) = A$ for $f(\lambda) = \lambda$. Moreover,

$$(3.25) \quad d\mu_\psi(\lambda) = \sum_{j=1}^n \|P_j\psi\|^2 d\Theta(\lambda - \lambda_j),$$

where $d\Theta(\lambda - \lambda_j)$ is the Dirac measure centered at λ_j. ◇

Next we want to define this operator for unbounded Borel functions. Since we expect the resulting operator to be unbounded, we need a suitable domain first. Motivated by (3.18), we set

$$(3.26) \quad \mathfrak{D}_f = \{\psi \in \mathfrak{H} \mid \int_\mathbb{R} |f(\lambda)|^2 d\mu_\psi(\lambda) < \infty\}.$$

This is clearly a linear subspace of $\mathfrak{H}$ since $\mu_{\alpha\psi}(\Omega) = |\alpha|^2 \mu_\psi(\Omega)$ and since $\mu_{\varphi+\psi}(\Omega) = \|P(\Omega)(\varphi+\psi)\|^2 \leq 2(\|P(\Omega)\varphi\|^2 + \|P(\Omega)\psi\|^2) = 2(\mu_\varphi(\Omega) + \mu_\psi(\Omega))$ (by the triangle inequality).

For every $\psi \in \mathfrak{D}_f$, the sequence of bounded Borel functions

$$(3.27) \quad f_n = \chi_{\Omega_n} f, \qquad \Omega_n = \{\lambda \mid |f(\lambda)| \leq n\},$$

is a Cauchy sequence converging to f in the sense of $L^2(\mathbb{R}, d\mu_\psi)$. Hence, by virtue of (3.18), the vectors $\psi_n = P(f_n)\psi$ form a Cauchy sequence in $\mathfrak{H}$ and

we can define
$$P(f)\psi = \lim_{n\to\infty} P(f_n)\psi, \qquad \psi \in \mathfrak{D}_f. \tag{3.28}$$

By construction, $P(f)$ is a linear operator such that (3.18) holds. Since $f \in L^1(\mathbb{R}, d\mu_\psi)$ (μ_ψ is finite), (3.17) also remains true at least for $\varphi = \psi$.

In addition, $\mathfrak{D}_f$ is dense. Indeed, let Ω_n be defined as in (3.27) and abbreviate $\psi_n = P(\Omega_n)\psi$. Now observe that $d\mu_{\psi_n} = \chi_{\Omega_n} d\mu_\psi$ and hence $\psi_n \in \mathfrak{D}_f$. Moreover, $\psi_n \to \psi$ by (3.18) since $\chi_{\Omega_n} \to 1$ in $L^2(\mathbb{R}, d\mu_\psi)$.

The operator $P(f)$ has some additional properties. One calls an unbounded operator A **normal** if $\mathfrak{D}(A) = \mathfrak{D}(A^*)$ and $\|A\psi\| = \|A^*\psi\|$ for all $\psi \in \mathfrak{D}(A)$. Note that normal operators are closed since the graph norms on $\mathfrak{D}(A) = \mathfrak{D}(A^*)$ are identical.

Theorem 3.2. *For every Borel function f, the operator*
$$P(f) \equiv \int_{\mathbb{R}} f(\lambda) dP(\lambda), \qquad \mathfrak{D}(P(f)) = \mathfrak{D}_f, \tag{3.29}$$

is normal and satisfies
$$\|P(f)\psi\|^2 = \int_{\mathbb{R}} |f(\lambda)|^2 d\mu_\psi(\lambda), \quad \langle \psi, P(f)\psi \rangle = \int_{\mathbb{R}} f(\lambda) d\mu_\psi(\lambda) \tag{3.30}$$

for $\psi \in \mathfrak{D}_f$.

Let f, g be Borel functions and $\alpha, \beta \in \mathbb{C}$. Then we have
$$P(f)^* = P(f^*), \tag{3.31}$$

$$\alpha P(f) + \beta P(g) \subseteq P(\alpha f + \beta g), \quad \mathfrak{D}(\alpha P(f) + \beta P(g)) = \mathfrak{D}_{|f|+|g|}, \tag{3.32}$$

and
$$P(f)P(g) \subseteq P(fg), \quad \mathfrak{D}(P(f)P(g)) = \mathfrak{D}_g \cap \mathfrak{D}_{fg}. \tag{3.33}$$

Proof. We begin by showing (3.31). Let f be given and define f_n, Ω_n as above. Since (3.31) holds for f_n by our previous theorem, we get
$$\langle \varphi, P(f)\psi \rangle = \langle P(f^*)\varphi, \psi \rangle$$
for every $\varphi, \psi \in \mathfrak{D}_f = \mathfrak{D}_{f^*}$ by continuity. Thus it remains to show that $\mathfrak{D}(P(f)^*) \subseteq \mathfrak{D}_f$. If $\psi \in \mathfrak{D}(P(f)^*)$, we have $\langle \psi, P(f)\varphi \rangle = \langle \tilde{\psi}, \varphi \rangle$ for all $\varphi \in \mathfrak{D}_f$ by definition. By construction of $P(f)$ we have $P(f_n) = P(f)P(\Omega_n)$ and thus
$$\langle P(f_n^*)\psi, \varphi \rangle = \langle \psi, P(f_n)\varphi \rangle = \langle \psi, P(f)P(\Omega_n)\varphi \rangle = \langle P(\Omega_n)\tilde{\psi}, \varphi \rangle$$
for every $\varphi \in \mathfrak{H}$ shows $P(f_n^*)\psi = P(\Omega_n)\tilde{\psi}$. This proves existence of the limit
$$\lim_{n\to\infty} \int_{\mathbb{R}} |f_n|^2 d\mu_\psi = \lim_{n\to\infty} \|P(f_n^*)\psi\|^2 = \lim_{n\to\infty} \|P(\Omega_n)\tilde{\psi}\|^2 = \|\tilde{\psi}\|^2,$$
which by monotone convergence implies $f \in L^2(\mathbb{R}, d\mu_\psi)$; that is, $\psi \in \mathfrak{D}_f$.

3.1. The spectral theorem

That $P(f)$ is normal follows from $\mathfrak{D}_f = \mathfrak{D}_{|f|} = \mathfrak{D}_{f^*}$ and (3.30), which implies $\|P(f)\psi\|^2 = \|P(|f|)\psi\|^2 = \|P(f^*)\psi\|^2$.

To show (3.32), note $\mathfrak{D}(\alpha P(f) + \beta P(g)) = \mathfrak{D}(P(f)) \cap \mathfrak{D}(P(g)) = \mathfrak{D}_f \cap \mathfrak{D}_g = \mathfrak{D}_{|f|+|g|}$ and set $f_n = \chi_{\Omega_n} f$, $g_n = \chi_{\Omega_n} g$, where $\Omega_n = \{\lambda | |f(\lambda)| + |g(\lambda)| \leq n\}$. Then $P(f_n)\psi \to P(f)\psi$, $P(g_n)\psi \to P(g)\psi$ and $\alpha P(f_n)\psi + \beta P(g_n)\psi = P(\alpha f_n + \beta g_n)\psi = P((\alpha f + \beta g)\chi_{\Omega_n})\psi \to P(\alpha f + \beta g)\psi$ for $\psi \in \mathfrak{D}_{|f|+|g|}$.

To show (3.33), we start with the case where g is bounded and define f_n, Ω_n as usual. Then $P(f_n) = P(f)P(\Omega_n)$ as noted above and $P(\Omega_n)P(g)\psi \to P(g)\psi$ plus $P(f)P(\Omega_n)P(g)\psi = P(f_n)P(g)\psi = P(f_n g)\psi \to P(fg)\psi$ for $\psi \in \mathfrak{D}_f$ shows $P(g)\psi \in \mathfrak{D}(P(f))$ and $P(f)P(g)\psi = P(fg)\psi$; that is, $P(f)P(g) = P(fg)$.

Finally, if g is unbounded, define g_n, Ω_n as usual. Then $P(g_n)\psi \to P(g)\psi$ and $P(f)P(g_n)\psi = P(fg_n)\psi \to P(fg)\psi$ for $\psi \in \mathfrak{D}_g \cap \mathfrak{D}_{fg}$; that is, $P(g)\psi \in \mathfrak{D}(P(f))$ and $P(f)P(g)\psi = P(fg)\psi$. □

These considerations seem to indicate some kind of correspondence between the operators $P(f)$ in $\mathfrak{H}$ and f in $L^2(\mathbb{R}, d\mu_\psi)$. Recall that $U : \mathfrak{H} \to \tilde{\mathfrak{H}}$ is called unitary if it is a bijection which preserves norms $\|U\psi\| = \|\psi\|$ (and hence scalar products). The operators A in $\mathfrak{H}$ and $\tilde{A}$ in $\tilde{\mathfrak{H}}$ are said to be **unitarily equivalent** if

$$(3.34) \qquad UA = \tilde{A}U, \qquad U\mathfrak{D}(A) = \mathfrak{D}(\tilde{A}).$$

Clearly, A is self-adjoint if and only if $\tilde{A}$ is and $\sigma(A) = \sigma(\tilde{A})$.

Now let us return to our original problem and consider the subspace

$$(3.35) \qquad \mathfrak{H}_\psi = \{P(g)\psi | g \in L^2(\mathbb{R}, d\mu_\psi)\} \subseteq \mathfrak{H}.$$

Note that $\mathfrak{H}_\psi$ is closed since L^2 is and $\psi_n = P(g_n)\psi$ converges in $\mathfrak{H}$ if and only if g_n converges in L^2. It even turns out that we can restrict $P(f)$ to $\mathfrak{H}_\psi$ (see Section 2.5).

Lemma 3.3. *The subspace $\mathfrak{H}_\psi$ reduces $P(f)$; that is, $P_\psi P(f) \subseteq P(f)P_\psi$, where P_ψ is the projection onto $\mathfrak{H}_\psi$.*

Proof. First, suppose f is bounded. Any $\varphi \in \mathfrak{H}$ can be decomposed as $\varphi = P(g)\psi + \varphi^\perp$. Moreover, $\langle P(h)\psi, P(f)\varphi^\perp \rangle = \langle P(f^*h)\psi, \varphi^\perp \rangle = 0$ for every bounded function h implies $P(f)\varphi^\perp \in \mathfrak{H}_\psi^\perp$. Hence $P_\psi P(f)\varphi = P_\psi P(f)P(g)\psi = P_\psi P(fg)\psi = P(f)P_\psi \varphi$, which by definition says that $\mathfrak{H}_\psi$ reduces $P(f)$.

If f is unbounded, we consider $f_n = f\chi_{\Omega_n}$ as before. Then, for every $\varphi \in \mathfrak{D}_f$, $P(f_n)P_\psi \varphi = P_\psi P(f_n)\varphi$. Letting $n \to \infty$, we have $P(\Omega_n)P_\psi \varphi \to P_\psi \varphi$ and $P(f_n)P_\psi \varphi = P(f)P(\Omega_n)P_\psi \varphi \to P_\psi P(f)\varphi$. Finally, closedness of $P(f)$ implies $P_\psi \varphi \in \mathfrak{D}_f$ and $P(f)P_\psi \varphi = P_\psi P(f)\varphi$. □

In particular, we can decompose $P(f) = P(f)|_{\mathfrak{H}_\psi} \oplus P(f)|_{\mathfrak{H}_\psi^\perp}$. Note that

(3.36) $$P_\psi \mathfrak{D}_f = \mathfrak{D}_f \cap \mathfrak{H}_\psi = \{P(g)\psi | g, fg \in L^2(\mathbb{R}, d\mu_\psi)\}$$

and $P(f)P(g)\psi = P(fg)\psi \in \mathfrak{H}_\psi$ in this case.

By (3.30), the operator

(3.37) $$\begin{aligned} U_\psi : \mathfrak{H}_\psi &\to L^2(\mathbb{R}, d\mu_\psi) \\ P(g)\psi &\mapsto g \end{aligned}$$

is unitary. Moreover, by (3.36) we have $U_\psi P_\psi \mathfrak{D}(P(f)) = U_\psi(\mathfrak{D}_f \cap \mathfrak{H}_\psi) = \{g \in L^2(\mathbb{R}, d\mu_\psi) | fg \in L^2(\mathbb{R}, d\mu_\psi)\} = \mathfrak{D}(f)$ and thus

(3.38) $$U_\psi P(f)|_{\mathfrak{H}_\psi} = f U_\psi,$$

where f is identified with its corresponding multiplication operator.

The vector ψ is called **cyclic** if $\mathfrak{H}_\psi = \mathfrak{H}$, and in this case, our picture is complete. Otherwise we need to extend this approach. A set $\{\psi_j\}_{j \in J}$ (J some index set) is called a set of spectral vectors if $\|\psi_j\| = 1$ and $\mathfrak{H}_{\psi_i} \perp \mathfrak{H}_{\psi_j}$ for all $i \neq j$. A set of spectral vectors is called a **spectral basis** if $\bigoplus_j \mathfrak{H}_{\psi_j} = \mathfrak{H}$. Luckily, a spectral basis always exists:

Lemma 3.4. *For every projection-valued measure P, there is a (at most countable) spectral basis $\{\psi_n\}$ such that*

(3.39) $$\mathfrak{H} = \bigoplus_n \mathfrak{H}_{\psi_n}$$

and a corresponding unitary operator

(3.40) $$U = \bigoplus_n U_{\psi_n} : \mathfrak{H} \to \bigoplus_n L^2(\mathbb{R}, d\mu_{\psi_n})$$

such that for every Borel function f,

(3.41) $$UP(f) = fU, \qquad U\mathfrak{D}_f = \mathfrak{D}(f).$$

Proof. It suffices to show that a spectral basis exists. This can be easily done using a Gram–Schmidt-type construction. First of all, observe that if $\{\psi_j\}_{j \in J}$ is a spectral set and $\psi \perp \mathfrak{H}_{\psi_j}$ for all j, we have $\mathfrak{H}_\psi \perp \mathfrak{H}_{\psi_j}$ for all j. Indeed, $\psi \perp \mathfrak{H}_{\psi_j}$ implies $P(g)\psi \perp \mathfrak{H}_{\psi_j}$ for every bounded function g since $\langle P(g)\psi, P(f)\psi_j \rangle = \langle \psi, P(g^*f)\psi_j \rangle = 0$. But $P(g)\psi$ with g bounded is dense in $\mathfrak{H}_\psi$, implying $\mathfrak{H}_\psi \perp \mathfrak{H}_{\psi_j}$.

Now start with some total set $\{\tilde\psi_j\}$. Normalize $\tilde\psi_1$ and choose this to be ψ_1. Move to the first $\tilde\psi_j$ which is not in $\mathfrak{H}_{\psi_1}$, project to the orthogonal complement of $\mathfrak{H}_{\psi_1}$ and normalize it. Choose the result to be ψ_2. Proceeding like this, we get a set of spectral vectors $\{\psi_j\}$ such that $\mathrm{span}\{\tilde\psi_j\} \subseteq \bigoplus_j \mathfrak{H}_{\psi_j}$. Hence $\mathfrak{H} = \overline{\mathrm{span}\{\tilde\psi_j\}} \subseteq \bigoplus_j \mathfrak{H}_{\psi_j}$. $\square$

3.1. The spectral theorem

It is important to observe that the cardinality of a spectral basis is *not* well-defined (in contradistinction to the cardinality of an ordinary basis of the Hilbert space). However, it can be at most equal to the cardinality of an ordinary basis. In particular, since $\mathfrak{H}$ is separable, it is at most countable. The minimal cardinality of a spectral basis is called the **spectral multiplicity** of P. If the spectral multiplicity is one, the spectrum is called **simple**.

Example. Let $\mathfrak{H} = \mathbb{C}^2$ and $A = \begin{pmatrix} 0 & 0 \\ 0 & 1 \end{pmatrix}$ and consider the associated projection-valued measure $P_A(\Omega)$ as before. Then $\psi_1 = (1,0)$ and $\psi_2 = (0,1)$ are a spectral basis. However, $\psi = (1,1)$ is cyclic and hence the spectrum of A is simple. If $A = \begin{pmatrix} 1 & 0 \\ 0 & 1 \end{pmatrix}$, there is no cyclic vector (why?) and hence the spectral multiplicity is two. ◇

Now observe that to every projection-valued measure P we can assign a self-adjoint operator $A = \int_{\mathbb{R}} \lambda dP(\lambda)$. The question is whether we can invert this map. To do this, we consider the resolvent $R_A(z) = \int_{\mathbb{R}} (\lambda - z)^{-1} dP(\lambda)$. From (3.17) the corresponding quadratic form is given by

$$(3.42) \qquad F_\psi(z) = \langle \psi, R_A(z)\psi \rangle = \int_{\mathbb{R}} \frac{1}{\lambda - z} d\mu_\psi(\lambda),$$

which is known as the **Borel transform** of the measure μ_ψ. By

$$(3.43) \qquad \text{Im}(F_\psi(z)) = \text{Im}(z) \int_{\mathbb{R}} \frac{1}{|\lambda - z|^2} d\mu_\psi(\lambda),$$

we infer that $F_\psi(z)$ is a holomorphic map from the upper half-plane into itself. Such functions are called **Herglotz** or **Nevanlinna functions** (see Section 3.4). Moreover, the measure μ_ψ can be reconstructed from $F_\psi(z)$ by the **Stieltjes inversion formula**

$$(3.44) \qquad \mu_\psi(\lambda) = \lim_{\delta \downarrow 0} \lim_{\varepsilon \downarrow 0} \frac{1}{\pi} \int_{-\infty}^{\lambda+\delta} \text{Im}(F_\psi(t + i\varepsilon)) dt.$$

(The limit with respect to δ is only here to ensure right continuity of $\mu_\psi(\lambda)$.) Conversely, if $F_\psi(z)$ is a Herglotz–Nevanlinna function satisfying $|F_\psi(z)| \leq \frac{M}{|\text{Im}(z)|}$, then it is the Borel transform of a unique measure μ_ψ (given by the Stieltjes inversion formula) satisfying $\mu_\psi(\mathbb{R}) \leq M$.

So let A be a given self-adjoint operator and consider the expectation of the resolvent of A,

$$(3.45) \qquad F_\psi(z) = \langle \psi, R_A(z)\psi \rangle.$$

This function is holomorphic for $z \in \rho(A)$ and satisfies

$$(3.46) \qquad F_\psi(z^*) = F_\psi(z)^* \quad \text{and} \quad |F_\psi(z)| \leq \frac{\|\psi\|^2}{|\text{Im}(z)|}$$

(see (2.72) and Theorem 2.19). Moreover, the first resolvent formula (2.84) shows that it maps the upper half-plane to itself:

$$\text{Im}(F_\psi(z)) = \text{Im}(z)\|R_A(z)\psi\|^2; \tag{3.47}$$

that is, it is a Herglotz–Nevanlinna function. So by our above remarks, there is a corresponding measure $\mu_\psi(\lambda)$ given by the Stieltjes inversion formula. It is called the **spectral measure** corresponding to ψ.

More generally, by polarization (2.17), for each $\varphi, \psi \in \mathfrak{H}$ we can find a corresponding complex measure $\mu_{\varphi,\psi}$ such that

$$\langle \varphi, R_A(z)\psi \rangle = \int_\mathbb{R} \frac{1}{\lambda - z} d\mu_{\varphi,\psi}(\lambda). \tag{3.48}$$

The measure $\mu_{\varphi,\psi}$ is conjugate linear in φ and linear in ψ since a complex measure is uniquely determined by its Borel transform (Problem 3.29). Moreover, a comparison with our previous considerations begs us to define a family of operators via the sesquilinear forms

$$s_\Omega(\varphi, \psi) = \int_\mathbb{R} \chi_\Omega(\lambda) d\mu_{\varphi,\psi}(\lambda). \tag{3.49}$$

Since the associated quadratic form is nonnegative, $q_\Omega(\psi) = s_\Omega(\psi, \psi) = \mu_\psi(\Omega) \geq 0$, the Cauchy–Schwarz inequality for sesquilinear forms (Problem 0.24) implies $|s_\Omega(\varphi, \psi)| \leq q_\Omega(\varphi)^{1/2} q_\Omega(\psi)^{1/2} = \mu_\varphi(\Omega)^{1/2} \mu_\psi(\Omega)^{1/2} \leq \mu_\varphi(\mathbb{R})^{1/2} \mu_\psi(\mathbb{R})^{1/2} \leq \|\varphi\| \|\psi\|$. Hence, Corollary 1.9 implies that there is indeed a family of nonnegative ($0 \leq \langle \psi, P_A(\Omega)\psi \rangle \leq \|\psi\|^2$) and hence self-adjoint operators $P_A(\Omega)$ such that

$$\langle \varphi, P_A(\Omega)\psi \rangle = \int_\mathbb{R} \chi_\Omega(\lambda) d\mu_{\varphi,\psi}(\lambda). \tag{3.50}$$

Lemma 3.5. *The family of operators $P_A(\Omega)$ forms a projection-valued measure.*

Proof. We first show $P_A(\Omega_1) P_A(\Omega_2) = P_A(\Omega_1 \cap \Omega_2)$ in two steps. First observe (using the first resolvent formula (2.84))

$$\int_\mathbb{R} \frac{1}{\lambda - \tilde{z}} d\mu_{R_A(z^*)\varphi, \psi}(\lambda) = \langle R_A(z^*)\varphi, R_A(\tilde{z})\psi \rangle = \langle \varphi, R_A(z)R_A(\tilde{z})\psi \rangle$$

$$= \frac{1}{z - \tilde{z}} (\langle \varphi, R_A(z)\psi \rangle - \langle \varphi, R_A(\tilde{z})\psi \rangle)$$

$$= \frac{1}{z - \tilde{z}} \int_\mathbb{R} \left(\frac{1}{\lambda - z} - \frac{1}{\lambda - \tilde{z}} \right) d\mu_{\varphi,\psi}(\lambda) = \int_\mathbb{R} \frac{1}{\lambda - \tilde{z}} \frac{d\mu_{\varphi,\psi}(\lambda)}{\lambda - z},$$

3.1. The spectral theorem

implying $d\mu_{R_A(z^*)\varphi,\psi}(\lambda) = (\lambda - z)^{-1} d\mu_{\varphi,\psi}(\lambda)$ by Problem 3.29. Secondly, we compute

$$\int_{\mathbb{R}} \frac{1}{\lambda - z} d\mu_{\varphi,P_A(\Omega)\psi}(\lambda) = \langle \varphi, R_A(z) P_A(\Omega) \psi \rangle = \langle R_A(z^*)\varphi, P_A(\Omega) \psi \rangle$$

$$= \int_{\mathbb{R}} \chi_\Omega(\lambda) d\mu_{R_A(z^*)\varphi,\psi}(\lambda) = \int_{\mathbb{R}} \frac{1}{\lambda - z} \chi_\Omega(\lambda) d\mu_{\varphi,\psi}(\lambda),$$

implying $d\mu_{\varphi,P_A(\Omega)\psi}(\lambda) = \chi_\Omega(\lambda) d\mu_{\varphi,\psi}(\lambda)$. Equivalently, we have

$$\langle \varphi, P_A(\Omega_1) P_A(\Omega_2) \psi \rangle = \langle \varphi, P_A(\Omega_1 \cap \Omega_2) \psi \rangle$$

since $\chi_{\Omega_1} \chi_{\Omega_2} = \chi_{\Omega_1 \cap \Omega_2}$. In particular, choosing $\Omega_1 = \Omega_2$, we see that $P_A(\Omega_1)$ is a projector.

To see $P_A(\mathbb{R}) = \mathbb{I}$, let $\psi \in \text{Ker}(P_A(\mathbb{R}))$. Then $0 = d\mu_{\varphi,P_A(\mathbb{R})\psi}(\lambda) = \chi_{\mathbb{R}}(\lambda) d\mu_{\varphi,\psi}(\lambda) = d\mu_{\varphi,\psi}(\lambda)$ implies $\langle \varphi, R_A(z) \psi \rangle = 0$ which implies $\psi = 0$.

Now let $\Omega = \bigcup_{n=1}^{\infty} \Omega_n$ with $\Omega_n \cap \Omega_m = \emptyset$ for $n \neq m$. Then

$$\sum_{j=1}^{n} \langle \psi, P_A(\Omega_j) \psi \rangle = \sum_{j=1}^{n} \mu_\psi(\Omega_j) \to \langle \psi, P_A(\Omega) \psi \rangle = \mu_\psi(\Omega)$$

by σ-additivity of μ_ψ. Hence P_A is weakly σ-additive which implies strong σ-additivity, as pointed out earlier. $\square$

Now we can prove the **spectral theorem** for self-adjoint operators.

Theorem 3.6 (Spectral theorem). *To every self-adjoint operator A there corresponds a unique projection-valued measure P_A such that*

$$(3.51) \qquad A = \int_{\mathbb{R}} \lambda dP_A(\lambda).$$

Proof. Existence has already been established. Moreover, Theorem 3.1 shows that $P_A((\lambda - z)^{-1}) = R_A(z)$, $z \in \mathbb{C} \setminus \mathbb{R}$. Since the measures $\mu_{\varphi,\psi}$ are uniquely determined by the resolvent and the projection-valued measure is uniquely determined by the measures $\mu_{\varphi,\psi}$, we are done. $\square$

The quadratic form of A is given by

$$(3.52) \qquad q_A(\psi) = \int_{\mathbb{R}} \lambda d\mu_\psi(\lambda)$$

and can be defined for every ψ in the **form domain**

$$(3.53) \qquad \mathfrak{Q}(A) = \mathfrak{D}(|A|^{1/2}) = \{\psi \in \mathfrak{H} | \int_{\mathbb{R}} |\lambda| d\mu_\psi(\lambda) < \infty\}$$

(which is larger than the domain $\mathfrak{D}(A) = \{\psi \in \mathfrak{H} | \int_{\mathbb{R}} \lambda^2 d\mu_\psi(\lambda) < \infty\}$). This extends our previous definition for nonnegative operators.

Note that if A and $\tilde{A}$ are unitarily equivalent as in (3.34), then $UR_A(z) = R_{\tilde{A}}(z)U$ and hence

(3.54) $$d\mu_\psi = d\tilde{\mu}_{U\psi}.$$

In particular, we have $UP_A(f) = P_{\tilde{A}}(f)U$, $U\mathfrak{D}(P_A(f)) = \mathfrak{D}(P_{\tilde{A}}(f))$.

Finally, let us give a characterization of the spectrum of A in terms of the associated projectors.

Theorem 3.7. *The spectrum of A is given by*

(3.55) $$\sigma(A) = \{\lambda \in \mathbb{R} | P_A((\lambda - \varepsilon, \lambda + \varepsilon)) \neq 0 \text{ for all } \varepsilon > 0\}.$$

Proof. Let $\Omega_n = (\lambda_0 - \frac{1}{n}, \lambda_0 + \frac{1}{n})$. Suppose $P_A(\Omega_n) \neq 0$. Then we can find a $\psi_n \in P_A(\Omega_n)\mathfrak{H}$ with $\|\psi_n\| = 1$. Since

$$\|(A - \lambda_0)\psi_n\|^2 = \|(A - \lambda_0)P_A(\Omega_n)\psi_n\|^2$$
$$= \int_\mathbb{R} (\lambda - \lambda_0)^2 \chi_{\Omega_n}(\lambda) d\mu_{\psi_n}(\lambda) \leq \frac{1}{n^2},$$

we conclude $\lambda_0 \in \sigma(A)$ by Lemma 2.17.

Conversely, if $P_A((\lambda_0 - \varepsilon, \lambda_0 + \varepsilon)) = 0$, set

$$f_\varepsilon(\lambda) = \chi_{\mathbb{R}\setminus(\lambda_0-\varepsilon,\lambda_0+\varepsilon)}(\lambda)(\lambda - \lambda_0)^{-1}.$$

Then

$$(A - \lambda_0)P_A(f_\varepsilon) = P_A((\lambda - \lambda_0)f_\varepsilon(\lambda)) = P_A(\mathbb{R}\setminus(\lambda_0 - \varepsilon, \lambda_0 + \varepsilon)) = \mathbb{I}.$$

Similarly, $P_A(f_\varepsilon)(A - \lambda_0) = \mathbb{I}|_{\mathfrak{D}(A)}$ and hence $\lambda_0 \in \rho(A)$. $\square$

In particular, $P_A((\lambda_1, \lambda_2)) = 0$ if and only if $(\lambda_1, \lambda_2) \subseteq \rho(A)$.

Corollary 3.8. *We have*

(3.56) $$P_A(\sigma(A)) = \mathbb{I} \quad \text{and} \quad P_A(\mathbb{R} \cap \rho(A)) = 0.$$

Proof. For every $\lambda \in \mathbb{R} \cap \rho(A)$ there is some open interval I_λ with $P_A(I_\lambda) = 0$. These intervals form an open cover for $\mathbb{R} \cap \rho(A)$, and there is a countable subcover J_n. Setting $\Omega_n = J_n \setminus \bigcup_{m<n} J_m$, we have disjoint Borel sets which cover $\mathbb{R} \cap \rho(A)$ and satisfy $P_A(\Omega_n) = 0$. Finally, strong σ-additivity shows $P_A(\mathbb{R} \cap \rho(A))\psi = \sum_n P_A(\Omega_n)\psi = 0$. $\square$

Consequently,

(3.57) $$P_A(f) = P_A(\sigma(A))P_A(f) = P_A(\chi_{\sigma(A)}f).$$

In other words, $P_A(f)$ is not affected by the values of f on $\mathbb{R}\setminus\sigma(A)$!

3.1. The spectral theorem

It is clearly more intuitive to write $P_A(f) = f(A)$ and we will do so from now on. This notation is justified by the elementary observation

$$(3.58) \qquad P_A(\sum_{j=0}^{n} \alpha_j \lambda^j) = \sum_{j=0}^{n} \alpha_j A^j.$$

Moreover, this also shows that if A is bounded and $f(A)$ can be defined via a convergent power series, then this agrees with our present definition by Theorem 3.1.

Problem 3.1. *Show that a self-adjoint operator P is a projection if and only if $\sigma(P) \subseteq \{0,1\}$.*

Problem 3.2. *Consider the **parity operator** $\Pi : L^2(\mathbb{R}^d) \to L^2(\mathbb{R}^d)$, $\psi(x) \mapsto \psi(-x)$. Show that Π is self-adjoint. Compute its spectrum $\sigma(\Pi)$ and the corresponding projection-valued measure P_Π.*

Problem 3.3. *Show that (3.7) is a projection-valued measure. What is the corresponding operator?*

Problem 3.4. *Show that $P(\lambda)$ defined in (3.14) satisfies properties (i)–(iv) stated there.*

Problem 3.5. *Show that for a self-adjoint operator A we have $\|R_A(z)\| = \operatorname{dist}(z, \sigma(A))^{-1}$.*

Problem 3.6. *Suppose A is self-adjoint and $\|B - z_0\| \leq r$. Show that $\sigma(A+B) \subseteq \sigma(A) + \overline{B_r(z_0)}$, where $B_r(z_0)$ is the ball of radius r around z_0. (Hint: Problem 2.18.)*

Problem 3.7. *Show that for a self-adjoint operator A we have $\|AR_A(z)\| \leq \frac{|z|}{|\operatorname{Im}(z)|}$. Find some A for which equality is attained.*

Conclude that for every $\psi \in \mathfrak{H}$ we have

$$(3.59) \qquad \lim_{z \to \infty} \|AR_A(z)\psi\| = 0,$$

where the limit is taken in any sector $\varepsilon|\operatorname{Re}(z)| \leq |\operatorname{Im}(z)|$, $\varepsilon > 0$.

Problem 3.8. *Suppose A is self-adjoint. Show that, if $\psi \in \mathfrak{D}(A^n)$, then*

$$(3.60) \qquad R_A(z)\psi = -\sum_{j=0}^{n} \frac{A^j \psi}{z^{j+1}} + O\left(\frac{\|A^n \psi\|}{|z|^n \operatorname{Im}(z)}\right), \quad \text{as} \quad z \to \infty.$$

(Hint: Proceed as in (2.90) and use the previous problem.)

Problem 3.9. *Suppose A is self-adjoint. Let λ_0 be an eigenvalue and ψ a corresponding normalized eigenvector. Compute μ_ψ.*

Problem 3.10. *Suppose A is self-adjoint. Show*
$$\mathrm{Ran}(P_A(\{\lambda_0\})) = \mathrm{Ker}(A - \lambda_0).$$
(Hint: Start by verifying $\mathrm{Ran}(P_A(\{\lambda_0\})) \subseteq \mathrm{Ker}(A-\lambda_0)$. To see the converse, let $\psi \in \mathrm{Ker}(A - \lambda_0)$ and use the previous example to compute $\langle \psi, P_A(\{\lambda_0\})\psi\rangle$.)

3.2. More on Borel measures

Section 3.1 showed that in order to understand self-adjoint operators, one needs to understand multiplication operators on $L^2(\mathbb{R}, d\mu)$, where $d\mu$ is a finite Borel measure. This is the purpose of the present section.

The set of all **growth points**, that is,

(3.61) $\quad\sigma(\mu) = \{\lambda \in \mathbb{R} | \mu((\lambda - \varepsilon, \lambda + \varepsilon)) > 0 \text{ for all } \varepsilon > 0\},$

is called the spectrum of μ. The same proof as for Corollary 3.8 shows that the spectrum $\sigma = \sigma(\mu)$ is a support for μ; that is, $\mu(\mathbb{R}\backslash\sigma) = 0$. In fact, it coincides with the (topological) support, $\sigma(\mu) = \mathrm{supp}(\mu)$, as defined in (A.7).

In the previous section we have already seen that the Borel transform of μ,

(3.62) $$F(z) = \int_\mathbb{R} \frac{1}{\lambda - z} d\mu(\lambda),$$

plays an important role.

Theorem 3.9. *The Borel transform of a finite Borel measure is a Herglotz–Nevanlinna function. It is holomorphic in $\mathbb{C}\backslash\sigma(\mu)$ and satisfies*

(3.63) $\quad F(z^*) = F(z)^*, \qquad |F(z)| \leq \dfrac{\mu(\mathbb{R})}{\mathrm{Im}(z)}, \qquad z \in \mathbb{C}_+.$

Proof. First of all, note
$$\mathrm{Im}(F(z)) = \int_\mathbb{R} \mathrm{Im}\left(\frac{1}{\lambda - z}\right) d\mu(\lambda) = \mathrm{Im}(z) \int_\mathbb{R} \frac{d\mu(\lambda)}{|\lambda - z|^2},$$
which shows that F maps $\mathbb{C}_+$ to $\mathbb{C}_+$. Moreover, $F(z^*) = F(z)^*$ is obvious and
$$|F(z)| \leq \int_\mathbb{R} \frac{d\mu(\lambda)}{|\lambda - z|} \leq \frac{1}{\mathrm{Im}(z)} \int_\mathbb{R} d\mu(\lambda)$$
establishes the bound. Moreover, since $\mu(\mathbb{R}\backslash\sigma) = 0$, we have
$$F(z) = \int_\sigma \frac{1}{\lambda - z} d\mu(\lambda),$$

which together with the bound

$$\frac{1}{|\lambda - z|} \le \frac{1}{\text{dist}(z, \sigma)}$$

allows the application of the dominated convergence theorem to conclude that F is continuous on $\mathbb{C}\backslash\sigma$. To show that F is holomorphic in $\mathbb{C}\backslash\sigma$, by Morera's theorem, it suffices to check $\int_\Gamma F(z)dz = 0$ for every triangle $\Gamma \subset \mathbb{C}\backslash\sigma$. Since $(\lambda - z)^{-1}$ is bounded for $(\lambda, z) \in \sigma \times \Gamma$, this follows from $\int_\Gamma (\lambda - z)^{-1}dz = 0$ by using Fubini, $\int_\Gamma F(z)dz = \int_\Gamma \int_\mathbb{R} (\lambda - z)^{-1}d\mu(\lambda)\, dz = \int_\mathbb{R} \int_\Gamma (\lambda - z)^{-1}dz\, d\mu(\lambda) = 0$. □

Note that F cannot be holomorphically extended to a larger domain. In fact, if F is holomorphic in a neighborhood of some $\lambda \in \mathbb{R}$, then $F(\lambda) = F(\lambda^*) = F(\lambda)^*$ implies $\text{Im}(F(\lambda)) = 0$, and the Stieltjes inversion formula (Theorem 3.23) shows that $\lambda \in \mathbb{R}\backslash\sigma(\mu)$.

Associated with this measure is the operator

(3.64) $\quad Af(\lambda) = \lambda f(\lambda), \qquad \mathfrak{D}(A) = \{f \in L^2(\mathbb{R}, d\mu) | \lambda f(\lambda) \in L^2(\mathbb{R}, d\mu)\}.$

By Theorem 3.7 the spectrum of A is precisely the spectrum of μ; that is,

(3.65) $\quad \sigma(A) = \sigma(\mu).$

Note that $1 \in L^2(\mathbb{R}, d\mu)$ is a cyclic vector for A and that

(3.66) $\quad d\mu_{g,f}(\lambda) = g(\lambda)^* f(\lambda)d\mu(\lambda).$

Now what can we say about the function $f(A)$ (which is precisely the multiplication operator by f) of A? We are only interested in the case where f is real-valued. Introduce the measure

(3.67) $\quad (f_\star\mu)(\Omega) = \mu(f^{-1}(\Omega)).$

Then

(3.68) $\quad \int_\mathbb{R} g(\lambda)d(f_\star\mu)(\lambda) = \int_\mathbb{R} g(f(\lambda))d\mu(\lambda).$

In fact, it suffices to check this formula for simple functions g, which follows since $\chi_\Omega \circ f = \chi_{f^{-1}(\Omega)}$. In particular, we have

(3.69) $\quad P_{f(A)}(\Omega) = \chi_{f^{-1}(\Omega)}.$

It is tempting to conjecture that $f(A)$ is unitarily equivalent to multiplication by λ in $L^2(\mathbb{R}, d(f_\star\mu))$ via the map

(3.70) $\quad L^2(\mathbb{R}, d(f_\star\mu)) \to L^2(\mathbb{R}, d\mu), \quad g \mapsto g \circ f.$

However, this map is only unitary if its range is $L^2(\mathbb{R}, d\mu)$.

Lemma 3.10. *Suppose f is injective. Then*
$$U : L^2(\mathbb{R}, d\mu) \to L^2(\mathbb{R}, d(f_\star\mu)), \quad g \mapsto g \circ f^{-1} \tag{3.71}$$
is a unitary map such that $Uf(\lambda) = \lambda$.

Example. Let $f(\lambda) = \lambda^2$. Then $(g \circ f)(\lambda) = g(\lambda^2)$ and the range of the above map is given by the symmetric functions. Note that we can still get a unitary map $L^2(\mathbb{R}, d(f_\star\mu)) \oplus L^2(\mathbb{R}, d(f_\star\mu)) \to L^2(\mathbb{R}, d\mu)$, $(g_1, g_2) \mapsto g_1(\lambda^2) + g_2(\lambda^2)(\chi_{(0,\infty)}(\lambda) - \chi_{(0,\infty)}(-\lambda))$. ◇

Lemma 3.11. *Let f be real-valued. The spectrum of $f(A)$ is given by*
$$\sigma(f(A)) = \sigma(f_\star\mu). \tag{3.72}$$
In particular,
$$\sigma(f(A)) \subseteq \overline{f(\sigma(A))}, \tag{3.73}$$
where equality holds if f is continuous and the closure can be dropped if, in addition, $\sigma(A)$ is bounded (i.e., compact) or $|f(\lambda)| \to \infty$ for $|\lambda| \to \infty$.

Proof. The first formula follows by comparing
$$\sigma(f_\star\mu) = \{\lambda \in \mathbb{R} \mid \mu(f^{-1}(\lambda - \varepsilon, \lambda + \varepsilon)) > 0 \text{ for all } \varepsilon > 0\}$$
with (2.77).

Next, let $\lambda \notin \overline{f(\sigma(A))}$. Then there exists an $\varepsilon > 0$ such that $(\lambda - \varepsilon, \lambda + \varepsilon) \not\subset \overline{f(\sigma(A))}$ and thus $\|(f(A) - \lambda)^{-1}\| \leq \varepsilon^{-1}$ which shows $\lambda \notin \sigma(f(A))$.

If f is continuous, $f^{-1}((f(\lambda) - \varepsilon, f(\lambda) + \varepsilon))$ contains an open interval around λ and hence $f(\lambda) \in \sigma(f(A))$ if $\lambda \in \sigma(A)$. If, in addition, $\sigma(A)$ is compact, then $f(\sigma(A))$ is compact and hence closed. Similarly, if $|f(\lambda)| \to \infty$ for $|\lambda| \to \infty$, suppose $\eta_n = f(\lambda_n) \to \eta$. Then, λ_n is bounded (otherwise η_n were also unbounded and could not be convergent), and we can pass to a convergent subsequence $\lambda_{n_m} \to \lambda \in \sigma(A)$ such that $\eta = f(\lambda) \in f(\sigma(A))$. □

Whether two operators with simple spectrum are unitarily equivalent can be read off from the corresponding measures:

Lemma 3.12. *Let A_1, A_2 be self-adjoint operators with simple spectrum and corresponding spectral measures μ_1 and μ_2 of cyclic vectors. Then A_1 and A_2 are unitarily equivalent if and only if μ_1 and μ_2 are mutually absolutely continuous.*

Proof. Without restriction, we can assume that A_j is multiplication by λ in $L^2(\mathbb{R}, d\mu_j)$. Let $U : L^2(\mathbb{R}, d\mu_1) \to L^2(\mathbb{R}, d\mu_2)$ be a unitary map such that $UA_1 = A_2 U$. Then we also have $Uf(A_1) = f(A_2)U$ for every bounded Borel function and hence
$$Uf(\lambda) = Uf(\lambda) \cdot 1 = f(\lambda)U(1)(\lambda)$$

3.2. More on Borel measures

and thus U is multiplication by $u(\lambda) = U(1)(\lambda)$. Moreover, since U is unitary, we have

$$\mu_1(\Omega) = \int_\mathbb{R} |\chi_\Omega|^2 d\mu_1 = \int_\mathbb{R} |u\,\chi_\Omega|^2 d\mu_2 = \int_\Omega |u|^2 d\mu_2;$$

that is, $d\mu_1 = |u|^2 d\mu_2$. Reversing the roles of A_1 and A_2, we obtain $d\mu_2 = |v|^2 d\mu_1$, where $v = U^{-1}1$.

The converse is left as an exercise (Problem 3.18). □

Next, we recall the unique decomposition of μ with respect to Lebesgue measure,

(3.74) $$d\mu = d\mu_{ac} + d\mu_s,$$

where μ_{ac} is **absolutely continuous** with respect to Lebesgue measure (i.e., we have $\mu_{ac}(B) = 0$ for all B with Lebesgue measure zero) and μ_s is **singular** with respect to Lebesgue measure (i.e., μ_s is supported, $\mu_s(\mathbb{R}\backslash B) = 0$, on a set B with Lebesgue measure zero). The singular part μ_s can be further decomposed into a **(singularly) continuous** and a **pure point** part,

(3.75) $$d\mu_s = d\mu_{sc} + d\mu_{pp},$$

where μ_{sc} is continuous on $\mathbb{R}$ and μ_{pp} is a step function. Since the measures $d\mu_{ac}$, $d\mu_{sc}$, and $d\mu_{pp}$ are mutually singular, they have mutually disjoint supports M_{ac}, M_{sc}, and M_{pp}. Note that these sets are *not* unique. We will choose them such that M_{pp} is the set of all jumps of $\mu(\lambda)$ and such that M_{sc} has Lebesgue measure zero.

To the sets M_{ac}, M_{sc}, and M_{pp} correspond projectors $P^{ac} = \chi_{M_{ac}}(A)$, $P^{sc} = \chi_{M_{sc}}(A)$, and $P^{pp} = \chi_{M_{pp}}(A)$ satisfying $P^{ac} + P^{sc} + P^{pp} = \mathbb{I}$. In other words, we have a corresponding direct sum decomposition of both our Hilbert space

(3.76) $$L^2(\mathbb{R}, d\mu) = L^2(\mathbb{R}, d\mu_{ac}) \oplus L^2(\mathbb{R}, d\mu_{sc}) \oplus L^2(\mathbb{R}, d\mu_{pp})$$

and our operator

(3.77) $$A = (AP^{ac}) \oplus (AP^{sc}) \oplus (AP^{pp}).$$

The corresponding spectra, $\sigma_{ac}(A) = \sigma(\mu_{ac})$, $\sigma_{sc}(A) = \sigma(\mu_{sc})$, and $\sigma_{pp}(A) = \sigma(\mu_{pp})$ are called the absolutely continuous, singularly continuous, and pure point spectrum of A, respectively.

It is important to observe that $\sigma_{pp}(A)$ is in general not equal to the set of eigenvalues

(3.78) $$\sigma_p(A) = \{\lambda \in \mathbb{R} | \lambda \text{ is an eigenvalue of } A\}$$

since we only have $\sigma_{pp}(A) = \overline{\sigma_p(A)}$.

Example. Let $\mathfrak{H} = \ell^2(\mathbb{N})$ and let A be given by $A\delta_n = \frac{1}{n}\delta_n$, where δ_n is the sequence which is 1 at the n'th place and zero otherwise (that is, A is a diagonal matrix with diagonal elements $\frac{1}{n}$). Then $\sigma_p(A) = \{\frac{1}{n}|n \in \mathbb{N}\}$ but $\sigma(A) = \sigma_{pp}(A) = \sigma_p(A) \cup \{0\}$. To see this, just observe that δ_n is the eigenvector corresponding to the eigenvalue $\frac{1}{n}$ and for $z \notin \sigma(A)$ we have $R_A(z)\delta_n = \frac{n}{1-nz}\delta_n$. At $z = 0$ this formula still gives the inverse of A, but it is unbounded and hence $0 \in \sigma(A)$ but $0 \notin \sigma_p(A)$. Since a continuous measure cannot live on a single point and hence also not on a countable set, we have $\sigma_{ac}(A) = \sigma_{sc}(A) = \emptyset$. ◇

Example. An example with purely absolutely continuous spectrum is given by taking μ to be the Lebesgue measure. An example with purely singularly continuous spectrum is given by taking μ to be the Cantor measure. ◇

Finally, we show how the spectrum can be read off from the boundary values of $\mathrm{Im}(F)$ toward the real line. We define the following sets:

$$M_{ac} = \{\lambda | 0 < \limsup_{\varepsilon \downarrow 0} \mathrm{Im}(F(\lambda + i\varepsilon)) < \infty\},$$

(3.79)
$$M_s = \{\lambda | \limsup_{\varepsilon \downarrow 0} \mathrm{Im}(F(\lambda + i\varepsilon)) = \infty\},$$

$$M_p = \{\lambda | \lim_{\varepsilon \downarrow 0} \varepsilon\, \mathrm{Im}(F(\lambda + i\varepsilon)) > 0\},$$

$$M = M_{ac} \cup M_s = \{\lambda | 0 < \limsup_{\varepsilon \downarrow 0} \mathrm{Im}(F(\lambda + i\varepsilon))\}.$$

Then, by Lemma 3.25 and Theorem 3.27, we conclude that these sets are minimal supports for μ_{ac}, μ_s, μ_{pp}, and μ, respectively. In fact, by Theorem 3.27, we could even restrict ourselves to values of λ, where the lim sup is a lim (finite or infinite).

Lemma 3.13. *The spectrum of μ is given by*

(3.80) $\quad \sigma(\mu) = \overline{M}, \quad M = \{\lambda | 0 < \liminf_{\varepsilon \downarrow 0} \mathrm{Im}(F(\lambda + i\varepsilon))\}.$

Proof. First, observe that F is real holomorphic near $\lambda \notin \sigma(\mu)$ and hence $\mathrm{Im}(F(\lambda)) = 0$ in this case. Thus $M \subseteq \sigma(\mu)$ and since $\sigma(\mu)$ is closed, we even have $\overline{M} \subseteq \sigma(\mu)$. To see the converse, note that by Theorem 3.27, the set M is a support for M. Thus, if $\lambda \in \sigma(\mu)$, then

$$0 < \mu((\lambda - \varepsilon, \lambda + \varepsilon)) = \mu((\lambda - \varepsilon, \lambda + \varepsilon) \cap M)$$

for all $\varepsilon > 0$ and we can find a sequence $\lambda_n \in (\lambda - 1/n, \lambda + 1/n) \cap M$ converging to λ from inside M. This shows the remaining part $\sigma(\mu) \subseteq \overline{M}$. □

By Lemma 3.25, the set M_p is precisely the set of point masses of μ and thus

(3.81) $$\sigma(\mu_{pp}) = \overline{M_p}.$$

3.2. More on Borel measures

To recover $\sigma(\mu_{ac})$ from M_{ac}, we need the **essential closure** of a Borel set $\Omega \subseteq \mathbb{R}$,

(3.82) $\quad \overline{\Omega}^{ess} = \{\lambda \in \mathbb{R} | |(\lambda - \varepsilon, \lambda + \varepsilon) \cap \Omega| > 0 \text{ for all } \varepsilon > 0\},$

where $|\Omega|$ denotes the Lebesgue measure of a Borel set Ω. Note that $\overline{\Omega}^{ess}$ is closed, whereas, in contradistinction to the ordinary closure, we might have $\Omega \not\subseteq \overline{\Omega}^{ess}$ (e.g., every isolated point of Ω will disappear).

Lemma 3.14. *The absolutely continuous spectrum of μ is given by*

(3.83) $\quad\quad\quad\quad\quad\quad \sigma(\mu_{ac}) = \overline{M}_{ac}^{ess}.$

Proof. We use that $0 < \mu_{ac}((\lambda - \varepsilon, \lambda + \varepsilon)) = \mu_{ac}((\lambda - \varepsilon, \lambda + \varepsilon) \cap M_{ac})$ is equivalent to $|(\lambda - \varepsilon, \lambda + \varepsilon) \cap M_{ac}| > 0$. One direction follows from the definition of absolute continuity and the other from minimality of M_{ac}. $\square$

Problem 3.11. Show that the set of eigenvalues of the operator A of multiplication by λ in $L^2(\mathbb{R}, d\mu)$ is precisely the set of point masses of μ: $\sigma_p(A) = M_p$.

Problem 3.12. Construct a multiplication operator on $L^2(\mathbb{R})$ which has dense point spectrum.

Problem 3.13. Let λ be Lebesgue measure on $\mathbb{R}$, and let f be a strictly increasing function. Show that

$$d(f_\star \lambda) = d(f^{-1}),$$

where f^{-1} is the inverse of f extended to all of $\mathbb{R}$ by setting $f^{-1}(y) = x$ for $y \in [f(x-), f(x+)]$.

Moreover, if $f \in AC(\mathbb{R})$ with $f' > 0$, then

$$d(f_\star \lambda) = \frac{1}{f'(f^{-1}(\lambda))} d\lambda.$$

Problem 3.14. Let $d\mu(\lambda) = \chi_{[0,1]}(\lambda) d\lambda$ and $f(\lambda) = \chi_{(-\infty, t]}(\lambda)$, $t \in \mathbb{R}$. Compute $f_\star \mu$.

Problem 3.15. Let A be the multiplication operator by the Cantor function in $L^2(0, 1)$. Compute the spectrum of A. Determine the spectral types.

Problem 3.16. Find a multiplication operator in $L^2(0, 1)$ with purely singular continuous spectrum. (Hint: Find a function whose inverse is the Cantor function and use Problem 3.13.)

Problem 3.17. Let A be multiplication by λ in $L^2(\mathbb{R})$. Find a function f such that $\sigma(f(A)) \neq \overline{f(\sigma(A))}$. Moreover, find a continuous function f such that $\sigma(f(A)) \neq f(\sigma(A))$.

Problem 3.18. Show the missing direction in the proof of Lemma 3.12.

Problem 3.19. Show $\overline{(\Omega_1 \cup \Omega_2)}^{ess} = \overline{\Omega_1}^{ess} \cup \overline{\Omega_2}^{ess}$ and $\overline{\Omega}^{ess} = \emptyset$ if $|\Omega| = 0$. Moreover, show that $\overline{\Omega}^{ess}$ is closed and satisfies $\overline{\Omega}^{ess} \subseteq \overline{\Omega}$.

3.3. Spectral types

Our next aim is to transfer the results of the previous section to arbitrary self-adjoint operators A using Lemma 3.4. To this end, we will need a spectral measure which contains the information from all measures in a spectral basis. This will be the case if there is a vector ψ such that for every $\varphi \in \mathfrak{H}$ its spectral measure μ_φ is absolutely continuous with respect to μ_ψ. Such a vector will be called a **maximal spectral vector** of A, and μ_ψ will be called a **maximal spectral measure** of A.

Lemma 3.15. *For every self-adjoint operator A there is a maximal spectral vector.*

Proof. Let $\{\psi_j\}_{j \in J}$ be a spectral basis and choose nonzero numbers ε_j with $\sum_{j \in J} |\varepsilon_j|^2 = 1$. Then I claim that

$$\psi = \sum_{j \in J} \varepsilon_j \psi_j$$

is a maximal spectral vector. Let φ be given. Then we can write it as $\varphi = \sum_j f_j(A)\psi_j$ and hence $d\mu_\varphi = \sum_j |f_j|^2 d\mu_{\psi_j}$. But $\mu_\psi(\Omega) = \sum_j |\varepsilon_j|^2 \mu_{\psi_j}(\Omega) = 0$ implies $\mu_{\psi_j}(\Omega) = 0$ for every $j \in J$ and thus $\mu_\varphi(\Omega) = 0$. $\square$

A set $\{\psi_j\}$ of spectral vectors is called **ordered** if ψ_k is a maximal spectral vector for A restricted to $(\bigoplus_{j=1}^{k-1} \mathfrak{H}_{\psi_j})^\perp$. As in the unordered case one can show

Theorem 3.16. *For every self-adjoint operator there is an ordered spectral basis.*

Observe that if $\{\psi_j\}$ is an ordered spectral basis, then $\mu_{\psi_{j+1}}$ is absolutely continuous with respect to μ_{ψ_j}.

If μ is a maximal spectral measure, we have $\sigma(A) = \sigma(\mu)$ and the following generalization of Lemma 3.11 holds.

Theorem 3.17 (Spectral mapping). *Let μ be a maximal spectral measure and let $f : \mathbb{R} \to \mathbb{C}$. Then the spectrum of $f(A)$ is given by*

(3.84) $\quad \sigma(f(A)) = \{z \in \mathbb{C} | \mu(f^{-1}(B_\varepsilon(z))) > 0 \text{ for all } \varepsilon > 0\}.$

In particular,

(3.85) $\quad \sigma(f(A)) \subseteq \overline{f(\sigma(A))},$

where equality holds if f is continuous and the closure can be dropped if, in addition, $\sigma(A)$ is bounded or $|f(\lambda)| \to \infty$ for $|\lambda| \to \infty$.

3.3. Spectral types

Next, we want to introduce the splitting (3.76) for arbitrary self-adjoint operators A. It is tempting to pick a spectral basis and treat each summand in the direct sum separately. However, since it is not clear that this approach is independent of the spectral basis chosen, we use the more sophisticated definition

$$\mathfrak{H}_{ac} = \{\psi \in \mathfrak{H} | \mu_\psi \text{ is absolutely continuous}\},$$
$$\mathfrak{H}_{sc} = \{\psi \in \mathfrak{H} | \mu_\psi \text{ is singularly continuous}\},$$
(3.86)
$$\mathfrak{H}_{pp} = \{\psi \in \mathfrak{H} | \mu_\psi \text{ is pure point}\}.$$

Lemma 3.18. *We have*

(3.87) $$\mathfrak{H} = \mathfrak{H}_{ac} \oplus \mathfrak{H}_{sc} \oplus \mathfrak{H}_{pp}.$$

There are Borel sets M_{xx} such that the projector onto $\mathfrak{H}_{xx}$ is given by $P^{xx} = \chi_{M_{xx}}(A)$, $xx \in \{ac, sc, pp\}$. In particular, the subspaces $\mathfrak{H}_{xx}$ reduce A. For the sets M_{xx} one can choose the corresponding supports of some maximal spectral measure μ.

Proof. We will use the unitary operator U of Lemma 3.4. Pick $\varphi \in \mathfrak{H}$ and write $\varphi = \sum_n \varphi_n$ with $\varphi_n \in \mathfrak{H}_{\psi_n}$. Let $f_n = U\varphi_n$. Then, by construction of the unitary operator U, $\varphi_n = f_n(A)\psi_n$ and hence $d\mu_{\varphi_n} = |f_n|^2 d\mu_{\psi_n}$. Moreover, since the subspaces $\mathfrak{H}_{\psi_n}$ are orthogonal, we have

$$d\mu_\varphi = \sum_n |f_n|^2 d\mu_{\psi_n}$$

and hence

$$d\mu_{\varphi,xx} = \sum_n |f_n|^2 d\mu_{\psi_n,xx}, \qquad xx \in \{ac, sc, pp\}.$$

This shows

$$U\mathfrak{H}_{xx} = \bigoplus_n L^2(\mathbb{R}, d\mu_{\psi_n,xx}), \qquad xx \in \{ac, sc, pp\},$$

and it reduces our problem to the considerations of the previous section.

Furthermore, note that if μ is a maximal spectral measure, then every support for μ_{xx} is also a support for $\mu_{\varphi,xx}$ for every $\varphi \in \mathfrak{H}$. $\square$

The **absolutely continuous**, **singularly continuous**, and **pure point spectrum** of A are defined as
(3.88)
$$\sigma_{ac}(A) = \sigma(A|_{\mathfrak{H}_{ac}}), \quad \sigma_{sc}(A) = \sigma(A|_{\mathfrak{H}_{sc}}), \quad \text{and} \quad \sigma_{pp}(A) = \sigma(A|_{\mathfrak{H}_{pp}}),$$

respectively. If μ is a maximal spectral measure, we have $\sigma_{ac}(A) = \sigma(\mu_{ac})$, $\sigma_{sc}(A) = \sigma(\mu_{sc})$, and $\sigma_{pp}(A) = \sigma(\mu_{pp})$.

If A and $\tilde{A}$ are unitarily equivalent via U, then so are $A|_{\mathfrak{H}_{xx}}$ and $\tilde{A}|_{\tilde{\mathfrak{H}}_{xx}}$ by (3.54). In particular, $\sigma_{xx}(A) = \sigma_{xx}(\tilde{A})$.

Problem 3.20. *Compute $\sigma(A)$, $\sigma_{ac}(A)$, $\sigma_{sc}(A)$, and $\sigma_{pp}(A)$ for the multiplication operator $A = \frac{1}{1+x^2}$ in $L^2(\mathbb{R})$. What is its spectral multiplicity?*

Problem 3.21. *Show Theorem 3.17. What needs to be changed in the proof of the second part of Lemma 3.11 to cover complex-valued functions?*

3.4. Appendix: Herglotz–Nevanlinna functions

Let $\mathbb{C}_{\pm} = \{z \in \mathbb{C} | \pm \operatorname{Im}(z) > 0\}$ be the upper, respectively, lower, half-plane. A holomorphic function $F : \mathbb{C}_+ \to \overline{\mathbb{C}_+}$ mapping the upper half-plane to itself is called a Herglotz–Nevanlinna function. The sum of two Herglotz–Nevanlinna functions is again one and so is the composition. We can define F on $\mathbb{C}_-$ using $F(z^*) = F(z)^*$.

Example. The following are examples of Herglotz–Nevanlinna functions: $F(z) = a + bz$ if $\operatorname{Im}(a), b \geq 0$ and $F(z) = \frac{1}{\lambda - z}$ if $\operatorname{Im}(\lambda) \leq 0$. Note that we have $F(z^*) = F(z)^*$ only if $a \in \mathbb{R}$ and $\lambda \in \mathbb{R}$ in the previous two examples. $F(z) = \log(z)$ and $F(z) = \sqrt{z}$ provided we use the standard branches with $\arg(z) \in (-\pi, \pi]$. ◇

In Theorem 3.9 we have seen that the Borel transform of a finite measure is a Herglotz–Nevanlinna function satisfying a growth estimate. Here we want to show that the converse is also true. Even though our interest is mainly in Herglotz–Nevanlinna functions, we will need the connection with their counterparts on the unit disc which will allow for simpler proofs below.

Recall that

$$(3.89) \qquad z \mapsto \frac{\mathrm{i} - z}{\mathrm{i} + z}$$

is a conformal bijection from $\mathbb{C}_+$ to the unit disc $\mathbb{D} = \{z \in \mathbb{C} | |z| < 1\}$ and hence

$$(3.90) \qquad F(z) = \mathrm{i}C\left(\frac{\mathrm{i} - z}{\mathrm{i} + z}\right)$$

is a Herglotz–Nevanlinna function if and only if $C(z)$ is a Carathéodory function; that is, $C : \mathbb{D} \to \{z \in \mathbb{C} | \operatorname{Re}(z) \geq 0\}$. Note that the converse is given by

$$(3.91) \qquad C(z) = -\mathrm{i}F\left(\mathrm{i}\frac{\mathrm{i} - z}{\mathrm{i} + z}\right).$$

Theorem 3.19 (Herglotz representation)**.** *Every Carathéodory function is of the form*

$$(3.92) \qquad C(z) = \mathrm{i}c + \int_{[-\pi,\pi]} \frac{\mathrm{e}^{\mathrm{i}\varphi} + z}{\mathrm{e}^{\mathrm{i}\varphi} - z} d\nu(\varphi)$$

3.4. Appendix: Herglotz–Nevanlinna functions

for some finite measure $d\nu$ and some real constant c. Explicitly, we have

$$(3.93) \qquad \int_{[-\pi,\pi]} d\nu = \text{Re}(C(0)), \qquad c = \text{Im}(C(0)).$$

Proof. Let $C : \mathbb{D} \to \mathbb{C}$ be a Carathéodory function and $0 < r < 1$. Then, for every z with $|z| < r$, we have

$$C(z) = \frac{1}{4\pi i} \int_{|\zeta|=r} \left(\frac{\zeta + z}{\zeta - z} + \frac{r^2/\zeta + z^*}{r^2/\zeta - z^*} \right) C(\zeta) \frac{d\zeta}{\zeta}$$

$$= \frac{1}{2\pi i} \int_{|\zeta|=r} \text{Re}\left(\frac{\zeta + z}{\zeta - z} \right) C(\zeta) \frac{d\zeta}{\zeta}.$$

Taking real parts and setting $\zeta = re^{i\varphi}$, we see

$$\text{Re}(C(z)) = \int_{-\pi}^{\pi} P_{|z|/r}(\arg(z) - \varphi) d\nu_r(\varphi),$$

where

$$P_r(\varphi) = \text{Re}\left(\frac{1 + re^{i\varphi}}{1 - re^{i\varphi}} \right) = \frac{1 - r^2}{1 - 2r\cos(\varphi) + r^2}, \qquad d\nu_r(\varphi) = \text{Re}(C(re^{i\varphi})) \frac{d\varphi}{2\pi}.$$

In particular, taking $z = 0$, we have

$$\int_{-\pi}^{\pi} d\nu_r(\varphi) = \text{Re}(C(0)) < \infty$$

and hence (since for $|z| < 1$ the functions $P_{|z|/r}$ converge uniformly to $P_{|z|}$ as $r \uparrow 1$)

$$\text{Re}(C(z)) = \lim_{r \uparrow 1} \int_{-\pi}^{\pi} P_{|z|/r}(\arg(z) - \varphi) d\nu_r(\varphi)$$

$$= \lim_{r \uparrow 1} \int_{-\pi}^{\pi} P_{|z|}(\arg(z) - \varphi) d\nu_r(\varphi).$$

Moreover, since the sequence of measures ν_r is bounded, Lemma A.35 implies that there is a subsequence which converges vaguely to some measure $d\nu$ which satisfies

$$\text{Re}(C(z)) = \int_{[-\pi,\pi]} P_{|z|}(\arg(z) - \varphi) d\nu(\varphi) = \int_{[-\pi,\pi]} \text{Re}\left(\frac{e^{i\varphi} + z}{e^{i\varphi} - z} \right) d\nu(\varphi).$$

Since a holomorphic function is determined by its real part up to an imaginary constant, the claim follows. $\square$

The corresponding result for Herglotz–Nevanlinna functions now follows via the connection (3.90).

Theorem 3.20. *Every Herglotz–Nevanlinna function is of the form*

$$F(z) = bz + a + \int_{\mathbb{R}} \left(\frac{1}{\lambda - z} - \frac{\lambda}{1 + \lambda^2} \right) d\mu(\lambda),$$

(3.94)
$$= bz + a + \int_{\mathbb{R}} \frac{1 + z\lambda}{\lambda - z} \frac{d\mu(\lambda)}{1 + \lambda^2}$$

with

(3.95) $$b \geq 0, a \in \mathbb{R}, \qquad \int_{\mathbb{R}} \frac{d\mu(\lambda)}{1 + \lambda^2} < \infty.$$

Conversely, given a, b, μ as above, $F(z)$ given by (3.94) is holomorphic in $\mathbb{C} \setminus \operatorname{supp}(\mu)$ where $\operatorname{supp}(\mu)$ is the (topological) support of μ and satisfies $F(z^*) = F(z)^*$. Moreover,

(3.96) $$\operatorname{Im}(F(z)) = \operatorname{Im}(z) \left(b + \int_{\mathbb{R}} \frac{d\mu(\lambda)}{|\lambda - z|^2} \right), \qquad z \in \mathbb{C} \setminus \operatorname{supp}(\mu),$$

as well as

(3.97) $$\frac{dF(z)}{dz} = b + \int_{\mathbb{R}} \frac{d\mu(\lambda)}{(\lambda - z)^2}, \qquad z \in \mathbb{C} \setminus \operatorname{supp}(\mu).$$

Proof. The homeomorphism $f : (-\pi, \pi) \to \mathbb{R}$, $\varphi \mapsto \mathrm{i}\frac{1-e^{\mathrm{i}\varphi}}{1+e^{\mathrm{i}\varphi}}$ maps our measure ν to a finite measure $\tilde{\mu} = f_* \mu$, defined via $\tilde{\mu}(A) = \nu(f^{-1}(A))$, such that (Theorem A.31)

$$\int_{\mathbb{R}} g(f^{-1}(\lambda)) d\tilde{\mu}(\lambda) = \int_{(-\pi, \pi)} g(\varphi) d\nu(\varphi).$$

In particular, with $g(\varphi) = \frac{e^{\mathrm{i}\varphi} + z}{e^{\mathrm{i}\varphi} - z}$, $f^{-1}(\lambda) = \arg(\frac{\mathrm{i}-\lambda}{\mathrm{i}+\lambda})$, we obtain that any Herglotz–Nevanlinna function $F(z) = \mathrm{i}C(\frac{\mathrm{i}-z}{\mathrm{i}+z})$ is of the form (3.94) with

$$d\mu(\lambda) = (1 + \lambda^2) d\tilde{\mu}(\lambda), \qquad a = -c, \qquad b = \nu(\{-\pi, \pi\}).$$

The last part follows as in Theorem 3.9, using the estimate

$$\left| \frac{1 + z\lambda}{\lambda - z} \right| \leq \frac{1 + |z|(|\operatorname{Re}(\lambda - z)| + |\operatorname{Re}(z)|)}{\sqrt{\operatorname{Re}(\lambda - z)^2 + \operatorname{Im}(z)^2}} \leq |z| + \frac{1 + |\operatorname{Re}(z)z|}{\operatorname{dist}(z, \operatorname{supp}(\mu))}$$

$\square$

3.4. Appendix: Herglotz–Nevanlinna functions

Example. The Herglotz representations of the functions from the previous examples are given by

$$\frac{1}{\lambda_0 - z} = \int_\mathbb{R} \frac{1}{\lambda - z} d\Theta(\lambda - \lambda_0) \, d\lambda, \quad z \in \mathbb{C}\setminus\{\lambda_0\},$$

$$\log(z) = \int_{-\infty}^0 \left(\frac{1}{\lambda - z} - \frac{\lambda}{1 + \lambda^2}\right) d\lambda, \quad z \in \mathbb{C}\setminus(-\infty, 0],$$

$$\sqrt{z} = \frac{1}{\sqrt{2}} + \int_{-\infty}^0 \left(\frac{1}{\lambda - z} - \frac{\lambda}{1 + \lambda^2}\right)\sqrt{-\lambda} \, d\lambda, \quad z \in \mathbb{C}\setminus(-\infty, 0].$$

◇

The general representation (3.94) can be simplified in case of a growth estimate as follows:

Corollary 3.21. *Let μ be the measure associated with a Herglotz–Nevanlinna function $F(z)$. Then*

$$(3.98) \qquad b = 0, \quad \int_\mathbb{R} d\mu \le M \quad \Leftrightarrow \quad \mathrm{Im}(F(iy)) \le \frac{M}{y}.$$

Moreover, in this case,

$$(3.99) \qquad F(z) = \tilde{a} + \int_\mathbb{R} \frac{d\mu(\lambda)}{\lambda - z}, \quad \tilde{a} = a - \int_\mathbb{R} \frac{\lambda}{1 + \lambda^2} d\mu(\lambda) \in \mathbb{R},$$

and

$$(3.100) \qquad F(z) = \tilde{a} - \frac{1}{z}\int_\mathbb{R} d\mu + o(z^{-1}),$$

as $z \to \infty$ in any sector in the upper (or lower) half-plane.

Proof. Note that the second condition reads

$$(3.101) \qquad y\,\mathrm{Im}(F(iy)) = by^2 + \int_\mathbb{R} \frac{y^2}{\lambda^2 + y^2} d\mu(\lambda) \le M$$

and implies $b = 0$ and $\int_\mathbb{R} d\mu \le M$ (use monotone convergence). The converse is straightforward.

For the last claim, observe

$$F(z) = \tilde{a} - \frac{\mu(\mathbb{R})}{z} + \frac{1}{z}\int_\mathbb{R} \frac{\lambda}{z - \lambda} d\mu(\lambda)$$

and use dominated convergence with the elementary estimate

$$\left|\frac{\lambda}{\lambda - z}\right| \le \frac{|z|}{|\mathrm{Im}(z)|}.$$

□

In particular, note the useful formula

(3.102) $$\lim_{\lambda \to \infty} \lambda \, \text{Im}(F(i\lambda)) = \mu(\mathbb{R}).$$

If we have the stronger condition $|F(iy)| \leq \frac{M}{y}$, then we must also have $\tilde{a} = 0$ in the last corollary.

Theorem 3.22. *A function F is a Herglotz–Nevanlinna function satisfying*

(3.103) $$|F(iy)| \leq \frac{M}{y}$$

if and only if it is the Borel transform of a finite measure μ.

Our next question is if μ is uniquely determined by F. This is answered by the following theorem:

Theorem 3.23. *Let F be a Herglotz–Nevanlinna function. Then the measure μ in (3.94) is unique and can be reconstructed via the Stieltjes inversion formula*

(3.104) $$\frac{1}{2}\left(\mu((\lambda_1, \lambda_2)) + \mu([\lambda_1, \lambda_2])\right) = \lim_{\varepsilon \downarrow 0} \frac{1}{\pi} \int_{\lambda_1}^{\lambda_2} \text{Im}(F(\lambda + i\varepsilon))d\lambda$$

for any $\lambda_1 < \lambda_2$. The constants a, b are given by

(3.105) $$a = \text{Re}(F(i)), \quad b = \text{Im}(F(i)) - \int_{\mathbb{R}} \frac{d\mu(\lambda)}{1 + \lambda^2}.$$

Proof. Without loss of generality, we can assume $b = 0$. By Fubini we have

$$\frac{1}{\pi} \int_{\lambda_1}^{\lambda_2} \text{Im}(F(\lambda + i\varepsilon))d\lambda = \frac{1}{\pi} \int_{\lambda_1}^{\lambda_2} \int_{\mathbb{R}} \frac{\varepsilon}{(x - \lambda)^2 + \varepsilon^2} d\mu(x) d\lambda$$

$$= \int_{\mathbb{R}} \frac{1}{\pi} \int_{\lambda_1}^{\lambda_2} \frac{\varepsilon}{(x - \lambda)^2 + \varepsilon^2} d\lambda \, d\mu(x),$$

where

$$\frac{1}{\pi} \int_{\lambda_1}^{\lambda_2} \frac{\varepsilon}{(x - \lambda)^2 + \varepsilon^2} d\lambda = \frac{1}{\pi}\left(\arctan\left(\frac{\lambda_2 - x}{\varepsilon}\right) - \arctan\left(\frac{\lambda_1 - x}{\varepsilon}\right)\right)$$

$$\to \frac{1}{2}\left(\chi_{[\lambda_1, \lambda_2]}(x) + \chi_{(\lambda_1, \lambda_2)}(x)\right)$$

pointwise. Hence the result follows from the dominated convergence theorem since $0 \leq \frac{1}{\pi}\left(\arctan(\frac{\lambda_2 - x}{\varepsilon}) - \arctan(\frac{\lambda_1 - x}{\varepsilon})\right) \leq 1$. □

Of course this implies that there is a one-to-one correspondence between Herglotz–Nevanlinna functions and triples (a, b, μ) satisfying $a \in \mathbb{R}$, $b \geq 0$, and $\int_{\mathbb{R}} \frac{d\mu(\lambda)}{1+\lambda^2} < \infty$ given by (3.94).

We can even strengthen the content of the last theorem a bit:

3.4. Appendix: Herglotz–Nevanlinna functions

Corollary 3.24. *Let F be a Herglotz–Nevanlinna function with associated measure μ. The measures $\pi^{-1}\operatorname{Im}(F(\lambda+i\varepsilon))d\lambda$ converge vaguely to $d\mu$ as $\varepsilon \downarrow 0$. In fact, we have*

$$\tag{3.106} \lim_{\varepsilon \downarrow 0} \frac{1}{\pi}\int_{\mathbb{R}} f(\lambda)\operatorname{Im}(F(\lambda+i\varepsilon))d\lambda = \int_{\mathbb{R}} f\, d\mu$$

for every continuous f satisfying $|f(\lambda)| \le \frac{C}{1+\lambda^2}$ and for any $f \in C_b(\mathbb{R})$ provided $b=0$ and $\mu(\mathbb{R}) < \infty$. Furthermore,

$$\tag{3.107} \lim_{\varepsilon \downarrow 0} \frac{1}{\pi}\int_{\lambda_0}^{\lambda_1} f(\lambda)\operatorname{Im}(F(\lambda+i\varepsilon))d\lambda = \frac{1}{2}\left(\int_{(\lambda_0,\lambda_1)} f\, d\mu + \int_{[\lambda_0,\lambda_1]} f\, d\mu\right)$$

for any $\lambda_1 < \lambda_2$ and every $f \in C[\lambda_0, \lambda_1]$.

Proof. As in the previous theorem, using Fubini, we have

$$\lim_{\varepsilon \downarrow 0} \frac{1}{\pi}\int_{\lambda_0}^{\lambda_1} f(\lambda)\operatorname{Im}(F(\lambda+i\varepsilon))d\lambda = \lim_{\varepsilon \downarrow 0}\int_{\mathbb{R}} f_\varepsilon d\mu,$$

where

$$f_\varepsilon(x) = \frac{1}{\pi}\int_{\lambda_0}^{\lambda_1} f(\lambda)\frac{\varepsilon}{(\lambda-x)^2+\varepsilon^2}d\lambda.$$

Now the claim follows from dominated convergence since

$$|f_\varepsilon(x)| \le \frac{1}{\pi}\int_{\mathbb{R}} \frac{C}{1+\lambda^2}\frac{\varepsilon}{(\lambda-x)^2+\varepsilon^2}d\lambda = \frac{(1+\varepsilon)C}{(1+\varepsilon)^2+x^2}$$

and

$$\lim_{\varepsilon \downarrow 0} f_\varepsilon(x) = \begin{cases} f(x), & x \in (\lambda_0, \lambda_1), \\ \frac{1}{2}f(\lambda_j), & x = \lambda_j,\ j=0,1, \\ 0, & \text{else.} \end{cases}$$

The last case follows using $|f_\varepsilon(x)| \le \frac{1}{\pi}\int_{\mathbb{R}} C\frac{\varepsilon}{(\lambda-x)^2+\varepsilon^2}d\lambda = C$ if $|f(\lambda)| \le C$. □

This last result raises questions of how properties of μ can be read off from the boundary behavior of $F(\lambda+i\varepsilon)$ as $\varepsilon \downarrow 0$. As a first result we show:

Lemma 3.25. *Let F be a Herglotz–Nevanlinna function with associated measure μ. Then*

$$\tag{3.108} \lim_{\varepsilon \downarrow 0} \varepsilon \operatorname{Im}(F(\lambda+i\varepsilon)) = \mu(\{\lambda\}).$$

Proof. This follows from

$$\lim_{\varepsilon \downarrow 0} \varepsilon \operatorname{Im}(F(\lambda+i\varepsilon)) = \lim_{\varepsilon \downarrow 0}\int_{\mathbb{R}} \frac{\varepsilon^2}{(x-\lambda)^2+\varepsilon^2}d\mu(x) = \int_{\mathbb{R}} \chi_{\{\lambda\}}d\mu = \mu(\{\lambda\})$$

by virtue of the dominated convergence theorem. □

Furthermore, the Radon–Nikodym derivative of μ can be obtained from the boundary values of F.

Theorem 3.26. *Let F be a Herglotz–Nevanlinna function with associated measure μ. Then*
(3.109)
$$(\underline{D}\mu)(\lambda) \le \liminf_{\varepsilon \downarrow 0} \frac{1}{\pi} \operatorname{Im}\left(F(\lambda + i\varepsilon)\right) \le \limsup_{\varepsilon \downarrow 0} \frac{1}{\pi} \operatorname{Im}\left(F(\lambda + i\varepsilon)\right) \le (\overline{D}\mu)(\lambda).$$

Proof. Without loss of generality, we can assume $b = 0$. We need to estimate
$$\operatorname{Im}(F(\lambda + i\varepsilon)) = \int_{\mathbb{R}} K_\varepsilon(t - \lambda) d\mu(t), \qquad K_\varepsilon(t) = \frac{\varepsilon}{t^2 + \varepsilon^2}.$$
We first split the integral into two parts:
$$\operatorname{Im}(F(\lambda + i\varepsilon)) = \int_{I_\delta} K_\varepsilon(t-\lambda) d\mu(t) + \int_{\mathbb{R} \setminus I_\delta} K_\varepsilon(t-\lambda) d\mu(t), \quad I_\delta = (\lambda - \delta, \lambda + \delta).$$
Clearly the second part can be estimated by
$$\int_{\mathbb{R} \setminus I_\delta} K_\varepsilon(t - \lambda) d\mu(t) \le K_\varepsilon(\delta) \mu(\mathbb{R}).$$
To estimate the first part, we integrate
$$K'_\varepsilon(s) \, ds \, d\mu(t)$$
over the triangle $\{(s,t) | \lambda - s < t < \lambda + s, \ 0 < s < \delta\} = \{(s,t) | \lambda - \delta < t < \lambda + \delta, \ |t - \lambda| < s < \delta\}$ and obtain
$$\int_0^\delta \mu(I_s) K'_\varepsilon(s) ds = \int_{I_\delta} (K_\varepsilon(\delta) - K_\varepsilon(t - \lambda)) d\mu(t).$$
Now suppose there are constants c and C such that $c \le \frac{\mu(I_s)}{2s} \le C$, $0 \le s \le \delta$. Then
$$2c \arctan(\frac{\delta}{\varepsilon}) \le \int_{I_\delta} K_\varepsilon(t - \lambda) d\mu(t) \le 2C \arctan(\frac{\delta}{\varepsilon})$$
since
$$\delta K_\varepsilon(\delta) + \int_0^\delta -s K'_\varepsilon(s) ds = \arctan(\frac{\delta}{\varepsilon}).$$
Thus the claim follows combining both estimates. □

As a consequence of Theorem A.45 and Theorem A.46, we obtain (cf. also Lemma A.47)

Theorem 3.27. *Let F be a Herglotz–Nevanlinna function with associated measure μ. Then the limit*
(3.110)
$$\operatorname{Im}(F(\lambda)) = \lim_{\varepsilon \downarrow 0} \operatorname{Im}(F(\lambda + i\varepsilon))$$

3.4. Appendix: Herglotz–Nevanlinna functions

exists a.e. with respect to both μ and Lebesgue measure (finite or infinite) and

$$(D\mu)(\lambda) = \frac{1}{\pi} \operatorname{Im}(F(\lambda)) \tag{3.111}$$

whenever $(D\mu)(\lambda)$ exists.

Moreover, the set $\{\lambda|\operatorname{Im}(F(\lambda)) = \infty\}$ is a support for the singularly continuous part and $\{\lambda|0 < \operatorname{Im}(F(\lambda)) < \infty\}$ is a minimal support for the absolutely continuous part.

In particular,

Corollary 3.28. *The measure μ is purely absolutely continuous on I if $\limsup_{\varepsilon \downarrow 0} \operatorname{Im}(F(\lambda + i\varepsilon)) < \infty$ for all $\lambda \in I$.*

The limit of the real part can be computed as well.

Corollary 3.29. *The limit*

$$\lim_{\varepsilon \downarrow 0} F(\lambda + i\varepsilon) \tag{3.112}$$

exists a.e. with respect to both μ and Lebesgue measure. It is finite a.e. with respect to Lebesgue measure.

Proof. If $F(z)$ is a Herglotz–Nevanlinna function, then so is $\sqrt{F(z)}$. Moreover, $\sqrt{F(z)}$ has values in the first quadrant; that is, both $\operatorname{Re}(\sqrt{F(z)})$ and $\operatorname{Im}(\sqrt{F(z)})$ are positive for $z \in \mathbb{C}_+$. Hence both $\sqrt{F(z)}$ and $i\sqrt{F(z)}$ are Herglotz–Nevanlinna functions, and by virtue of Theorem 3.27, both limits $\lim_{\varepsilon \downarrow 0} \operatorname{Re}(\sqrt{F(\lambda + i\varepsilon)})$ and $\lim_{\varepsilon \downarrow 0} \operatorname{Im}(\sqrt{F(\lambda + i\varepsilon)})$ exist and are finite a.e. with respect to Lebesgue measure. By taking squares, the same is true for $F(z)$ and hence $\lim_{\varepsilon \downarrow 0} F(\lambda + i\varepsilon)$ exists and is finite a.e. with respect to Lebesgue measure. Since $\lim_{\varepsilon \downarrow 0} \operatorname{Im}(F(\lambda + i\varepsilon)) = \infty$ implies $\lim_{\varepsilon \downarrow 0} F(\lambda + i\varepsilon) = \infty$, the result follows. $\square$

Finally, we note that the growth of F along the imaginary axis is related to the growth of its associated measure.

Lemma 3.30. *Let F be a Herglotz–Nevanlinna function with associated measure μ and $b = 0$. Then, for every $0 < \gamma < 2$, we have*

$$\int_{\mathbb{R}} \frac{d\mu(\lambda)}{1 + |\lambda|^\gamma} < \infty \iff \int_1^\infty \frac{\operatorname{Im}(F(iy))}{y^\gamma} dy < \infty. \tag{3.113}$$

Proof. First of all, note that we can split $F(z) = F_1(z) + F_2(z)$ according to $d\mu = \chi_{[-1,1]} d\mu + (1 + \chi_{[-1,1]}) d\mu$. The part $F_1(z)$ corresponds to a finite measure and does not contribute by (3.103). Hence we can assume that μ is not supported near 0. Then Fubini shows

$$\int_0^\infty \frac{\mathrm{Im}(F(iy))}{y^\gamma} dy = \int_0^\infty \int_\mathbb{R} \frac{y^{1-\gamma}}{\lambda^2 + y^2} d\mu(\lambda) dy = \frac{\pi/2}{\sin(\gamma\pi/2)} \int_\mathbb{R} \frac{1}{|\lambda|^\gamma} d\mu(\lambda),$$

which proves the claim. Here we have used (Problem 3.36)

$$\int_0^\infty \frac{y^{1-\gamma}}{\lambda^2 + y^2} dy = \frac{\pi/2}{|\lambda|^\gamma \sin(\gamma\pi/2)}.$$

$\square$

The case $\gamma = 0$ is the content of Theorem 3.22, and for the case $\gamma = 2$, see Problem 3.37.

Problem 3.22. *Let μ be a finite Borel measure and F its Borel transform. Show that (3.103) holds with $M = \mu(\mathbb{R})$.*

Problem 3.23. *Show*

(3.114) $$\left|\frac{1}{\lambda - z}\right| \leq \frac{1}{1+|\lambda|} \frac{1+|z|}{|\mathrm{Im}(z)|}$$

and

(3.115) $$\left|\frac{1}{\lambda - z} - \frac{\lambda}{1+\lambda^2}\right| \leq \frac{1}{1+\lambda^2} \frac{1+|z|^2}{|\mathrm{Im}(z)|}$$

for every $\lambda \in \mathbb{R}$. (Hint: First search for the maximum of $|\lambda - z|^{-1}$ and $|\lambda||\lambda - z|^{-1}$ as a function of λ (cf. also Problem 3.7).)

Problem 3.24. *Let F be a Herglotz–Nevanlinna function. Show that*

$$F(z) = bz + o(z)$$

as $z \to \infty$ in any sector in the upper (or lower) half-plane. If in addition $\int_\mathbb{R} |\lambda| d\mu(\lambda) < \infty$, show that

$$F(z) = bz + \tilde{a} + \int_\mathbb{R} \frac{d\mu(\lambda)}{\lambda - z} = bz + \tilde{a} + o(1)$$

as $z \to \infty$ in any sector in the upper (or lower) half-plane. (Hint: Use Problem 3.23.)

Problem 3.25. *Let F be a Herglotz–Nevanlinna function with associated measure μ. Show the estimate*

$$\mu([\lambda - \varepsilon, \lambda + \varepsilon]) \leq 2\varepsilon \, \mathrm{Im}(F(\lambda + i\varepsilon)), \qquad \varepsilon > 0.$$

Conclude that μ is Hölder continuous of exponent $\alpha \in (0, 1]$ if $C(\lambda) = \sup_{\varepsilon \in (0,1]} \varepsilon^{1-\alpha} \mathrm{Im}(F(\lambda + i\varepsilon))$ is locally bounded.

3.4. Appendix: Herglotz–Nevanlinna functions

Problem 3.26. *Find all rational Herglotz functions $F : \mathbb{C} \to \mathbb{C}$ satisfying $F(z^*) = F(z)^*$ and $\lim_{|z| \to \infty} |zF(z)| = M < \infty$. What can you say about the zeros of F?*

Problem 3.27. *Show that every non-constant Herglotz–Nevanlinna function maps $\mathbb{C}_+$ to $\mathbb{C}_+$.*

Problem 3.28. *Let F be a Herglotz–Nevanlinna function corresponding to a discrete measure (i.e., μ is a pure point measure with isolated masses). Show that F has only first order zeros and poles on the real line which are interlacing. (Hint: Note that F is strictly increasing on the real line, away from its poles.)*

Problem 3.29. *A complex measure $d\mu$ is a measure which can be written as a complex linear combination of positive measures $d\mu_j$:*

$$d\mu = d\mu_1 - d\mu_2 + \mathrm{i}(d\mu_3 - d\mu_4).$$

Let

$$F(z) = \int_\mathbb{R} \frac{d\mu}{\lambda - z}$$

be the Borel transform of a complex measure. Show that μ is uniquely determined by F via the Stieltjes inversion formula

$$\frac{1}{2}\left(\mu((\lambda_1, \lambda_2)) + \mu([\lambda_1, \lambda_2])\right) = \lim_{\varepsilon \downarrow 0} \frac{1}{2\pi\mathrm{i}} \int_{\lambda_1}^{\lambda_2} (F(\lambda + \mathrm{i}\varepsilon) - F(\lambda - \mathrm{i}\varepsilon))d\lambda.$$

Problem 3.30. *Compute the Borel transform of the complex measure given by $d\mu(\lambda) = \frac{d\lambda}{(\lambda - \mathrm{i})^2}$.*

Problem 3.31. *Show that the second part of Corollary 3.24 extends to the case where $\lambda_1 = -\infty$ or $\lambda_2 = \infty$ under the conditions used in the first part.*

Problem 3.32. *Let F be a Herglotz–Nevanlinna function. Then*

$$\lim_{\varepsilon \downarrow 0} \varepsilon \operatorname{Re}\left(F(\lambda + \mathrm{i}\varepsilon)\right) = 0.$$

Problem 3.33 (Exponential Herglotz representation)**.** *Show that every Herglotz–Nevanlinna function can be written as*

$$F(z) = \exp\left(c + \int_\mathbb{R} \left(\frac{1}{\lambda - z} - \frac{\lambda}{1 + \lambda^2}\right)\xi(\lambda)d\lambda\right),$$

where

$$\xi(\lambda) = \lim_{\varepsilon \downarrow 0} \frac{1}{\pi} \arg(F(\lambda + \mathrm{i}\varepsilon))$$

for a.e. λ and $c = \log|F(\mathrm{i})|$. Moreover, $0 \le \xi(\lambda) \le 1$. (Hint: $\log(F(z))$ is also a Herglotz–Nevanlinna function.)

Problem 3.34. Show that
$$F(z) = 1 + \int_{\mathbb{R}} \frac{d\mu(\lambda)}{\lambda - z}, \qquad \int_{\mathbb{R}} d\mu < \infty,$$
if and only if
$$F(z) = \exp\left(\int_{\mathbb{R}} \frac{\xi(\lambda)d\lambda}{\lambda - z}\right), \qquad \int_{\mathbb{R}} \xi(\lambda)d\lambda < \infty.$$
Moreover, in this case $\int_{\mathbb{R}} d\mu = \int_{\mathbb{R}} \xi(\lambda)d\lambda$. (Hint: $\log(1+z) = z + O(z^{-2})$.)

Problem 3.35. Suppose $\|\operatorname{Im}(F(.+i\varepsilon))\|_p \leq \pi C$ for some $1 < p \leq \infty$. Then $d\mu(\lambda) = g(\lambda)d\lambda$ with $\|g\|_p \leq C$. (Hint: Use Theorem 3.23 and Hölder's inequality to show $\mu((\lambda_1, \lambda_2)) \leq C|\lambda_2 - \lambda_1|^{1/q}$. Now use outer regularity to conclude that μ is absolutely continuous.)

Problem 3.36. Show
$$\int_0^\infty \frac{y^{1-\gamma}}{1+y^2} dy = \frac{\pi/2}{\sin(\gamma\pi/2)}, \qquad \gamma \in (0, 2),$$
by proving
$$\int_{-\infty}^\infty \frac{e^{\alpha x}}{1+e^x} dx = \frac{\pi}{\sin(\alpha\pi)}, \qquad \alpha \in (0, 1).$$
(Hint: To compute the last integral, use a contour consisting of the straight lines connecting the points $-R$, R, $R+2\pi i$, $-R+2\pi i$. Evaluate the contour integral using the residue theorem and let $R \to \infty$. Show that the contributions from the vertical lines vanish in the limit and relate the integrals along the horizontal lines.)

Problem 3.37. In Lemma 3.30 we assumed $0 < \gamma < 2$. Show that in the case $\gamma = 2$ we have
$$\int_{\mathbb{R}} \frac{\log(1+\lambda^2)}{1+\lambda^2} d\mu(\lambda) < \infty \quad \Longleftrightarrow \quad \int_1^\infty \frac{\operatorname{Im}(F(iy))}{y^2} dy < \infty.$$
(Hint: $\int_1^\infty \frac{y^{-1}}{\lambda^2 + y^2} dy = \frac{\log(1+\lambda^2)}{2\lambda^2}$.)

Chapter 4

Applications of the spectral theorem

This chapter can be mostly skipped on first reading. You might want to have a look at the first section and then come back to the remaining ones later.

Now let us show how the spectral theorem can be used. We will give a few typical applications:

First, we will derive an operator-valued version of the Stieltjes inversion formula. To do this, we need to show how to integrate a family of functions of A with respect to a parameter. Moreover, we will show that these integrals can be evaluated by computing the corresponding integrals of the complex-valued functions.

Secondly, we will consider commuting operators and show how certain facts, which are known to hold for the resolvent of an operator A, can be established for a larger class of functions.

Then we will show how the eigenvalues below the essential spectrum and dimension of $\operatorname{Ran} P_A(\Omega)$ can be estimated using the quadratic form.

Finally, we will investigate tensor products of operators.

4.1. Integral formulas

We begin with the first task by having a closer look at the projections $P_A(\Omega)$. They project onto subspaces corresponding to expectation values in the set Ω. In particular, the number

(4.1) $$\langle \psi, \chi_\Omega(A)\psi \rangle$$

is the probability for a measurement of a to lie in Ω. In addition, we have

$$(4.2) \quad \langle \psi, A\psi \rangle = \int_\Omega \lambda\, d\mu_\psi(\lambda) \in \mathrm{hull}(\Omega), \quad \psi \in P_A(\Omega)\mathfrak{H},\ \|\psi\|=1,$$

where $\mathrm{hull}(\Omega)$ is the convex hull of Ω.

The space $\mathrm{Ran}\,\chi_{\{\lambda_0\}}(A)$ is called the **eigenspace** corresponding to λ_0 since we have

$$(4.3) \quad \langle \varphi, A\psi \rangle = \int_\mathbb{R} \lambda \chi_{\{\lambda_0\}}(\lambda) d\mu_{\varphi,\psi}(\lambda) = \lambda_0 \int_\mathbb{R} d\mu_{\varphi,\psi}(\lambda) = \lambda_0 \langle \varphi, \psi \rangle$$

and hence $A\psi = \lambda_0 \psi$ for all $\psi \in \mathrm{Ran}\,\chi_{\{\lambda_0\}}(A)$. The dimension of the eigenspace is called the **multiplicity** of the eigenvalue.

Moreover, since

$$(4.4) \quad \lim_{\varepsilon \downarrow 0} \frac{-i\varepsilon}{\lambda - \lambda_0 - i\varepsilon} = \chi_{\{\lambda_0\}}(\lambda),$$

we infer from Theorem 3.1 that

$$(4.5) \quad \lim_{\varepsilon \downarrow 0} -i\varepsilon R_A(\lambda_0 + i\varepsilon)\psi = \chi_{\{\lambda_0\}}(A)\psi.$$

Similarly, we can obtain an operator-valued version of the Stieltjes inversion formula. But first we need to recall a few facts from integration in Banach spaces.

We will consider the case of mappings $f: I \to X$ where $I = [t_0, t_1] \subset \mathbb{R}$ is a compact interval and X is a Banach space. As before, a function $f: I \to X$ is called simple if the image of f is finite, $f(I) = \{x_i\}_{i=1}^n$, and if each inverse image $f^{-1}(x_i)$, $1 \le i \le n$, is a Borel set. The set of simple functions $S(I,X)$ forms a vector space and can be equipped with the sup norm

$$(4.6) \quad \|f\|_\infty = \sup_{t \in I} \|f(t)\|.$$

The corresponding Banach space obtained after completion is called the set of regulated functions $R(I,X)$.

Observe that $C(I,X) \subset R(I,X)$. In fact, consider the simple function $f_n = \sum_{i=0}^{n-1} f(s_i) \chi_{[s_i, s_{i+1})}$, where $s_i = t_0 + i\frac{t_1-t_0}{n}$. Since $f \in C(I,X)$ is uniformly continuous, we infer that f_n converges uniformly to f.

For $f \in S(I,X)$ we can define a linear map $\int : S(I,X) \to X$ by

$$(4.7) \quad \int_I f(t) dt = \sum_{i=1}^n x_i |f^{-1}(x_i)|,$$

where $|\Omega|$ denotes the Lebesgue measure of Ω. This map satisfies

$$(4.8) \quad \left\| \int_I f(t) dt \right\| \le \|f\|_\infty (t_1 - t_0)$$

4.1. Integral formulas

and hence it can be extended uniquely to a linear map $\int : R(I, X) \to X$ with the same norm $(t_1 - t_0)$ by Theorem 0.29. We even have

$$\text{(4.9)} \qquad \left\| \int_I f(t)dt \right\| \leq \int_I \|f(t)\|dt,$$

which clearly holds for $f \in S(I, X)$ and thus for all $f \in R(I, X)$ by continuity. In addition, if $A \in \mathfrak{L}(X, Y)$, then

$$\text{(4.10)} \qquad A \int_I f(t)dt = \int_I A f(t)dt, \qquad f \in R(I, X).$$

Again, this holds for simple functions and thus extends to all regulated functions by continuity. In particular, if $\ell \in X^*$ is a continuous linear functional, then

$$\text{(4.11)} \qquad \ell\left(\int_I f(t)dt\right) = \int_I \ell(f(t))dt, \qquad f \in R(I, X),$$

and if $A(t) \in R(I, \mathfrak{L}(\mathfrak{H}))$, then

$$\text{(4.12)} \qquad \left(\int_I A(t)dt\right)\psi = \int_I (A(t)\psi)dt.$$

If $I = \mathbb{R}$, we say that $f : I \to X$ is integrable if $f \in R([-r, r], X)$ for all $r > 0$ and if $\|f(t)\|$ is integrable. In this case, we can set

$$\text{(4.13)} \qquad \int_\mathbb{R} f(t)dt = \lim_{r \to \infty} \int_{[-r,r]} f(t)dt$$

and (4.9) and (4.11) still hold.

We will use the standard notation $\int_{t_2}^{t_3} f(s)ds = \int_I \chi_{(t_2,t_3)}(s)f(s)ds$ and $\int_{t_3}^{t_2} f(s)ds = -\int_{t_2}^{t_3} f(s)ds$.

We write $f \in C^1(I, X)$ if

$$\text{(4.14)} \qquad \frac{d}{dt}f(t) = \lim_{\varepsilon \to 0} \frac{f(t + \varepsilon) - f(t)}{\varepsilon}$$

exists for all $t \in I$ and defines a continuous function $I \to X$. In particular, if $f \in C(I, X)$, then $F(t) = \int_{t_0}^t f(s)ds \in C^1(I, X)$ and $dF/dt = f$ as can be seen from
(4.15)
$$\|F(t + \varepsilon) - F(t) - f(t)\varepsilon\| = \left\|\int_t^{t+\varepsilon} (f(s) - f(t))ds\right\| \leq |\varepsilon| \sup_{s \in [t, t+\varepsilon]} \|f(s) - f(t)\|.$$

The important facts for us are the following two results.

Lemma 4.1. *Suppose $f : I \times \mathbb{R} \to \mathbb{C}$ is a bounded uniformly continuous function and set $F(\lambda) = \int_I f(t, \lambda) dt$. Let A be self-adjoint. Then $f(t, A) \in R(I, \mathfrak{L}(\mathfrak{H}))$ and*

$$\text{(4.16)} \quad F(A) = \int_I f(t, A) dt, \quad \text{respectively,} \quad F(A)\psi = \int_I f(t, A)\psi \, dt.$$

Proof. That $f(t, A) \in R(I, \mathfrak{L}(\mathfrak{H}))$ follows from the spectral theorem, since it is no restriction to assume that A is multiplication by λ in some L^2 space. We compute

$$\langle \varphi, (\int_I f(t, A) dt) \psi \rangle = \int_I \langle \varphi, f(t, A)\psi \rangle dt$$

$$= \int_I \int_{\mathbb{R}} f(t, \lambda) d\mu_{\varphi,\psi}(\lambda) dt$$

$$= \int_{\mathbb{R}} \int_I f(t, \lambda) dt \, d\mu_{\varphi,\psi}(\lambda)$$

$$= \int_{\mathbb{R}} F(\lambda) d\mu_{\varphi,\psi}(\lambda) = \langle \varphi, F(A)\psi \rangle$$

by Fubini's theorem and hence the first claim follows. $\square$

Lemma 4.2. *Suppose $f : \mathbb{R} \to \mathfrak{L}(\mathfrak{H})$ is integrable and $A \in \mathfrak{L}(\mathfrak{H})$. Then*

$$\text{(4.17)} \quad A \int_{\mathbb{R}} f(t) dt = \int_{\mathbb{R}} Af(t) dt.$$

Proof. It suffices to prove the case where f is simple and of compact support. But for such functions the claim is straightforward. $\square$

Now we can prove an operator-valued version of the Stieltjes inversion formula.

Theorem 4.3 (Stone's formula). *Let A be self-adjoint. Then*
(4.18)
$$\frac{1}{2\pi \mathrm{i}} \int_{\lambda_1}^{\lambda_2} \left(R_A(\lambda + \mathrm{i}\varepsilon) - R_A(\lambda - \mathrm{i}\varepsilon) \right) d\lambda \xrightarrow{s} \frac{1}{2} \left(P_A([\lambda_1, \lambda_2]) + P_A((\lambda_1, \lambda_2)) \right)$$

strongly.

Proof. By

$$\frac{1}{2\pi \mathrm{i}} \int_{\lambda_1}^{\lambda_2} \left(\frac{1}{x - \lambda - \mathrm{i}\varepsilon} - \frac{1}{x - \lambda + \mathrm{i}\varepsilon} \right) d\lambda = \frac{1}{\pi} \int_{\lambda_1}^{\lambda_2} \frac{\varepsilon}{(x - \lambda)^2 + \varepsilon^2} d\lambda$$

$$= \frac{1}{\pi} \left(\arctan\left(\frac{\lambda_2 - x}{\varepsilon}\right) - \arctan\left(\frac{\lambda_1 - x}{\varepsilon}\right) \right)$$

$$\to \frac{1}{2} \left(\chi_{[\lambda_1, \lambda_2]}(x) + \chi_{(\lambda_1, \lambda_2)}(x) \right),$$

4.2. Commuting operators

the result follows combining the last part of Theorem 3.1 with Lemma 4.1. $\square$

Note that by using the first resolvent formula, Stone's formula can also be written in the form

$$\langle \psi, \frac{1}{2}(P_A([\lambda_1, \lambda_2]) + P_A((\lambda_1, \lambda_2)))\psi \rangle = \lim_{\varepsilon \downarrow 0} \frac{1}{\pi} \int_{\lambda_1}^{\lambda_2} \operatorname{Im}\langle \psi, R_A(\lambda + i\varepsilon)\psi \rangle d\lambda$$

(4.19)
$$= \lim_{\varepsilon \downarrow 0} \frac{\varepsilon}{\pi} \int_{\lambda_1}^{\lambda_2} \|R_A(\lambda + i\varepsilon)\psi\|^2 d\lambda.$$

Problem 4.1. *Let Γ be a differentiable Jordan curve in $\rho(A)$. Show*

$$\chi_\Omega(A) = \int_\Gamma R_A(z) dz,$$

where Ω is the intersection of the interior of Γ with $\mathbb{R}$.

Problem 4.2. *Suppose $f \in C^1(I, X)$. Show that $\frac{d}{dt} f(t) = 0$ if and only if f is constant. In particular, the fundamental theorem of calculus holds:*

$$f(t) = f(t_0) + \int_{t_0}^t f'(s) ds,$$

where $f'(t) = \frac{d}{dt} f(t)$.

Problem 4.3. *Suppose $f \in C(\mathbb{R})$ is bounded. Show that*

$$\frac{1}{2\pi i} \int_\mathbb{R} f(\lambda)(R_A(\lambda + i\varepsilon) - R_A(\lambda - i\varepsilon)) d\lambda \xrightarrow{s} f(A)$$

and

$$\frac{1}{2\pi i} \int_{\lambda_1}^{\lambda_2} f(\lambda)(R_A(\lambda + i\varepsilon) - R_A(\lambda - i\varepsilon)) d\lambda \xrightarrow{s} \frac{1}{2}(P_A([\lambda_1, \lambda_2]) + P_A((\lambda_1, \lambda_2))) f(A).$$

(Hint: For the first part, show $\frac{1}{\pi} \int f(\lambda) \operatorname{Im}(\frac{1}{x - \lambda - \varepsilon}) d\lambda \to f(x)$. Now the second part follows from the first if f vanishes at the boundary points and it remains to show the cases $f(\lambda) = 1$ and $f(\lambda) = \lambda$.)

4.2. Commuting operators

Now we come to commuting operators. As a preparation we first prove

Lemma 4.4. *Let $K \subseteq \mathbb{R}$ be closed and let $C_\infty(K)$ be the set of all continuous functions on K which vanish at ∞ (if K is unbounded) with the sup norm. The $*$-subalgebra generated by the function*

(4.20)
$$\lambda \mapsto \frac{1}{\lambda - z}$$

for one $z \in \mathbb{C} \backslash K$ is dense in $C_\infty(K)$.

Proof. If K is compact, the claim follows directly from the complex Stone–Weierstraß theorem since $(\lambda_1-z)^{-1} = (\lambda_2-z)^{-1}$ implies $\lambda_1 = \lambda_2$. Otherwise, replace K by $\tilde{K} = K \cup \{\infty\}$, which is compact, and set $(\infty - z)^{-1} = 0$. Then we can again apply the complex Stone–Weierstraß theorem to conclude that our $*$-subalgebra is equal to $\{f \in C(\tilde{K}) | f(\infty) = 0\}$ which is equivalent to $C_\infty(K)$. □

We say that two bounded operators A, B **commute** if

(4.21) $$[A, B] = AB - BA = 0.$$

If A or B is unbounded, we soon run into trouble with this definition since the above expression might not even make sense for any nonzero vector (e.g., take $B = \langle \varphi, . \rangle \psi$ with $\psi \notin \mathfrak{D}(A)$). To avoid this nuisance, we will replace A by a bounded function of A. A good candidate is the resolvent. Hence if A is self-adjoint and B is bounded, we will say that A and B commute if

(4.22) $$[R_A(z), B] = [R_A(z^*), B] = 0$$

for one $z \in \rho(A)$.

Lemma 4.5. *Suppose A is self-adjoint and B is bounded. Then B commutes with A if and only if*

(4.23) $$BA \subseteq AB.$$

Moreover, in this case,

(4.24) $$[f(A), B] = 0$$

for every bounded Borel function f. If f is unbounded, the claim holds for every $\psi \in \mathfrak{D}(f(A))$ in the sense that $Bf(A) \subseteq f(A)B$.

Proof. First of all, (4.23) implies $B(A - z) \subseteq (A - z)B$ for all $z \in \mathbb{C}$. Multiplying with $R_A(z)$ from the right, we get $B = (A - z)BR_A(z)$ for $z \in \rho(A)$, and multiplying this last relation with $R_A(z)$ from the left, we finally get (4.22) for all $z \in \rho(A)$.

Conversely, (4.22) tells us that (4.24) holds for every f in the $*$-subalgebra generated by $R_A(z)$. Since this subalgebra is dense in $C_\infty(\sigma(A))$, the claim follows for all such $f \in C_\infty(\sigma(A))$. Next, fix $\psi \in \mathfrak{H}$ and let f be bounded. Choose a sequence $f_n \in C_\infty(\sigma(A))$ converging to f in $L^2(\mathbb{R}, d\mu_\psi)$. Then
$$Bf(A)\psi = \lim_{n \to \infty} Bf_n(A)\psi = \lim_{n \to \infty} f_n(A)B\psi = f(A)B\psi.$$
If f is unbounded, let $\psi \in \mathfrak{D}(f(A))$ and choose f_n as in (3.27). Then
$$f(A)B\psi = \lim_{n \to \infty} f_n(A)B\psi = \lim_{n \to \infty} Bf_n(A)\psi$$
shows $f \in L^2(\mathbb{R}, d\mu_{B\psi})$ (i.e., $B\psi \in \mathfrak{D}(f(A))$) and $f(A)B\psi = Bf(A)\psi$. □

4.2. Commuting operators

In the special case where B is an orthogonal projection, we obtain

Corollary 4.6. *Let A be self-adjoint and $\mathfrak{H}_1$ a closed subspace with corresponding projector P_1. Then $\mathfrak{H}_1$ reduces A if and only if P_1 and A commute.*

Furthermore, note

Corollary 4.7. *If A is self-adjoint and bounded, then (4.22) holds if and only if (4.21) holds.*

Proof. Since $\sigma(A)$ is compact, we have $\lambda \in C_\infty(\sigma(A))$, and hence (4.21) follows from (4.24) by our lemma. Conversely, since B commutes with every polynomial of A, the claim follows from the Neumann series. $\square$

As another consequence, we obtain

Theorem 4.8. *Suppose A is self-adjoint and has simple spectrum. A bounded operator B commutes with A if and only if $B = f(A)$ for some bounded Borel function.*

Proof. Let ψ be a cyclic vector for A. By our unitary equivalence it is no restriction to assume $\mathfrak{H} = L^2(\mathbb{R}, d\mu_\psi)$. Then
$$Bg(\lambda) = Bg(\lambda) \cdot 1 = g(\lambda)(B1)(\lambda)$$
since B commutes with the multiplication operator $g(\lambda)$. Hence B is multiplication by $f(\lambda) = (B1)(\lambda)$. $\square$

The assumption that the spectrum of A is simple is crucial as the example $A = \mathbb{I}$ shows. Note also that the functions $\exp(-itA)$ can also be used instead of resolvents.

Lemma 4.9. *Suppose A is self-adjoint and B is bounded. Then B commutes with A if and only if*

(4.25) $$[e^{-iAt}, B] = 0$$

for all $t \in \mathbb{R}$.

Proof. It suffices to show $[\hat{f}(A), B] = 0$ for $f \in \mathcal{S}(\mathbb{R})$, since these functions are dense in $C_\infty(\mathbb{R})$ by the complex Stone–Weierstraß theorem. Here $\hat{f}$ denotes the Fourier transform of f; see Section 7.1. But for such f we have

$$[\hat{f}(A), B] = \frac{1}{\sqrt{2\pi}}[\int_\mathbb{R} f(t)e^{-iAt}dt, B] = \frac{1}{\sqrt{2\pi}} \int_\mathbb{R} f(t)[e^{-iAt}, B]dt = 0$$

by Lemma 4.2. $\square$

The extension to the case where B is self-adjoint and unbounded is straightforward. We say that A and B commute in this case if

$$[R_A(z_1), R_B(z_2)] = [R_A(z_1^*), R_B(z_2)] = 0 \qquad (4.26)$$

for one $z_1 \in \rho(A)$ and one $z_2 \in \rho(B)$ (the claim for z_2^* follows by taking adjoints). From our above analysis it follows that this is equivalent to

$$[e^{-iAt}, e^{-iBs}] = 0, \quad t, s \in \mathbb{R}, \qquad (4.27)$$

respectively,

$$[f(A), g(B)] = 0, \qquad (4.28)$$

for arbitrary bounded Borel functions f and g. Note that in this case e^{-iAt} will leave the domain of A invariant and vice versa (show this).

Given a quantum mechanical system with Hamiltonian H, every observable A commuting with H corresponds to a **conserved quantity**. In fact, for $\psi \in \mathfrak{D}(A)$ and $\psi(t) = e^{-iHt}\psi$, we see that the expectations

$$\langle \psi(t), A\psi(t) \rangle = \langle e^{-iHt}\psi, Ae^{-iHt}\psi \rangle = \langle e^{-iHt}\psi, e^{-iHt}A\psi \rangle = \langle \psi, A\psi \rangle \qquad (4.29)$$

are independent of time.

Problem 4.4. *Let A and B be self-adjoint. Show that A and B commute if and only if the corresponding spectral projections $P_A(\Omega)$ and $P_B(\Omega)$ commute for every Borel set Ω. In particular, $\mathrm{Ran}(P_B(\Omega))$ reduces A and vice versa.*

Problem 4.5. *Let A and B be self-adjoint operators with pure point spectrum. Show that A and B commute if and only if they have a common orthonormal basis of eigenfunctions.*

Problem 4.6. *Let A_1, A_2 be self-adjoint, and let B be bounded. Show that*

$$BA_1 \subseteq A_2 B$$

implies

$$Bf(A_1) \subseteq f(A_2)B$$

for every Borel function f.

4.3. Polar decomposition

Let A be a closed operator. Recall that, by Problem 2.12, A^*A is self-adjoint and $\mathfrak{Q}(A^*A) = \mathfrak{D}(A)$. Hence we can define the absolute value of an operator via

$$|A| = \sqrt{A^*A}. \qquad (4.30)$$

A straightforward calculation shows

$$\||A|\psi\| = \langle \psi, |A|^2 \psi \rangle = \langle \psi, A^*A\psi \rangle = \|A\psi\|, \quad \psi \in \mathfrak{D}(|A|) = \mathfrak{D}(A), \qquad (4.31)$$

4.3. Polar decomposition

which in particular implies

(4.32) $$\mathrm{Ker}(A) = \mathrm{Ker}(|A|) = \mathrm{Ran}(|A|)^\perp.$$

As a consequence, the operator

(4.33) $$U = \begin{cases} \varphi = |A|\psi \mapsto A\psi & \text{if } \varphi \in \mathrm{Ran}(|A|), \\ \varphi \mapsto 0 & \text{if } \varphi \in \mathrm{Ker}(|A|) \end{cases}$$

extends to a well-defined **partial isometry**; that is, $U : \mathrm{Ker}(U)^\perp \to \mathrm{Ran}(U)$ is unitary, where $\mathrm{Ker}(U) = \mathrm{Ker}(A)$ and $\mathrm{Ran}(U) = \mathrm{Ker}(A^*)^\perp$. Equivalently,

(4.34) $$U^*U = P_{\mathrm{Ker}(A)}, \qquad UU^* = P_{\mathrm{Ker}(A^*)},$$

where P_M is the projection onto the subspace M. In particular, we have

Theorem 4.10 (Polar decomposition). *Every closed operator A admits a* **polar decomposition**

(4.35) $$A = U|A| = |A^*|U,$$

where U is a unitary from $\mathrm{Ker}(U) = \mathrm{Ker}(A)$ to $\mathrm{Ran}(U) = \mathrm{Ker}(A^*)^\perp$.

Proof. To get the second equality, use (Problem 2.3) $A^* = |A|U^*$, implying $|A^*|^2 = AA^* = (U|A|U^*)(U|A|U^*)$, which shows $|A^*| = U|A|U^*$. □

As a simple consequence, we also obtain

Corollary 4.11. *Let A be a closed operator with polar decomposition $A = U|A|$. Then*

$$A = U|A| = |A^*|U = UA^*U,$$
$$A^* = U^*|A^*| = |A|U^* = U^*AU^*,$$
$$|A| = U^*A = A^*U = U^*|A^*|U,$$
$$|A^*| = UA^* = AU^* = U|A|U^*.$$

Problem 4.7. *Show $\mathrm{Ran}(A) = \mathrm{Ran}(|A^*|)$.*

Problem 4.8. *Compute $|A|$ for the rank one operator $A = \alpha\langle\varphi,.\rangle\psi$. Compute $|A^*|$ also.*

Problem 4.9. *Let $f(\lambda), g(\lambda)$ be two Borel functions such that $f(\lambda)g(\lambda) = \lambda$ (for a.e. $\lambda \in \mathbb{R}$). Then*

$$A = g(|A^*|)Uf(|A|).$$

(Hint: Begin with $A = g(|A^*|)f(|A^*|)U$ and use Problem 4.6.)

Problem 4.10. *Let A be closed. Show that A is normal if and only if $|A| = |A^*|$. (Hint: Problem 2.13.)*

4.4. The min-max theorem

In many applications, a self-adjoint operator has a number of eigenvalues below the bottom of the essential spectrum. The essential spectrum is obtained from the spectrum by removing all discrete eigenvalues with finite multiplicity (we will have a closer look at this in Section 6.2). In general, there is no way of computing the lowest eigenvalues and their corresponding eigenfunctions explicitly. However, one often has some idea about how the eigenfunctions might approximately look.

So suppose we have a normalized function ψ_1 which is an approximation for the eigenfunction φ_1 of the lowest eigenvalue E_1. Then by Theorem 2.20 we know that

$$\langle \psi_1, A\psi_1 \rangle \geq \langle \varphi_1, A\varphi_1 \rangle = E_1. \tag{4.36}$$

If we add some free parameters to ψ_1, one can optimize them and obtain quite good upper bounds for the first eigenvalue. This is known as the **Rayleigh–Ritz method**.

But is there also something one can say about the next eigenvalues? Suppose we know the first eigenfunction φ_1. Then we can restrict A to the orthogonal complement of φ_1 and proceed as before: E_2 will be the infimum over all expectations restricted to this subspace. If we restrict to the orthogonal complement of an approximating eigenfunction ψ_1, there will still be a component in the direction of φ_1 left and hence the infimum of the expectations will be lower than E_2. Thus the optimal choice $\psi_1 = \varphi_1$ will give the maximal value E_2.

More precisely, let $\{\varphi_j\}_{j=1}^N$ be an orthonormal basis for the space spanned by the eigenfunctions corresponding to eigenvalues below the essential spectrum. Here the essential spectrum $\sigma_{ess}(A)$ is given by precisely those values in the spectrum which are *not* isolated eigenvalues of finite multiplicity (see Section 6.2). Assume they satisfy $(A - E_j)\varphi_j = 0$, where $E_j \leq E_{j+1}$ are the eigenvalues (counted according to their multiplicity). If the number of eigenvalues N is finite, we set $E_j = \inf \sigma_{ess}(A)$ for $j > N$ and choose φ_j orthonormal such that $\|(A - E_j)\varphi_j\| \leq \varepsilon$.

Define

$$U(\psi_1, \ldots, \psi_n) = \{\psi \in \mathfrak{D}(A) \mid \|\psi\| = 1,\ \psi \in \operatorname{span}\{\psi_1, \ldots, \psi_n\}^\perp\}. \tag{4.37}$$

(i) We have

$$\inf_{\psi \in U(\psi_1, \ldots, \psi_{n-1})} \langle \psi, A\psi \rangle \leq E_n + O(\varepsilon). \tag{4.38}$$

4.4. The min-max theorem

In fact, set $\psi = \sum_{j=1}^n \alpha_j \varphi_j$ and choose α_j such that $\psi \in U(\psi_1, \ldots, \psi_{n-1})$. Then

$$(4.39) \qquad \langle \psi, A\psi \rangle = \sum_{j=1}^n |\alpha_j|^2 E_j + O(\varepsilon) \leq E_n + O(\varepsilon)$$

and the claim follows.

(ii) We have

$$(4.40) \qquad \inf_{\psi \in U(\varphi_1,\ldots,\varphi_{n-1})} \langle \psi, A\psi \rangle \geq E_n - O(\varepsilon).$$

In fact, set $\psi = \varphi_n$.

Since ε can be chosen arbitrarily small, we have proven the following.

Theorem 4.12 (Max-min). *Let A be self-adjoint and let $E_1 \leq E_2 \leq E_3 \cdots$ be the eigenvalues of A below the essential spectrum, respectively, the infimum of the essential spectrum, once there are no more eigenvalues left. Then*

$$(4.41) \qquad E_n = \sup_{\psi_1,\ldots,\psi_{n-1}} \inf_{\psi \in U(\psi_1,\ldots,\psi_{n-1})} \langle \psi, A\psi \rangle,$$

where $U(\psi_1, \ldots, \psi_{n-1})$ is defined in (4.37).

Clearly the same result holds if $\mathfrak{D}(A)$ is replaced by the quadratic form domain $\mathfrak{Q}(A)$ in the definition of U. In addition, as long as E_n is an eigenvalue, the sup and inf are in fact max and min, explaining the name.

Corollary 4.13. *Suppose A and B are self-adjoint operators with $A \geq B$ (i.e., $A - B \geq 0$). Then $E_n(A) \geq E_n(B)$.*

Similarly we obtain (Problem 4.11):

Theorem 4.14 (Min-max). *Let A be self-adjoint and let $E_1 \leq E_2 \leq E_3 \cdots$ be the eigenvalues of A below the essential spectrum, respectively, the infimum of the essential spectrum, once there are no more eigenvalues left. Then*

$$(4.42) \qquad E_n = \inf_{\psi_1,\ldots,\psi_{n-1}} \sup_{\psi \in V(\psi_1,\ldots,\psi_{n-1})} \langle \psi, A\psi \rangle,$$

where

$$(4.43) \qquad V(\psi_1, \ldots, \psi_n) = \{\psi \in \mathfrak{D}(A)|\, \|\psi\| = 1,\, \psi \in \text{span}\{\psi_1, \ldots, \psi_n\}\}.$$

Problem 4.11. *Prove Theorem 4.14.*

Problem 4.12. *Suppose A, A_n are self-adjoint, bounded and $A_n \to A$. Then $E_k(A_n) \to E_k(A)$. (Hint: $\|A - A_n\| \leq \varepsilon$ is equivalent to $A - \varepsilon \leq A_n \leq A + \varepsilon$.)*

4.5. Estimating eigenspaces

Next, we show that the dimension of the range of $P_A(\Omega)$ can be estimated if we have some functions which lie approximately in this space.

Theorem 4.15. *Suppose A is a self-adjoint operator and ψ_j, $1 \leq j \leq k$, are linearly independent elements of $\mathfrak{H}$.*

(i) *Let $\lambda \in \mathbb{R}$, $\psi_j \in \mathfrak{Q}(A)$. If*

(4.44) $$\langle \psi, A\psi \rangle < \lambda \|\psi\|^2$$

for any nonzero linear combination $\psi = \sum_{j=1}^{k} c_j \psi_j$, then

(4.45) $$\dim \operatorname{Ran} P_A((-\infty, \lambda)) \geq k.$$

Similarly, $\langle \psi, A\psi \rangle > \lambda \|\psi\|^2$ implies $\dim \operatorname{Ran} P_A((\lambda, \infty)) \geq k$.

(ii) *Let $\lambda_1 < \lambda_2$, $\psi_j \in \mathfrak{D}(A)$. If*

(4.46) $$\left\|\left(A - \frac{\lambda_2 + \lambda_1}{2}\right)\psi\right\| < \frac{\lambda_2 - \lambda_1}{2} \|\psi\|$$

for any nonzero linear combination $\psi = \sum_{j=1}^{k} c_j \psi_j$, then

(4.47) $$\dim \operatorname{Ran} P_A((\lambda_1, \lambda_2)) \geq k.$$

Proof. (i) Let $M = \operatorname{span}\{\psi_j\} \subseteq \mathfrak{H}$. We claim $\dim P_A((-\infty, \lambda))M = \dim M = k$. For this it suffices to show $\operatorname{Ker} P_A((-\infty, \lambda))|_M = \{0\}$. Suppose $P_A((-\infty, \lambda))\psi = 0$, $\psi \neq 0$. Then we see that for any nonzero linear combination ψ,

$$\langle \psi, A\psi \rangle = \int_{\mathbb{R}} \eta \, d\mu_\psi(\eta) = \int_{[\lambda, \infty)} \eta \, d\mu_\psi(\eta)$$

$$\geq \lambda \int_{[\lambda, \infty)} d\mu_\psi(\eta) = \lambda \|\psi\|^2.$$

This contradicts our assumption (4.44).

(ii) This is just the previous case (i) applied to $(A - (\lambda_2 + \lambda_1)/2)^2$ with $\lambda = (\lambda_2 - \lambda_1)^2/4$. □

Another useful estimate is

Theorem 4.16 (Temple's inequality). *Let $\lambda_1 < \lambda_2$ and $\psi \in \mathfrak{D}(A)$ with $\|\psi\| = 1$ such that*

(4.48) $$\lambda = \langle \psi, A\psi \rangle \in (\lambda_1, \lambda_2).$$

4.6. Tensor products of operators

If there is one isolated eigenvalue E between λ_1 and λ_2, that is, $\sigma(A) \cap (\lambda_1, \lambda_2) = \{E\}$, then

$$\lambda - \frac{\|(A-\lambda)\psi\|^2}{\lambda_2 - \lambda} \leq E \leq \lambda + \frac{\|(A-\lambda)\psi\|^2}{\lambda - \lambda_1}. \tag{4.49}$$

Proof. First of all, we can assume $\lambda = 0$ if we replace A by $A - \lambda$. To prove the first inequality, observe that by assumption $(E, \lambda_2) \subset \rho(A)$ and hence the spectral theorem implies $(A - \lambda_2)(A - E) \geq 0$. Thus $\langle \psi, (A - \lambda_2)(A - E)\rangle = \|A\psi\|^2 + \lambda_2 E \geq 0$ and the first inequality follows after dividing by $\lambda_2 > 0$. Similarly, $(A - \lambda_1)(A - E) \geq 0$ implies the second inequality. $\square$

Note that the last inequality only provides additional information if $\|(A - \lambda)\psi\|^2 \leq (\lambda_2 - \lambda)(\lambda - \lambda_1)$.

A typical application is if $E = E_0$ is the lowest eigenvalue. In this case any normalized trial function ψ will give the bound $E_0 \leq \langle \psi, A\psi \rangle$. If, in addition, we also have some estimate $\lambda_2 \leq E_1$ for the second eigenvalue E_1, then Temple's inequality can give a bound from below. For λ_1 we can choose any value $\lambda_1 < E_0$; in fact, if we let $\lambda_1 \to -\infty$, we just recover the bound we already know.

4.6. Tensor products of operators

Recall the definition of the tensor product of Hilbert space from Section 1.4. Suppose A_j, $1 \leq j \leq n$, are (essentially) self-adjoint operators on $\mathfrak{H}_j$. For every monomial $\lambda_1^{n_1} \cdots \lambda_n^{n_n}$, we can define

$$(A_1^{n_1} \otimes \cdots \otimes A_n^{n_n})\psi_1 \otimes \cdots \otimes \psi_n = (A_1^{n_1}\psi_1) \otimes \cdots \otimes (A_n^{n_n}\psi_n), \quad \psi_j \in \mathfrak{D}(A_j^{n_j}), \tag{4.50}$$

and extend this definition by linearity to the span of all such functions (check that this definition is well-defined by showing that the corresponding operator on $\mathcal{F}(\mathfrak{H}_1, \ldots, \mathfrak{H}_n)$ vanishes on $\mathcal{N}(\mathfrak{H}_1, \ldots, \mathfrak{H}_n)$). Hence for every polynomial $P(\lambda_1, \ldots, \lambda_n)$ of degree N, we obtain an operator

$$P(A_1, \ldots, A_n)\psi_1 \otimes \cdots \otimes \psi_n, \quad \psi_j \in \mathfrak{D}(A_j^N), \tag{4.51}$$

defined on the set

$$\mathfrak{D} = \text{span}\{\psi_1 \otimes \cdots \otimes \psi_n \mid \psi_j \in \mathfrak{D}(A_j^N)\}. \tag{4.52}$$

Moreover, if P is real-valued, then the operator $P(A_1, \ldots, A_n)$ on $\mathfrak{D}$ is symmetric and we can consider its closure, which will again be denoted by $P(A_1, \ldots, A_n)$.

Theorem 4.17. *Suppose A_j, $1 \leq j \leq n$, are self-adjoint operators on $\mathfrak{H}_j$ and let $P(\lambda_1, \ldots, \lambda_n)$ be a real-valued polynomial and define $P(A_1, \ldots, A_n)$ as above.*

Then $P(A_1, \ldots, A_n)$ is self-adjoint and its spectrum is the closure of the range of P on the product of the spectra of the A_j; that is,

(4.53) $\qquad \sigma(P(A_1, \ldots, A_n)) = \overline{P(\sigma(A_1), \ldots, \sigma(A_n))}.$

Proof. By the spectral theorem, it is no restriction to assume that A_j is multiplication by λ_j on $L^2(\mathbb{R}, d\mu_j)$ and $P(A_1, \ldots, A_n)$ is hence multiplication by $P(\lambda_1, \ldots, \lambda_n)$ on $L^2(\mathbb{R}^n, d\mu_1 \times \cdots \times d\mu_n)$. Since $\mathfrak{D}$ contains the set of all functions $\psi_1(\lambda_1) \cdots \psi_n(\lambda_n)$ for which $\psi_j \in L_c^2(\mathbb{R}, d\mu_j)$, it follows that the domain of the closure of P contains $L_c^2(\mathbb{R}^n, d\mu_1 \times \cdots \times d\mu_n)$. Hence P is the maximally defined multiplication operator by $P(\lambda_1, \ldots, \lambda_n)$, which is self-adjoint.

Now let $\lambda = P(\lambda_1, \ldots, \lambda_n)$ with $\lambda_j \in \sigma(A_j)$. Then there exist Weyl sequences $\psi_{j,k} \in \mathfrak{D}(A_j^N)$ with $(A_j - \lambda_j)\psi_{j,k} \to 0$ as $k \to \infty$. Consequently, $(P - \lambda)\psi_k \to 0$, where $\psi_k = \psi_{1,k} \otimes \cdots \otimes \psi_{1,k}$ and hence $\lambda \in \sigma(P)$. Conversely, if $\lambda \notin \overline{P(\sigma(A_1), \ldots, \sigma(A_n))}$, then $|P(\lambda_1, \ldots, \lambda_n) - \lambda| \geq \varepsilon$ for a.e. λ_j with respect to μ_j and hence $(P - \lambda)^{-1}$ exists and is bounded; that is, $\lambda \in \rho(P)$. $\square$

The two main cases of interest are $A_1 \otimes A_2$, in which case

(4.54) $\qquad \sigma(A_1 \otimes A_2) = \overline{\sigma(A_1)\sigma(A_2)} = \overline{\{\lambda_1 \lambda_2 | \lambda_j \in \sigma(A_j)\}},$

and $A_1 \otimes \mathbb{I} + \mathbb{I} \otimes A_2$, in which case

(4.55) $\qquad \sigma(A_1 \otimes \mathbb{I} + \mathbb{I} \otimes A_2) = \overline{\sigma(A_1) + \sigma(A_2)} = \overline{\{\lambda_1 + \lambda_2 | \lambda_j \in \sigma(A_j)\}}.$

Problem 4.13. *Show that the closure can be omitted in (4.55) if at least one operator is bounded and in (4.54) if both operators are bounded.*

Chapter 5

Quantum dynamics

As in the finite dimensional case, the solution of the Schrödinger equation

(5.1) $$i\frac{d}{dt}\psi(t) = H\psi(t)$$

is given by

(5.2) $$\psi(t) = \exp(-itH)\psi(0).$$

A detailed investigation of this formula will be our first task. Moreover, in the finite dimensional case, the dynamics is understood once the eigenvalues are known and the same is true in our case once we know the spectrum. Note that, like any Hamiltonian system from classical mechanics, our system is not hyperbolic (i.e., the spectrum is not away from the real axis) and hence simple results such as all solutions tend to the equilibrium position cannot be expected.

5.1. The time evolution and Stone's theorem

In this section, we want to have a look at the initial value problem associated with the Schrödinger equation (2.12) in the Hilbert space $\mathfrak{H}$. If $\mathfrak{H}$ is one-dimensional (and hence A is a real number), the solution is given by

(5.3) $$\psi(t) = e^{-itA}\psi(0).$$

Our hope is that this formula also applies in the general case and that we can reconstruct a one-parameter unitary group $U(t)$ from its generator A (compare (2.11)) via $U(t) = \exp(-itA)$. We first investigate the family of operators $\exp(-itA)$.

Theorem 5.1. *Let A be self-adjoint and let $U(t) = \exp(-itA)$.*

(i) $U(t)$ is a strongly continuous one-parameter unitary group.
(ii) The limit $\lim_{t\to 0} \frac{1}{t}(U(t)\psi - \psi)$ exists if and only if $\psi \in \mathfrak{D}(A)$, in which case $\lim_{t\to 0} \frac{1}{t}(U(t)\psi - \psi) = -\mathrm{i}A\psi$.
(iii) $U(t)\mathfrak{D}(A) = \mathfrak{D}(A)$ and $AU(t) = U(t)A$.

Proof. The group property (i) follows directly from Theorem 3.1 and the corresponding statements for the function $\exp(-\mathrm{i}t\lambda)$. To prove strong continuity, observe that

$$\lim_{t\to t_0} \|\mathrm{e}^{-\mathrm{i}tA}\psi - \mathrm{e}^{-\mathrm{i}t_0 A}\psi\|^2 = \lim_{t\to t_0} \int_\mathbb{R} |\mathrm{e}^{-\mathrm{i}t\lambda} - \mathrm{e}^{-\mathrm{i}t_0\lambda}|^2 d\mu_\psi(\lambda)$$

$$= \int_\mathbb{R} \lim_{t\to t_0} |\mathrm{e}^{-\mathrm{i}t\lambda} - \mathrm{e}^{-\mathrm{i}t_0\lambda}|^2 d\mu_\psi(\lambda) = 0$$

by the dominated convergence theorem.

Similarly, if $\psi \in \mathfrak{D}(A)$, we obtain

$$\lim_{t\to 0} \|\frac{1}{t}(\mathrm{e}^{-\mathrm{i}tA}\psi - \psi) + \mathrm{i}A\psi\|^2 = \lim_{t\to 0} \int_\mathbb{R} |\frac{1}{t}(\mathrm{e}^{-\mathrm{i}t\lambda} - 1) + \mathrm{i}\lambda|^2 d\mu_\psi(\lambda) = 0$$

since $|\mathrm{e}^{-\mathrm{i}t\lambda} - 1| \leq |t\lambda|$. Now let $\tilde{A}$ be the generator defined as in (2.11). Then $\tilde{A}$ is a symmetric extension of A since we have

$$\langle \varphi, \tilde{A}\psi \rangle = \lim_{t\to 0}\langle \varphi, \frac{\mathrm{i}}{t}(U(t)-1)\psi \rangle = \lim_{t\to 0}\langle \frac{\mathrm{i}}{-t}(U(-t)-1)\varphi, \psi \rangle = \langle \tilde{A}\varphi, \psi \rangle$$

and hence $\tilde{A} = A$ by Corollary 2.2. This settles (ii).

To see (iii), replace $\psi \to U(s)\psi$ in (ii). $\square$

For our original problem, this implies that formula (5.3) is indeed the solution to the initial value problem of the Schrödinger equation

(5.4) $$\mathrm{i}\frac{d}{dt}\psi(t) = A\psi(t), \qquad \psi(0) = \psi_0 \in \mathfrak{D}(A).$$

Here $\psi : \mathbb{R} \to \mathfrak{H}$ is said to be differentiable if

(5.5) $$\frac{d}{dt}\psi(t) = \lim_{\varepsilon\to 0} \frac{\psi(t+\varepsilon) - \psi(t)}{\varepsilon}$$

exists for all $t \in \mathbb{R}$. In fact, it is the only solution.

Lemma 5.2. *The function $\psi(t) = U(t)\psi_0$ is the only solution of (5.4).*

Proof. Let $\psi(t)$ be a solution and consider $\varphi(t) = U(-t)\psi(t)$. Then a straightforward calculation shows that $\varphi(t)$ is differentiable with

$$\frac{d}{dt}\varphi(t) = \lim_{\varepsilon\to 0} \frac{1}{\varepsilon}\Big((U(-t+\varepsilon) - U(-t))(\psi(t) - \varepsilon\mathrm{i}A\psi(t) + o(\varepsilon))$$
$$+ U(-t)(\psi(t+\varepsilon) - \psi(t))\Big) = 0.$$

5.1. The time evolution and Stone's theorem

Hence (Problem 5.2), $\varphi(t) = \varphi(0) = \psi_0$, implying $\psi(t) = U(t)\psi_0$. □

Moreover,

(5.6) $\qquad \langle U(t)\psi, AU(t)\psi \rangle = \langle U(t)\psi, U(t)A\psi \rangle = \langle \psi, A\psi \rangle$

shows that the expectations of A are time independent. This corresponds to conservation of energy.

On the other hand, the generator of the time evolution of a quantum mechanical system should always be a self-adjoint operator since it corresponds to an observable (energy). Moreover, there should be a one-to-one correspondence between the unitary group and its generator. This is ensured by Stone's theorem.

Theorem 5.3 (Stone). *Let $U(t)$ be a weakly continuous one-parameter unitary group. Then its generator A is self-adjoint and $U(t) = \exp(-itA)$.*

Proof. First of all, observe that weak continuity together with item (iv) of Lemma 1.12 shows that $U(t)$ is in fact strongly continuous.

Next we show that A is densely defined. Pick $\psi \in \mathfrak{H}$ and set

$$\psi_\tau = \int_0^\tau U(t)\psi\, dt$$

(the integral is defined as in Section 4.1), implying $\lim_{\tau \to 0} \tau^{-1}\psi_\tau = \psi$. Moreover,

$$\frac{1}{t}(U(t)\psi_\tau - \psi_\tau) = \frac{1}{t}\int_t^{t+\tau} U(s)\psi\, ds - \frac{1}{t}\int_0^\tau U(s)\psi\, ds$$

$$= \frac{1}{t}\int_\tau^{\tau+t} U(s)\psi\, ds - \frac{1}{t}\int_0^t U(s)\psi\, ds$$

$$= \frac{1}{t}U(\tau)\int_0^t U(s)\psi\, ds - \frac{1}{t}\int_0^t U(s)\psi\, ds \to U(\tau)\psi - \psi$$

as $t \to 0$ shows $\psi_\tau \in \mathfrak{D}(A)$. As in the proof of the previous theorem, we can show that A is symmetric and that $U(t)\mathfrak{D}(A) = \mathfrak{D}(A)$.

Next, let us prove that A is essentially self-adjoint. By Corollary 2.8 it suffices to prove $\operatorname{Ker}(A^* - z^*) = \{0\}$ for $z \in \mathbb{C}\backslash\mathbb{R}$. Suppose $A^*\varphi = z^*\varphi$. Then for each $\psi \in \mathfrak{D}(A)$ we have

$$\frac{d}{dt}\langle \varphi, U(t)\psi \rangle = \langle \varphi, -iAU(t)\psi \rangle = -i\langle A^*\varphi, U(t)\psi \rangle = -iz\langle \varphi, U(t)\psi \rangle$$

and hence $\langle \varphi, U(t)\psi \rangle = \exp(-izt)\langle \varphi, \psi \rangle$. Since the left-hand side is bounded for all $t \in \mathbb{R}$ and the exponential on the right-hand side is not, we must have $\langle \varphi, \psi \rangle = 0$ implying $\varphi = 0$ since $\mathfrak{D}(A)$ is dense.

So A is essentially self-adjoint and we can introduce $V(t) = \exp(-it\overline{A})$. We are done if we can show $U(t) = V(t)$.

Let $\psi \in \mathfrak{D}(A)$ and abbreviate $\psi(t) = (U(t) - V(t))\psi$. Then
$$\lim_{s \to 0} \frac{\psi(t+s) - \psi(t)}{s} = \mathrm{i}\overline{A}\psi(t)$$
and hence $\frac{d}{dt}\|\psi(t)\|^2 = 2\operatorname{Re}\langle\psi(t), \mathrm{i}A\psi(t)\rangle = 0$. Since $\psi(0) = 0$, we have $\psi(t) = 0$ and hence $U(t)$ and $V(t)$ coincide on $\mathfrak{D}(A)$. Furthermore, since $\mathfrak{D}(A)$ is dense, we have $U(t) = V(t)$ by continuity. □

As an immediate consequence of the proof, we also note the following useful criterion.

Corollary 5.4. *Suppose $\mathfrak{D} \subseteq \mathfrak{D}(A)$ is dense and invariant under $U(t)$. Then A is essentially self-adjoint on $\mathfrak{D}$.*

Proof. As in the above proof, it follows that $\langle\varphi, \psi\rangle = 0$ for every $\psi \in \mathfrak{D}$ and $\varphi \in \operatorname{Ker}(A^* - z^*)$. □

Note that by Lemma 4.9 two strongly continuous one-parameter groups commute,
$$(5.7) \qquad [\mathrm{e}^{-\mathrm{i}tA}, \mathrm{e}^{-\mathrm{i}sB}] = 0,$$
if and only if the generators commute.

Clearly, for a physicist, one of the goals must be to understand the time evolution of a quantum mechanical system. We have seen that the time evolution is generated by a self-adjoint operator, the Hamiltonian, and is given by a linear first order differential equation, the Schrödinger equation. To understand the dynamics of such a first order differential equation, one must understand the spectrum of the generator. Some general tools for this endeavor will be provided in the following sections.

Problem 5.1. *Let $\mathfrak{H} = L^2(0, 2\pi)$ and consider the one-parameter unitary group given by $U(t)f(x) = f(x - t \mod 2\pi)$. What is the generator of U?*

Problem 5.2. *Suppose $\psi(t)$ is differentiable on $\mathbb{R}$. Show that*
$$\|\psi(t) - \psi(s)\| \leq M|t - s|, \qquad M = \sup_{\tau \in [s,t]} \|\frac{d\psi}{dt}(\tau)\|.$$

(Hint: Consider $f(\tau) = \|\psi(\tau) - \psi(s)\| - \tilde{M}(\tau - s)$ for $\tau \in [s, t]$. Suppose τ_0 is the largest τ for which the claim holds with $\tilde{M} > M$ and find a contradiction if $\tau_0 < t$.)

Problem 5.3. *Consider the **abstract wave equation***
$$(5.8) \qquad \frac{d^2}{dt^2}\psi(t) + H\psi(t) = 0,$$

5.1. The time evolution and Stone's theorem

where $H \geq 0$ is some nonnegative operator. If $H = -\Delta$ is the free Schrödinger operator (cf. Section 7.3), then (5.8) is the usual wave equation in $\mathbb{R}^d$.

Introducing $\varphi = (\psi, \frac{d}{dt}\psi)$, equation (5.8) can be written as an equivalent first order system

$$(5.9) \qquad i\frac{d}{dt}\varphi(t) = \mathcal{H}\varphi(t), \qquad \mathcal{H} = \begin{pmatrix} 0 & 1 \\ -H & 0 \end{pmatrix}.$$

Show that

$$(5.10) \qquad \varphi(t) = V(t)\varphi(0)$$

solves (5.9) provided $\varphi(0) = \mathfrak{D}(H) \oplus \mathfrak{D}(H^{1/2})$ with

$$V(t) = c(tH^{1/2}) \begin{pmatrix} 1 & 0 \\ 0 & 1 \end{pmatrix} + t\, s(tH^{1/2}) \begin{pmatrix} 0 & 1 \\ -H & 0 \end{pmatrix},$$

$$\mathfrak{D}(V(t)) = \mathfrak{D}(H^{1/2}) \oplus \mathfrak{H},$$

where $c(\lambda) = \cos(\lambda)$ and $s(\lambda) = \frac{\sin(\lambda)}{\lambda}$.

In particular, show that $V(t)$ is a strongly continuous one-parameter group (which, however, is not unitary).

Problem 5.4. *To make $V(t)$ from the previous example unitary, we will change the underlying Hilbert space.*

Let A be some closed operator such that $H = A^*A$ (cf. Problem 2.12). (For example, one could choose $A = |H|^{1/2}$ but for given H a factorization $H = A^*A$ might be easier to find than $|H|^{1/2}$.)

Suppose $\mathrm{Ker}(A) = \{0\}$ and let $\mathfrak{H}_0 = \mathrm{Ker}(A^*)^\perp = \overline{\mathrm{Ran}(A)} \subseteq \mathfrak{H}$. Moreover, let $\mathfrak{H}_1$ be the Hilbert space obtained from $\mathfrak{D}(A)$ by taking the completion with respect to the norm $\|\psi\|_1 = \|A\psi\| = \|H^{1/2}\psi\|$ (note that $\|.\|_1$ will be equivalent to the graph norm if $0 \in \rho(A)$ and $\mathfrak{H}_1$ will be just $\mathfrak{D}(A)$ in this case).

Show that $\mathcal{H}$ is self-adjoint in $\mathfrak{H}_1 \oplus \mathfrak{H}$ when defined on $\mathfrak{D}(\mathcal{H}) = \mathfrak{D}(A^*A) \oplus \mathfrak{D}(A)$ by showing that it is unitarily equivalent to a supersymmetric Dirac operator (cf. Problem 8.4).

More precisely, show that

$$U = \begin{pmatrix} 0 & 1 \\ iA & 0 \end{pmatrix}$$

extends to a unitary map $U : \mathfrak{H}_1 \otimes \mathfrak{H} \to \mathfrak{H} \otimes \mathfrak{H}_0$ and show that $U\mathcal{H}U^{-1} = D$, where D is given by

$$D = \begin{pmatrix} 0 & A^* \\ A & 0 \end{pmatrix}, \qquad \mathfrak{D}(D) = (\mathfrak{D}(A^*) \cap \mathfrak{H}_0) \oplus \mathfrak{D}(A) \subseteq \mathfrak{H}_0 \oplus \mathfrak{H}.$$

In particular, conclude that $V(t)$ defined in (5.10) extends to a unitary group in $\mathfrak{H}_1 \oplus \mathfrak{H}$ and hence the solution of (5.8) preserves the energy

$$\|A\psi(t)\|^2 + \|\frac{d}{dt}\psi(t)\|^2.$$

Problem 5.5. *How can the case* $\mathrm{Ker}(A) = \mathrm{Ker}(H) \neq \{0\}$ *be reduced to the case* $\mathrm{Ker}(A) = \mathrm{Ker}(H) = \{0\}$ *in the previous problem?*

5.2. The RAGE theorem

Now, let us discuss why the decomposition of the spectrum introduced in Section 3.3 is of physical relevance. Let $\|\varphi\| = \|\psi\| = 1$. The vector $\langle \varphi, \psi \rangle \varphi$ is the projection of ψ onto the (one-dimensional) subspace spanned by φ. Hence $|\langle \varphi, \psi \rangle|^2$ can be viewed as the part of ψ which is in the state φ. The first question one might raise is, how does

(5.11) $$|\langle \varphi, U(t)\psi \rangle|^2, \qquad U(t) = e^{-itA},$$

behave as $t \to \infty$? By the spectral theorem,

(5.12) $$\hat{\mu}_{\varphi,\psi}(t) = \langle \varphi, U(t)\psi \rangle = \int_{\mathbb{R}} e^{-it\lambda} d\mu_{\varphi,\psi}(\lambda)$$

is the **Fourier transform** of the measure $\mu_{\varphi,\psi}$. Thus our question is answered by Wiener's theorem.

Theorem 5.5 (Wiener)**.** *Let μ be a finite complex Borel measure on $\mathbb{R}$ and let*

(5.13) $$\hat{\mu}(t) = \int_{\mathbb{R}} e^{-it\lambda} d\mu(\lambda)$$

be its Fourier transform. Then the Cesàro time average of $\hat{\mu}(t)$ has the limit

(5.14) $$\lim_{T \to \infty} \frac{1}{T} \int_0^T |\hat{\mu}(t)|^2 dt = \sum_{\lambda \in \mathbb{R}} |\mu(\{\lambda\})|^2,$$

where the sum on the right-hand side is finite.

Proof. By Fubini, we have

$$\frac{1}{T} \int_0^T |\hat{\mu}(t)|^2 dt = \frac{1}{T} \int_0^T \int_{\mathbb{R}} \int_{\mathbb{R}} e^{-i(x-y)t} d\mu(x) d\mu^*(y) dt$$

$$= \int_{\mathbb{R}} \int_{\mathbb{R}} \left(\frac{1}{T} \int_0^T e^{-i(x-y)t} dt \right) d\mu(x) d\mu^*(y).$$

The function in parentheses is bounded by one and converges pointwise to $\chi_{\{0\}}(x-y)$ as $T \to \infty$. Thus, by the dominated convergence theorem, the

limit of the above expression is given by
$$\int_{\mathbb{R}}\int_{\mathbb{R}} \chi_{\{0\}}(x-y) d\mu(x) d\mu^*(y) = \int_{\mathbb{R}} \mu(\{y\}) d\mu^*(y) = \sum_{y \in \mathbb{R}} |\mu(\{y\})|^2,$$
which finishes the proof. □

To apply this result to our situation, observe that the subspaces $\mathfrak{H}_{ac}$, $\mathfrak{H}_{sc}$, and $\mathfrak{H}_{pp}$ are invariant with respect to time evolution since $P^{xx}U(t) = \chi_{M_{xx}}(A)\exp(-itA) = \exp(-itA)\chi_{M_{xx}}(A) = U(t)P^{xx}$, $xx \in \{ac, sc, pp\}$. Moreover, if $\psi \in \mathfrak{H}_{xx}$, we have $P^{xx}\psi = \psi$, which shows $\langle \varphi, f(A)\psi \rangle = \langle \varphi, P^{xx}f(A)\psi \rangle = \langle P^{xx}\varphi, f(A)\psi \rangle$, implying $d\mu_{\varphi,\psi} = d\mu_{P^{xx}\varphi,\psi}$. Thus if μ_ψ is ac, sc, or pp, so is $\mu_{\varphi,\psi}$ for every $\varphi \in \mathfrak{H}$.

That is, if $\psi \in \mathfrak{H}_c = \mathfrak{H}_{ac} \oplus \mathfrak{H}_{sc}$, then the Cesàro mean of $|\langle \varphi, U(t)\psi \rangle|^2$ tends to zero. In other words, the average of the probability of finding the system in any prescribed state tends to zero if we start in the continuous subspace $\mathfrak{H}_c$ of A.

If $\psi \in \mathfrak{H}_{ac}$, then $d\mu_{\varphi,\psi}$ is absolutely continuous with respect to Lebesgue measure and thus $\hat{\mu}_{\varphi,\psi}(t)$ is continuous and tends to zero as $|t| \to \infty$. In fact, this follows from the Riemann-Lebesgue lemma (see Lemma 7.7 below).

Now we want to draw some additional consequences from Wiener's theorem. This will eventually yield a dynamical characterization of the continuous and pure point spectrum due to Ruelle, Amrein, Georgescu, and Enß. But first we need a few definitions.

An operator $K \in \mathfrak{L}(\mathfrak{H})$ is called a **finite rank operator** if its range is finite dimensional. The dimension
$$\mathrm{rank}(K) = \dim \mathrm{Ran}(K)$$
is called the **rank** of K. If $\{\varphi_j\}_{j=1}^n$ is an orthonormal basis for $\mathrm{Ran}(K)$, we have

(5.15) $$K\psi = \sum_{j=1}^n \langle \varphi_j, K\psi \rangle \varphi_j = \sum_{j=1}^n \langle \psi_j, \psi \rangle \varphi_j,$$

where $\psi_j = K^*\varphi_j$. The elements ψ_j are linearly independent since $\mathrm{Ran}(K) = \mathrm{Ker}(K^*)^\perp$. Hence every finite rank operator is of the form (5.15). In addition, the adjoint of K is also finite rank and is given by

(5.16) $$K^*\psi = \sum_{j=1}^n \langle \varphi_j, \psi \rangle \psi_j.$$

The closure of the set of all finite rank operators in $\mathfrak{L}(\mathfrak{H})$ is called the set of **compact operators** $\mathfrak{C}(\mathfrak{H})$. It is straightforward to verify (Problem 5.6)

Lemma 5.6. *The set of all compact operators $\mathfrak{C}(\mathfrak{H})$ is a closed $*$-ideal in $\mathfrak{L}(\mathfrak{H})$.*

There is also a weaker version of compactness which is useful for us. The operator K is called **relatively compact** with respect to A if

(5.17) $$KR_A(z) \in \mathfrak{C}(\mathfrak{H})$$

for one $z \in \rho(A)$. By the first resolvent formula this then follows for all $z \in \rho(A)$. In particular, we have $\mathfrak{D}(A) \subseteq \mathfrak{D}(K)$.

Now let us return to our original problem.

Theorem 5.7. *Let A be self-adjoint and suppose K is relatively compact. Then*

(5.18)
$$\lim_{T \to \infty} \frac{1}{T} \int_0^T \|Ke^{-itA} P^c \psi\|^2 dt = 0 \quad \text{and} \quad \lim_{t \to \infty} \|Ke^{-itA} P^{ac} \psi\| = 0$$

for every $\psi \in \mathfrak{D}(A)$. If, in addition, K is bounded, then the result holds for every $\psi \in \mathfrak{H}$.

Proof. Let $\psi \in \mathfrak{H}_c$, respectively, $\psi \in \mathfrak{H}_{ac}$, and drop the projectors. Abbreviate $\psi(t) = e^{-itA}\psi$. Then, given a finite rank operator K as in (5.15), the claim follows from

$$\|K\psi(t)\|^2 = \left\|\sum_{j=1}^n \langle \psi_j, \psi(t)\rangle \varphi_j\right\|^2 = \sum_{j=1}^n |\langle \psi_j, \psi(t)\rangle|^2$$

together with Wiener's theorem, respectively, the Riemann-Lebesgue lemma. Hence it holds for every finite rank operator K.

If K is compact, there is a sequence K_n of finite rank operators such that $\|K - K_n\| \leq 1/n$ and hence

$$\|K\psi(t)\|^2 \leq \left(\|K_n\psi(t)\| + \frac{1}{n}\|\psi\|\right)^2 \leq 2\|K_n\psi(t)\|^2 + \frac{2}{n^2}\|\psi\|^2.$$

Thus the claim holds for every compact operator K.

If $\psi \in \mathfrak{D}(A)$, we can set $\psi = (A-i)^{-1}\varphi$, where $\varphi \in \mathfrak{H}_c$ if and only if $\psi \in \mathfrak{H}_c$ (since $\mathfrak{H}_c$ reduces A). Since $K(A+i)^{-1}$ is compact by assumption, the claim can be reduced to the previous situation. If K is also bounded, we can find a sequence $\psi_n \in \mathfrak{D}(A)$ such that $\|\psi - \psi_n\| \leq 1/n$ and hence

$$\|Ke^{-itA}\psi\| \leq \|Ke^{-itA}\psi_n\| + \frac{1}{n}\|K\|,$$

concluding the proof. $\square$

With the help of this result we can now prove an abstract version of the RAGE theorem.

5.2. The RAGE theorem

Theorem 5.8 (RAGE). *Let A be self-adjoint. Suppose $K_n \in \mathfrak{L}(\mathfrak{H})$ is a sequence of relatively compact operators which converges strongly to the identity. Then*

$$\mathfrak{H}_c = \{\psi \in \mathfrak{H} | \lim_{n\to\infty} \lim_{T\to\infty} \frac{1}{T}\int_0^T \|K_n e^{-itA}\psi\| dt = 0\},$$

(5.19) $$\mathfrak{H}_{pp} = \{\psi \in \mathfrak{H} | \lim_{n\to\infty} \sup_{t\geq 0} \|(\mathbb{I} - K_n)e^{-itA}\psi\| = 0\}.$$

Proof. Abbreviate $\psi(t) = \exp(-itA)\psi$. We begin with the first equation. Let $\psi \in \mathfrak{H}_c$. Then

$$\frac{1}{T}\int_0^T \|K_n \psi(t)\| dt \leq \left(\frac{1}{T}\int_0^T \|K_n \psi(t)\|^2 dt\right)^{1/2} \to 0$$

by Cauchy–Schwarz and the previous theorem. Conversely, if $\psi \notin \mathfrak{H}_c$, we can write $\psi = \psi^c + \psi^{pp}$. By our previous estimate it suffices to show $\|K_n \psi^{pp}(t)\| \geq \varepsilon > 0$ for n large. In fact, we even claim

(5.20) $$\lim_{n\to\infty} \sup_{t\geq 0} \|K_n \psi^{pp}(t) - \psi^{pp}(t)\| = 0.$$

By the spectral theorem, we can write $\psi^{pp}(t) = \sum_j \alpha_j(t)\psi_j$, where the ψ_j are orthonormal eigenfunctions and $\alpha_j(t) = \exp(-it\lambda_j)\alpha_j$. Truncate this expansion after N terms. Then this part converges uniformly to the desired limit by strong convergence of K_n. Moreover, by Lemma 1.14, we have $\|K_n\| \leq M$, and hence the error can be made arbitrarily small by choosing N large.

Now let us turn to the second equation. If $\psi \in \mathfrak{H}_{pp}$, the claim follows by (5.20). Conversely, if $\psi \notin \mathfrak{H}_{pp}$, we can write $\psi = \psi^c + \psi^{pp}$ and by our previous estimate it suffices to show that $\|(\mathbb{I} - K_n)\psi^c(t)\|$ does not tend to 0 as $n \to \infty$. If it did, we would have

$$0 = \lim_{T\to\infty} \frac{1}{T}\int_0^T \|(\mathbb{I} - K_n)\psi^c(t)\| dt$$

$$\geq \|\psi^c(t)\| - \lim_{T\to\infty} \frac{1}{T}\int_0^T \|K_n \psi^c(t)\| dt = \|\psi^c(t)\|,$$

a contradiction. □

In summary, regularity properties of spectral measures are related to the long-time behavior of the corresponding quantum mechanical system. However, a more detailed investigation of this topic is beyond the scope of this manuscript. For a survey containing several recent results, see [**35**].

It is often convenient to treat the observables as time dependent rather than the states. We set

(5.21) $$K(t) = e^{itA} K e^{-itA}$$

and note

(5.22) $$\langle \psi(t), K\psi(t) \rangle = \langle \psi, K(t)\psi \rangle, \qquad \psi(t) = e^{-itA}\psi.$$

This point of view is often referred to as the **Heisenberg picture** in the physics literature. If K is unbounded, we will assume $\mathfrak{D}(A) \subseteq \mathfrak{D}(K)$ such that the above equations make sense at least for $\psi \in \mathfrak{D}(A)$. The main interest is the behavior of $K(t)$ for large t. The strong limits are called **asymptotic observables** if they exist.

Theorem 5.9. *Suppose A is self-adjoint and K is relatively compact. Then*

(5.23) $$\lim_{T \to \infty} \frac{1}{T} \int_0^T e^{itA} K e^{-itA} \psi \, dt = \sum_{\lambda \in \sigma_p(A)} P_A(\{\lambda\}) K P_A(\{\lambda\}) \psi, \quad \psi \in \mathfrak{D}(A).$$

If K is in addition bounded, the result holds for every $\psi \in \mathfrak{H}$.

Proof. We will assume that K is bounded. To obtain the general result, use the same trick as before and replace K by $K R_A(z)$. Write $\psi = \psi^c + \psi^{pp}$. Then
$$\lim_{T \to \infty} \frac{1}{T} \left\| \int_0^T K(t)\psi^c dt \right\| \le \lim_{T \to \infty} \frac{1}{T} \int_0^T \| K(t)\psi^c dt \| = 0$$
by Theorem 5.7. As in the proof of the previous theorem, we can write $\psi^{pp} = \sum_j \alpha_j \psi_j$ and hence
$$\sum_j \alpha_j \frac{1}{T} \int_0^T K(t)\psi_j dt = \sum_j \alpha_j \left(\frac{1}{T} \int_0^T e^{it(A-\lambda_j)} dt \right) K\psi_j.$$

As in the proof of Wiener's theorem, we see that the operator in parentheses tends to $P_A(\{\lambda_j\})$ strongly as $T \to \infty$. Since this operator is also bounded by 1 for all T, we can interchange the limit with the summation and the claim follows. □

We also note the following corollary.

Corollary 5.10. *Under the same assumptions as in the RAGE theorem, we have*

(5.24) $$\lim_{n \to \infty} \lim_{T \to \infty} \frac{1}{T} \int_0^T e^{itA} K_n e^{-itA} \psi \, dt = P^{pp}\psi,$$

respectively,

(5.25) $$\lim_{n \to \infty} \lim_{T \to \infty} \frac{1}{T} \int_0^T e^{itA} (\mathbb{I} - K_n) e^{-itA} \psi \, dt = P^c\psi.$$

Problem 5.6. *Prove Lemma 5.6.*

Problem 5.7 (Mean ergodic theorem)**.** *Show*
$$\lim_{T\to\infty} \frac{1}{T} \int_0^T \langle \varphi, \mathrm{e}^{\mathrm{i}tA}\psi\rangle = \langle \varphi, P_A(\{0\})\psi\rangle$$
and conclude
$$\operatorname*{s-lim}_{T\to\infty} \frac{1}{T} \int_0^T \mathrm{e}^{\mathrm{i}tA} dt = P_A(\{0\}).$$
(Hint: Lemma 1.12 (iv).)

Problem 5.8. *Prove Corollary 5.10.*

Problem 5.9. *A finite measure is called a* **Rajchman measure** *if it satisfies*
$$\lim_{t\to\infty} \hat{\mu}(t) = 0. \tag{5.26}$$
Note that by the Riemann–Lebesgue lemma (cf. Lemma 7.7), every absolutely continuous measure is Rajchman. Moreover, by the Wiener theorem (cf. Theorem 5.5), every Rajchman measure is continuous (however, there are examples which show that the converse of both claims is not true).

Let A be self-adjoint. Show that the set of all vectors for which the spectral measure is a Rajchman measure,
$$\mathfrak{H}_{rc} = \{\psi \in \mathfrak{H} | \lim_{t\to\infty} \langle \psi, \mathrm{e}^{-\mathrm{i}tA}\psi\rangle = 0\} \subseteq \mathfrak{H}_{ac}, \tag{5.27}$$
is a closed subspace which is invariant under $\mathrm{e}^{-\mathrm{i}sA}$.

(Hint: First show $\psi \in \mathfrak{H}_{rc}$ if and only if $\lim_{t\to\infty}\langle\varphi, \mathrm{e}^{-\mathrm{i}tA}\psi\rangle = 0$ for every $\varphi \in \mathfrak{H}$. In fact, first show this for $\varphi \in \overline{\{\mathrm{e}^{-\mathrm{i}sA}\psi | s \in \mathbb{R}\}}$ and then extend it to the general case.)

Problem 5.10. *Under the same assumptions as for Theorem 5.7, show*
$$\lim_{t\to\infty} \|K\mathrm{e}^{-\mathrm{i}tA} P^{rc}\psi\| = 0,$$
where P^{rc} is the projector onto $\mathfrak{H}_{rc}$ from the previous problem.

5.3. The Trotter product formula

In many situations the operator is of the form $A + B$, where $\mathrm{e}^{\mathrm{i}tA}$ and $\mathrm{e}^{\mathrm{i}tB}$ can be computed explicitly. Since A and B will not commute in general, we cannot obtain $\mathrm{e}^{\mathrm{i}t(A+B)}$ from $\mathrm{e}^{\mathrm{i}tA}\mathrm{e}^{\mathrm{i}tB}$. However, we at least have

Theorem 5.11 (Trotter product formula)**.** *Suppose A, B, and $A+B$ are self-adjoint. Then*
$$\mathrm{e}^{\mathrm{i}t(A+B)} = \operatorname*{s-lim}_{n\to\infty} \left(\mathrm{e}^{\mathrm{i}\frac{t}{n}A} \mathrm{e}^{\mathrm{i}\frac{t}{n}B}\right)^n, \qquad t \in \mathbb{R}. \tag{5.28}$$

Proof. First of all, note that we have
$$\left(e^{i\tau A}e^{i\tau B}\right)^n - e^{it(A+B)}$$
$$= \sum_{j=0}^{n-1} \left(e^{i\tau A} e^{i\tau B}\right)^{n-1-j} \left(e^{i\tau A} e^{i\tau B} - e^{i\tau(A+B)}\right) \left(e^{i\tau(A+B)}\right)^j,$$
where $\tau = \frac{t}{n}$, and hence
$$\|(e^{i\tau A}e^{i\tau B})^n - e^{it(A+B)}\psi\| \le |t| \max_{|s| \le |t|} F_\tau(s),$$
where
$$F_\tau(s) = \|\frac{1}{\tau}(e^{i\tau A} e^{i\tau B} - e^{i\tau(A+B)})e^{is(A+B)}\psi\|.$$
Now for $\psi \in \mathfrak{D}(A+B) = \mathfrak{D}(A) \cap \mathfrak{D}(B)$, we have
$$\frac{1}{\tau}(e^{i\tau A} e^{i\tau B} - e^{i\tau(A+B)})\psi \to iA\psi + iB\psi - i(A+B)\psi = 0$$
as $\tau \to 0$. So $\lim_{\tau \to 0} F_\tau(s) = 0$ at least pointwise, but we need this uniformly with respect to $s \in [-|t|, |t|]$.

Pointwise convergence implies
$$\|\frac{1}{\tau}(e^{i\tau A} e^{i\tau B} - e^{i\tau(A+B)})\psi\| \le C(\psi)$$
and, since $\mathfrak{D}(A+B)$ is a Hilbert space when equipped with the graph norm $\|\psi\|_{\Gamma(A+B)}^2 = \|\psi\|^2 + \|(A+B)\psi\|^2$, we can invoke the uniform boundedness principle to obtain
$$\|\frac{1}{\tau}(e^{i\tau A} e^{i\tau B} - e^{i\tau(A+B)})\psi\| \le C\|\psi\|_{\Gamma(A+B)}.$$
Now
$$|F_\tau(s) - F_\tau(r)| \le \|\frac{1}{\tau}(e^{i\tau A} e^{i\tau B} - e^{i\tau(A+B)})(e^{is(A+B)} - e^{ir(A+B)})\psi\|$$
$$\le C\|(e^{is(A+B)} - e^{ir(A+B)})\psi\|_{\Gamma(A+B)}$$
shows that $F_\tau(.)$ is uniformly continuous and the claim follows by a standard $\frac{\varepsilon}{2}$ argument. $\square$

If the operators are semi-bounded from below, the same proof shows

Theorem 5.12 (Trotter product formula). *Suppose A, B, and $A+B$ are self-adjoint and semi-bounded from below. Then*
$$(5.29) \qquad e^{-t(A+B)} = \operatorname*{s-lim}_{n \to \infty} \left(e^{-\frac{t}{n}A} e^{-\frac{t}{n}B}\right)^n, \quad t \ge 0.$$

Problem 5.11. *Prove Theorem 5.12.*

Chapter 6

Perturbation theory for self-adjoint operators

The Hamiltonian of a quantum mechanical system is usually the sum of the kinetic energy H_0 (free Schrödinger operator) plus an operator V corresponding to the potential energy. Since H_0 is easy to investigate, one usually tries to consider V as a perturbation of H_0. This will only work if V is *small* with respect to H_0. Hence we study such perturbations of self-adjoint operators next.

6.1. Relatively bounded operators and the Kato–Rellich theorem

An operator B is called A **bounded** or **relatively bounded** with respect to A if $\mathfrak{D}(A) \subseteq \mathfrak{D}(B)$ and if there are constants $a, b \geq 0$ such that

(6.1) $$\|B\psi\| \leq a\|A\psi\| + b\|\psi\|, \qquad \psi \in \mathfrak{D}(A).$$

The infimum of all constants a for which a corresponding b exists such that (6.1) holds is called the A-**bound** of B.

The triangle inequality implies

Lemma 6.1. *Suppose B_j, $j = 1, 2$, are A bounded with respective A-bounds a_i, $i = 1, 2$. Then $\alpha_1 B_1 + \alpha_2 B_2$ is also A bounded with A-bound less than $|\alpha_1|a_1 + |\alpha_2|a_2$. In particular, the set of all A bounded operators forms a vector space.*

There are also the following equivalent characterizations:

Lemma 6.2. *Suppose A is closed with nonempty resolvent set and B is closable. Then the following are equivalent:*

(i) *B is A bounded.*

(ii) *$\mathfrak{D}(A) \subseteq \mathfrak{D}(B)$.*

(iii) *$BR_A(z)$ is bounded for one (and hence for all) $z \in \rho(A)$.*

Moreover, the A-bound of B is no larger than $\inf_{z \in \rho(A)} \|BR_A(z)\|$.

Proof. (i) $\Rightarrow$ (ii) is true by definition. (ii) $\Rightarrow$ (iii) since $BR_A(z)$ is a closed (Problem 2.9) operator defined on all of $\mathfrak{H}$ and hence bounded by the closed graph theorem (Theorem 2.9). To see (iii) $\Rightarrow$ (i), let $\psi \in \mathfrak{D}(A)$. Then

$$\|B\psi\| = \|BR_A(z)(A-z)\psi\| \le a\|(A-z)\psi\| \le a\|A\psi\| + (a|z|)\|\psi\|,$$

where $a = \|BR_A(z)\|$. Finally, note that if $BR_A(z)$ is bounded for one $z \in \rho(A)$, it is bounded for all $z \in \rho(A)$ by the first resolvent formula. $\square$

Example. Let A be the self-adjoint operator $A = -\frac{d^2}{dx^2}$, $\mathfrak{D}(A) = \{f \in H^2[0,1] | f(0) = f(1) = 0\}$ in the Hilbert space $L^2(0,1)$. If we want to add a potential represented by a multiplication operator with a real-valued (measurable) function q, then q will be relatively bounded if $q \in L^2(0,1)$: Indeed, since all functions in $\mathfrak{D}(A)$ are continuous on $[0,1]$ and hence bounded, we clearly have $\mathfrak{D}(A) \subset \mathfrak{D}(q)$ in this case. $\diamond$

We are mainly interested in the situation where A is self-adjoint and B is symmetric. Hence we will restrict our attention to this case.

Lemma 6.3. *Suppose A is self-adjoint and B is relatively bounded. The A-bound of B is given by*

(6.2) $$\lim_{\lambda \to \infty} \|BR_A(\pm i\lambda)\|.$$

If A is bounded from below, we can also replace $\pm i\lambda$ by $-\lambda$.

Proof. Let $\varphi = R_A(\pm i\lambda)\psi$, $\lambda > 0$, and let a_∞ be the A-bound of B. Then (use the spectral theorem to estimate the norms)

$$\|BR_A(\pm i\lambda)\psi\| \le a\|AR_A(\pm i\lambda)\psi\| + b\|R_A(\pm i\lambda)\psi\| \le (a + \frac{b}{\lambda})\|\psi\|.$$

Hence $\limsup_\lambda \|BR_A(\pm i\lambda)\| \le a_\infty$ which, together with the inequality $a_\infty \le \inf_\lambda \|BR_A(\pm i\lambda)\|$ from the previous lemma, proves the claim.

The case where A is bounded from below is similar, using

(6.3) $$\|BR_A(-\lambda)\psi\| \le \left(a \max\left(1, \frac{|\gamma|}{\lambda + \gamma}\right) + \frac{b}{\lambda + \gamma}\right) \|\psi\|,$$

for $-\lambda < \gamma$. $\square$

6.1. Relatively bounded operators and the Kato–Rellich theorem

Now we will show the basic perturbation result due to Kato and Rellich.

Theorem 6.4 (Kato–Rellich)**.** *Suppose A is (essentially) self-adjoint and B is symmetric with A-bound less than one. Then $A + B$, $\mathfrak{D}(A + B) = \mathfrak{D}(A)$, is (essentially) self-adjoint. If A is essentially self-adjoint, we have $\mathfrak{D}(\overline{A}) \subseteq \mathfrak{D}(\overline{B})$ and $\overline{A + B} = \overline{A} + B$.*

If A is bounded from below by γ, then $A + B$ is bounded from below by

$$(6.4) \qquad \gamma - \max\left(a|\gamma| + b, \frac{b}{a-1}\right).$$

Proof. Since by (6.1) the graph norm of A dominates those of B and $A+B$, we obtain $\mathfrak{D}(\overline{A}) \subseteq \mathfrak{D}(\overline{B})$ and $\mathfrak{D}(\overline{A}) \subseteq \mathfrak{D}(\overline{A+B})$. Thus we can assume that A is closed (i.e., self-adjoint). It suffices to show that $\operatorname{Ran}(A+B \pm i\lambda) = \mathfrak{H}$. By the above lemma, we can find a $\lambda > 0$ such that $\|BR_A(\pm i\lambda)\| < 1$. Hence $-1 \in \rho(BR_A(\pm i\lambda))$ and thus $\mathbb{I} + BR_A(\pm i\lambda)$ is onto. Thus

$$(A + B \pm i\lambda) = (\mathbb{I} + BR_A(\pm i\lambda))(A \pm i\lambda)$$

is onto and the proof of the first part is complete.

If A is bounded from below, we can replace $\pm i\lambda$ by $-\lambda$, and the above equation shows that R_{A+B} exists for λ sufficiently large. The explicit bound (6.4) follows after solving (6.3) from the proof of the previous lemma for λ. $\square$

Example. In our previous example, we have seen that $q \in L^2(0,1)$ is relatively bounded by checking $\mathfrak{D}(A) \subset \mathfrak{D}(q)$. However, working a bit harder (Problem 6.2), one can even show that the relative bound is 0 and hence $A + q$ is self-adjoint by the Kato–Rellich theorem. $\diamond$

Finally, let us show that there is also a connection between the resolvents.

Lemma 6.5. *Let A, B be two given operators with $\mathfrak{D}(A) \subseteq \mathfrak{D}(B)$ such that A and $A + B$ are closed. Then we have the* **second resolvent formula**

$$(6.5) \qquad R_{A+B}(z) - R_A(z) = -R_A(z)BR_{A+B}(z) = -R_{A+B}(z)BR_A(z)$$

for $z \in \rho(A) \cap \rho(A+B)$. The same conclusion holds if $A+B$ is replaced by a closed operator C with $\mathfrak{D}(C) = \mathfrak{D}(A)$ and $B = C - A$.

Proof. We compute

$$R_C(z) + R_A(z)BR_C(z) = R_A(z)(C - z)R_C(z) = R_A(z).$$

The second identity follows by interchanging the roles of A and C. $\square$

Problem 6.1. *Show that (6.1) implies*

$$\|B\psi\|^2 \leq \tilde{a}^2\|A\psi\|^2 + \tilde{b}^2\|\psi\|^2$$

with $\tilde{a} = a(1+\varepsilon^2)$ and $\tilde{b} = b(1+\varepsilon^{-2})$ for every $\varepsilon > 0$. Conversely, show that this inequality implies (6.1) with $a = \tilde{a}$ and $b = \tilde{b}$.

Problem 6.2. Let A be the self-adjoint operator $A = -\frac{d^2}{dx^2}$, $\mathfrak{D}(A) = \{f \in H^2[0,1] | f(0) = f(1) = 0\}$ in the Hilbert space $L^2(0,1)$ and $q \in L^2(0,1)$.

Show that for every $f \in \mathfrak{D}(A)$ we have

$$\|f\|_\infty^2 \leq \frac{\varepsilon}{2}\|f''\|^2 + \frac{1}{2\varepsilon}\|f\|^2$$

for every $\varepsilon > 0$. Conclude that the relative bound of q with respect to A is zero. (Hint: $|f(x)|^2 \leq |\int_0^1 f'(t)dt|^2 \leq \int_0^1 |f'(t)|^2 dt = -\int_0^1 f(t)^* f''(t) dt$.)

Problem 6.3. Let A be as in the previous example. Show that q is relatively bounded if and only if $x(1-x)q(x) \in L^2(0,1)$.

Problem 6.4. Compute the resolvent of $A + \alpha \langle \psi, . \rangle \psi$. (Hint: Show

$$(\mathbb{I} + \alpha \langle \varphi, . \rangle \psi)^{-1} = \mathbb{I} - \frac{\alpha}{1 + \alpha \langle \varphi, \psi \rangle} \langle \varphi, . \rangle \psi$$

and use the second resolvent formula.)

6.2. More on compact operators

Recall from Section 5.2 that we have introduced the set of compact operators $\mathfrak{C}(\mathfrak{H})$ as the closure of the set of all finite rank operators in $\mathfrak{L}(\mathfrak{H})$. Before we can proceed, we need to establish some further results for such operators. We begin by investigating the spectrum of self-adjoint compact operators and show that the spectral theorem takes a particularly simple form in this case.

Theorem 6.6 (Spectral theorem for compact operators). *Suppose the operator K is self-adjoint and compact. Then the spectrum of K consists of an at most countable number of eigenvalues which can only cluster at 0. Moreover, the eigenspace to each nonzero eigenvalue is finite dimensional.*

In addition, we have

(6.6) $$K = \sum_{\lambda \in \sigma(K)} \lambda P_K(\{\lambda\}).$$

Proof. It suffices to show $\operatorname{rank}(P_K((\lambda - \varepsilon, \lambda + \varepsilon))) < \infty$ for $0 < \varepsilon < |\lambda|$. Let K_n be a sequence of finite rank operators such that $\|K - K_n\| \leq 1/n$. If $\operatorname{Ran} P_K((\lambda - \varepsilon, \lambda + \varepsilon))$ is infinite dimensional, we can find a vector ψ_n in this range such that $\|\psi_n\| = 1$ and $K_n \psi_n = 0$. But this yields a contradiction since

$$\frac{1}{n} \geq |\langle \psi_n, (K - K_n)\psi_n \rangle| = |\langle \psi_n, K\psi_n \rangle| \geq |\lambda| - \varepsilon > 0$$

by (4.2). $\square$

6.2. More on compact operators

As a consequence, we obtain the canonical form of a general compact operator.

Theorem 6.7 (Canonical form of compact operators). *Let K be compact. There exist orthonormal sets $\{\hat{\phi}_j\}$, $\{\phi_j\}$ and positive numbers $s_j = s_j(K)$ such that*

(6.7) $$K = \sum_j s_j \langle \phi_j, . \rangle \hat{\phi}_j, \qquad K^* = \sum_j s_j \langle \hat{\phi}_j, . \rangle \phi_j.$$

Note $K\phi_j = s_j \hat{\phi}_j$ and $K^* \hat{\phi}_j = s_j \phi_j$, and hence $K^*K\phi_j = s_j^2 \phi_j$ and $KK^* \hat{\phi}_j = s_j^2 \hat{\phi}_j$.

The numbers $s_j(K)^2 > 0$ are the nonzero eigenvalues of KK^*, respectively, K^*K (counted with multiplicity), and $s_j(K) = s_j(K^*) = s_j$ are called **singular values** *of K. There are either finitely many singular values (if K is finite rank) or they converge to zero.*

Proof. By Lemma 5.6, K^*K is compact and hence Theorem 6.6 applies. Let $\{\phi_j\}$ be an orthonormal basis of eigenvectors for $P_{K^*K}((0,\infty))\mathfrak{H}$ and let s_j^2 be the eigenvalue corresponding to ϕ_j. Then, for any $\psi \in \mathfrak{H}$, we can write

$$\psi = \sum_j \langle \phi_j, \psi \rangle \phi_j + \tilde{\psi}$$

with $\tilde{\psi} \in \text{Ker}(K^*K) = \text{Ker}(K)$. Then

$$K\psi = \sum_j s_j \langle \phi_j, \psi \rangle \hat{\phi}_j,$$

where $\hat{\phi}_j = s_j^{-1} K\phi_j$, since $\|K\tilde{\psi}\|^2 = \langle \tilde{\psi}, K^*K\tilde{\psi} \rangle = 0$. By $\langle \hat{\phi}_j, \hat{\phi}_k \rangle = (s_j s_k)^{-1} \langle K\phi_j, K\phi_k \rangle = (s_j s_k)^{-1} \langle K^*K\phi_j, \phi_k \rangle = s_j s_k^{-1} \langle \phi_j, \phi_k \rangle$, we see that the $\{\hat{\phi}_j\}$ are orthonormal and the formula for K^* follows by taking the adjoint of the formula for K (Problem 6.5). The rest is straightforward. $\square$

If K is self-adjoint, we can choose $\phi_j = \sigma_j \hat{\phi}_j$, $\sigma_j^2 = 1$, to be the eigenvectors of K, and $\sigma_j s_j$ are the corresponding eigenvalues.

Moreover, note that we have (Problem 6.6)

(6.8) $$\|K\| = \max_j s_j(K).$$

From the max-min theorem (Theorem 4.12) we obtain:

Lemma 6.8. *Let K be compact; then*

(6.9) $$s_j(K) = \min_{\psi_1, \ldots, \psi_{n-1}} \sup_{\psi \in U(\psi_1, \ldots, \psi_{n-1})} \|K\psi\|,$$

where $U(\psi_1, \ldots, \psi_n) = \{\psi \in \mathfrak{H} \mid \|\psi\| = 1, \psi \in \text{span}\{\psi_1, \ldots, \psi_n\}^\perp\}$.

In particular, note

(6.10) $$s_j(AK) \le \|A\| s_j(K), \qquad s_j(KA) \le \|A\| s_j(K)$$

whenever K is compact and A is bounded (the second estimate follows from the first by taking adjoints).

Finally, let me remark that there are a number of other equivalent definitions for compact operators.

Lemma 6.9. *For $K \in \mathfrak{L}(\mathfrak{H})$, the following statements are equivalent:*

(i) K *is compact.*

(i') K^* *is compact.*

(ii) $A_n \in \mathfrak{L}(\mathfrak{H})$ *and* $A_n \xrightarrow{s} A$ *strongly implies* $A_n K \to A K$.

(iii) $\psi_n \rightharpoonup \psi$ *weakly implies* $K\psi_n \to K\psi$ *in norm.*

(iv) ψ_n *bounded implies that* $K\psi_n$ *has a (norm) convergent subsequence.*

Proof. (i) $\Leftrightarrow$ (i'). This is immediate from Theorem 6.7.

(i) $\Rightarrow$ (ii). Translating $A_n \to A_n - A$, it is no restriction to assume that $A = 0$. Since $\|A_n\| \le M$, it suffices to consider the case where K is finite rank. Then using (6.7) and applying the triangle plus Cauchy–Schwarz inequalities yields

$$\|A_n K\|^2 \le \sup_{\|\psi\|=1} \left(\sum_{j=1}^N s_j |\langle \phi_j, \psi \rangle| \|A_n \hat\phi_j\| \right)^2 \le \sum_{j=1}^N s_j^2 \|A_n \hat\phi_j\|^2 \to 0.$$

(ii) $\Rightarrow$ (iii). Again, replace $\psi_n \to \psi_n - \psi$ and assume $\psi = 0$. Choose $A_n = \langle \psi_n, . \rangle \varphi$, $\|\varphi\| = 1$. Then $\|K\psi_n\| = \|A_n K^*\| \to 0$.

(iii) $\Rightarrow$ (iv). If ψ_n is bounded, it has a weakly convergent subsequence by Lemma 1.13. Now apply (iii) to this subsequence.

(iv) $\Rightarrow$ (i). Let φ_j be an orthonormal basis and set

$$K_n = \sum_{j=1}^n \langle \varphi_j, . \rangle K \varphi_j.$$

Then

$$\gamma_n = \|K - K_n\| = \sup_{\|\psi\|=1} \|K(\psi - \sum_{j=1}^n \langle \varphi_j, \psi \rangle K \varphi_j)\|$$

$$= \sup_{\psi \in \operatorname{span}\{\varphi_j\}_{j=n}^\infty, \|\psi\|=1} \|K\psi\|$$

is a decreasing sequence tending to a limit $\varepsilon \ge 0$. Moreover, we can find a sequence of unit vectors $\psi_n \in \operatorname{span}\{\varphi_j\}_{j=n}^\infty$ for which $\|K\psi_n\| \ge \varepsilon/2$. By assumption, $K\psi_n$ has a convergent subsequence which, since ψ_n converges weakly to 0, converges to 0. Hence ε must be 0 and we are done. $\square$

The last condition explains the name compact. Moreover, note that one cannot replace $A_n K \to AK$ by $KA_n \to KA$ in (ii) unless one additionally requires A_n to be normal (then this follows by taking adjoints — recall that only for normal operators is taking adjoints continuous with respect to strong convergence). Without the requirement that A_n be normal, the claim is wrong as the following example shows.

Example. Let $\mathfrak{H} = \ell^2(\mathbb{N})$, let A_n be the operator which shifts each sequence n places to the left, and let $K = \langle \delta_1, .\rangle \delta_1$, where $\delta_1 = (1, 0, \dots)$. Then s-lim $A_n = 0$ but $\|KA_n\| = 1$. $\diamond$

Problem 6.5. *Deduce the formula for K^* from the one for K in (6.7).*

Problem 6.6. *Prove (6.8).*

6.3. Hilbert–Schmidt and trace class operators

Among the compact operators, two special classes are of particular importance. The first ones are integral operators

$$(6.11) \qquad K\psi(x) = \int_M K(x,y)\psi(y)d\mu(y), \qquad \psi \in L^2(M, d\mu),$$

where $K(x,y) \in L^2(M \times M, d\mu \otimes d\mu)$. Such an operator is called a **Hilbert–Schmidt operator**. Using Cauchy–Schwarz,

$$\int_M |K\psi(x)|^2 d\mu(x) = \int_M \left| \int_M |K(x,y)\psi(y)|d\mu(y) \right|^2 d\mu(x)$$
$$\leq \int_M \left(\int_M |K(x,y)|^2 d\mu(y) \right) \left(\int_M |\psi(y)|^2 d\mu(y) \right) d\mu(x)$$
$$(6.12) \qquad = \left(\int_M \int_M |K(x,y)|^2 d\mu(y)\, d\mu(x) \right) \left(\int_M |\psi(y)|^2 d\mu(y) \right),$$

we see that K is bounded. Next, pick an orthonormal basis $\varphi_j(x)$ for $L^2(M, d\mu)$. Then, by Lemma 1.10, $\varphi_i(x)\varphi_j(y)$ is an orthonormal basis for $L^2(M \times M, d\mu \otimes d\mu)$ and

$$(6.13) \qquad K(x,y) = \sum_{i,j} c_{i,j} \varphi_i(x) \varphi_j(y), \qquad c_{i,j} = \langle \varphi_i, K\varphi_j^* \rangle,$$

where

$$(6.14) \qquad \sum_{i,j} |c_{i,j}|^2 = \int_M \int_M |K(x,y)|^2 d\mu(y)\, d\mu(x) < \infty.$$

In particular,

(6.15) $$K\psi(x) = \sum_{i,j} c_{i,j} \langle \varphi_j^*, \psi \rangle \varphi_i(x)$$

shows that K can be approximated by finite rank operators (take finitely many terms in the sum) and is hence compact.

Using (6.7), we can also give a different characterization of Hilbert–Schmidt operators.

Lemma 6.10. *If $\mathfrak{H} = L^2(M, d\mu)$, then a compact operator K is Hilbert–Schmidt if and only if $\sum_j s_j(K)^2 < \infty$ and*

(6.16) $$\sum_j s_j(K)^2 = \int_M \int_M |K(x,y)|^2 d\mu(x) d\mu(y),$$

in this case.

Proof. If K is compact, we can define approximating finite rank operators K_n by considering only finitely many terms in (6.7):

$$K_n = \sum_{j=1}^n s_j \langle \phi_j, . \rangle \hat{\phi}_j.$$

Then K_n has the kernel $K_n(x,y) = \sum_{j=1}^n s_j \phi_j(y)^* \hat{\phi}_j(x)$ and

$$\int_M \int_M |K_n(x,y)|^2 d\mu(x) d\mu(y) = \sum_{j=1}^n s_j(K)^2.$$

Now if one side converges, so does the other and, in particular, (6.16) holds in this case.

Conversely, choose $\varphi_i = \hat{\varphi}_i$ in (6.15). Then a comparison with (6.7) shows

$$\sum_j c_{i,j}^* \hat{\phi}_j^* = s_i(K) \phi_i$$

and thus

$$s_i(K)^2 = \sum_j |c_{i,j}|^2.$$

$\square$

Hence, we will call a bounded operator Hilbert–Schmidt if it is compact and its singular values satisfy

(6.17) $$\sum_j s_j(K)^2 < \infty.$$

By our lemma, this coincides with our previous definition if $\mathfrak{H} = L^2(M, d\mu)$.

6.3. Hilbert–Schmidt and trace class operators

Since every Hilbert space is isomorphic to some $L^2(M, d\mu)$, we see that the Hilbert–Schmidt operators together with the norm

$$\|K\|_2 = \left(\sum_j s_j(K)^2\right)^{1/2} \tag{6.18}$$

form a Hilbert space (isomorphic to $L^2(M \times M, d\mu \otimes d\mu)$). Note that $\|K\|_2 = \|K^*\|_2$ (since $s_j(K) = s_j(K^*)$). There is another useful characterization for identifying Hilbert–Schmidt operators:

Lemma 6.11. *A bounded operator K is Hilbert–Schmidt if and only if*

$$\sum_j \|K\varphi_j\|^2 < \infty \tag{6.19}$$

for some orthonormal basis and

$$\sum_j \|K\varphi_j\|^2 = \|K\|_2^2 \tag{6.20}$$

for every orthonormal basis in this case.

Proof. First of all, note that (6.19) implies that K is compact. To see this, let P_n be the projection onto the space spanned by the first n elements of the orthonormal basis $\{\varphi_j\}$. Then $K_n = KP_n$ is finite rank and converges to K since

$$\|(K - K_n)\psi\| = \|\sum_{j>n} c_j K\varphi_j\| \leq \sum_{j>n} |c_j| \|K\varphi_j\| \leq \left(\sum_{j>n} \|K\varphi_j\|^2\right)^{1/2} \|\psi\|,$$

where $\psi = \sum_j c_j \varphi_j$.

The rest follows from (6.7) and

$$\sum_j \|K\varphi_j\|^2 = \sum_{j,k} |\langle \hat{\phi}_k, K\varphi_j\rangle|^2 = \sum_{j,k} |\langle K^*\hat{\phi}_k, \varphi_j\rangle|^2$$
$$= \sum_k \|K^*\hat{\phi}_k\|^2 = \sum_k s_k(K)^2.$$

Here we have used $\overline{\text{span}\{\hat{\phi}_k\}} = \text{Ker}(K^*)^\perp = \overline{\text{Ran}(K)}$ in the first step. $\square$

This approach can be generalized by defining

$$\|K\|_p = \left(\sum_j s_j(K)^p\right)^{1/p} \tag{6.21}$$

plus corresponding spaces

$$\mathcal{J}_p(\mathfrak{H}) = \{K \in \mathfrak{C}(\mathfrak{H}) | \|K\|_p < \infty\}, \tag{6.22}$$

which are known as **Schatten p-classes**. Note that by (6.8)

$$\|K\| \leq \|K\|_p \tag{6.23}$$

and that by $s_j(K) = s_j(K^*)$ we have

(6.24) $$\|K\|_p = \|K^*\|_p.$$

Lemma 6.12. *The spaces $\mathcal{J}_p(\mathfrak{H})$ together with the norm $\|.\|_p$ are Banach spaces. Moreover,*

(6.25) $$\|K\|_p = \sup\left\{\left(\sum_j |\langle\psi_j, K\varphi_j\rangle|^p\right)^{1/p} \Big| \{\psi_j\}, \{\varphi_j\} \text{ ONS}\right\},$$

where the sup is taken over all orthonormal sets.

Proof. The hard part is to prove (6.25): Choose q such that $\frac{1}{p} + \frac{1}{q} = 1$ and use Hölder's inequality to obtain $(s_j|...|^2 = (s_j^p|...|^2)^{1/p}|...|^{2/q})$

$$\sum_j s_j |\langle\varphi_n, \phi_j\rangle|^2 \leq \left(\sum_j s_j^p |\langle\varphi_n, \phi_j\rangle|^2\right)^{1/p}\left(\sum_j |\langle\varphi_n, \phi_j\rangle|^2\right)^{1/q}$$

$$\leq \left(\sum_j s_j^p |\langle\varphi_n, \phi_j\rangle|^2\right)^{1/p}.$$

Clearly the analogous equation holds for $\hat{\phi}_j, \psi_n$. Now using Cauchy–Schwarz and the above inequality, we have

$$|\langle\psi_n, K\varphi_n\rangle|^p = \left|\sum_j s_j^{1/2}\langle\varphi_n, \phi_j\rangle s_j^{1/2}\langle\hat{\phi}_j, \psi_n\rangle\right|^p$$

$$\leq \left(\sum_j s_j^p |\langle\varphi_n, \phi_j\rangle|^2\right)^{1/2}\left(\sum_j s_j^p |\langle\psi_n, \hat{\phi}_j\rangle|^2\right)^{1/2}.$$

Summing over n, a second appeal to Cauchy–Schwarz and interchanging the order of summation finally gives

$$\sum_n |\langle\psi_n, K\varphi_n\rangle|^p \leq \left(\sum_{n,j} s_j^p |\langle\varphi_n, \phi_j\rangle|^2\right)^{1/2}\left(\sum_{n,j} s_j^p |\langle\psi_n, \hat{\phi}_j\rangle|^2\right)^{1/2}$$

$$\leq \left(\sum_j s_j^p\right)^{1/2}\left(\sum_j s_j^p\right)^{1/2} = \sum_j s_j^p.$$

Since equality is attained for $\varphi_n = \phi_n$ and $\psi_n = \hat{\phi}_n$, equation (6.25) holds.

Now the rest is straightforward. From

$$\left(\sum_j |\langle\psi_j, (K_1 + K_2)\varphi_j\rangle|^p\right)^{1/p}$$

$$\leq \left(\sum_j |\langle\psi_j, K_1\varphi_j\rangle|^p\right)^{1/p} + \left(\sum_j |\langle\psi_j, K_2\varphi_j\rangle|^p\right)^{1/p}$$

$$\leq \|K_1\|_p + \|K_2\|_p,$$

6.3. Hilbert–Schmidt and trace class operators

we infer that $\mathcal{J}_p(\mathfrak{H})$ is a vector space and the triangle inequality. The other requirements for a norm are obvious and it remains to check completeness. If K_n is a Cauchy sequence with respect to $\|.\|_p$, it is also a Cauchy sequence with respect to $\|.\|$ ($\|K\| \leq \|K\|_p$). Since $\mathfrak{C}(\mathfrak{H})$ is closed, there is a compact K with $\|K - K_n\| \to 0$, and by $\|K_n\|_p \leq C$, we have

$$\left(\sum_j |\langle \psi_j, K\varphi_j\rangle|^p\right)^{1/p} \leq C$$

for every finite ONS. Since the right-hand side is independent of the ONS (and in particular on the number of vectors), K is in $\mathcal{J}_p(\mathfrak{H})$. □

In combination with (6.10), we also obtain:

Corollary 6.13. *The set of $\mathcal{J}$ operators forms a $*$-ideal in $\mathfrak{L}(\mathfrak{H})$ and*

(6.26) $\qquad \|KA\|_p \leq \|A\|\|K\|_p, \quad \text{respectively}, \quad \|AK\|_p \leq \|A\|\|K\|_p.$

The two most important cases are $p = 1$ and $p = 2$: $\mathcal{J}_2(\mathfrak{H})$ is the space of Hilbert–Schmidt operators investigated at the beginning of this section and $\mathcal{J}_1(\mathfrak{H})$ is the space of **trace class** operators. Since Hilbert–Schmidt operators are easy to identify, it is important to relate $\mathcal{J}_1(\mathfrak{H})$ with $\mathcal{J}_2(\mathfrak{H})$:

Lemma 6.14. *An operator is trace class if and only if it can be written as the product of two Hilbert–Schmidt operators, $K = K_1 K_2$, and in this case we have*

(6.27) $\qquad \|K\|_1 \leq \|K_1\|_2 \|K_2\|_2.$

Proof. By Cauchy–Schwarz we have

$$\sum_n |\langle \varphi_n, K\psi_n\rangle| = \sum_n |\langle K_1^*\varphi_n, K_2\psi_n\rangle| \leq \left(\sum_n \|K_1^*\varphi_n\|^2 \sum_n \|K_2\psi_n\|^2\right)^{1/2}$$
$$= \|K_1\|_2 \|K_2\|_2,$$

and hence $K = K_1 K_2$ is trace class if both K_1 and K_2 are Hilbert–Schmidt operators. To see the converse, let K be given by (6.7) and choose $K_1 = \sum_j \sqrt{s_j(K)} \langle \phi_j, .\rangle \hat{\phi}_j$, respectively, $K_2 = \sum_j \sqrt{s_j(K)} \langle \phi_j, .\rangle \phi_j$. □

Now we can also explain the name trace class:

Lemma 6.15. *If K is trace class, then for every orthonormal basis $\{\varphi_n\}$ the* **trace**

(6.28) $\qquad \operatorname{tr}(K) = \sum_n \langle \varphi_n, K\varphi_n\rangle$

is finite and independent of the orthonormal basis.

Proof. Let $\{\psi_n\}$ be another ONB. If we write $K = K_1 K_2$ with K_1, K_2 Hilbert–Schmidt, we have

$$\sum_n \langle \varphi_n, K_1 K_2 \varphi_n \rangle = \sum_n \langle K_1^* \varphi_n, K_2 \varphi_n \rangle = \sum_{n,m} \langle K_1^* \varphi_n, \psi_m \rangle \langle \psi_m, K_2 \varphi_n \rangle$$

$$= \sum_{m,n} \langle K_2^* \psi_m, \varphi_n \rangle \langle \varphi_n, K_1 \psi_m \rangle = \sum_m \langle K_2^* \psi_m, K_1 \psi_m \rangle$$

$$= \sum_m \langle \psi_m, K_2 K_1 \psi_m \rangle.$$

Hence the trace is independent of the ONB and we even have $\operatorname{tr}(K_1 K_2) = \operatorname{tr}(K_2 K_1)$. □

Clearly for self-adjoint trace class operators, the trace is the sum over all eigenvalues (counted with their multiplicity). To see this, one just has to choose the orthonormal basis to consist of eigenfunctions. This is even true for all trace class operators and is known as the Lidskij trace theorem (see [53] or [24] for an easy-to-read introduction).

Finally we note the following elementary properties of the trace:

Lemma 6.16. *Suppose K, K_1, K_2 are trace class and A is bounded.*

(i) *The trace is linear.*
(ii) $\operatorname{tr}(K^*) = \operatorname{tr}(K)^*$.
(iii) *If $K_1 \leq K_2$, then $\operatorname{tr}(K_1) \leq \operatorname{tr}(K_2)$.*
(iv) $\operatorname{tr}(AK) = \operatorname{tr}(KA)$.

Proof. (i) and (ii) are straightforward. (iii) follows from $K_1 \leq K_2$ if and only if $\langle \varphi, K_1 \varphi \rangle \leq \langle \varphi, K_2 \varphi \rangle$ for every $\varphi \in \mathfrak{H}$. (iv) By Problem 6.7 and (i), it is no restriction to assume that A is unitary. Let $\{\varphi_n\}$ be some ONB and note that $\{\psi_n = A\varphi_n\}$ is also an ONB. Then

$$\operatorname{tr}(AK) = \sum_n \langle \psi_n, AK\psi_n \rangle = \sum_n \langle A\varphi_n, AKA\varphi_n \rangle$$

$$= \sum_n \langle \varphi_n, KA\varphi_n \rangle = \operatorname{tr}(KA)$$

and the claim follows. □

Problem 6.7. *Show that every bounded operator can be written as a linear combination of two self-adjoint operators. Furthermore, show that every bounded self-adjoint operator can be written as a linear combination of two unitary operators. (Hint: $x \pm i\sqrt{1 - x^2}$ has absolute value one for $x \in [-1, 1]$.)*

6.3. Hilbert–Schmidt and trace class operators

Problem 6.8. Let $\mathfrak{H} = \ell^2(\mathbb{N})$ and let A be multiplication by a sequence $a(n)$. Show that $A \in \mathcal{J}_p(\ell^2(\mathbb{N}))$ if and only if $a \in \ell^p(\mathbb{N})$. Furthermore, show that $\|A\|_p = \|a\|_p$ in this case.

Problem 6.9. Show that $A \geq 0$ is trace class if (6.28) is finite for one (and hence all) ONB. (Hint: A is self-adjoint (why?) and $A = \sqrt{A}\sqrt{A}$.)

Problem 6.10. Show that, for an orthogonal projection P, we have

$$\dim \operatorname{Ran}(P) = \operatorname{tr}(P),$$

where we set $\operatorname{tr}(P) = \infty$ if (6.28) is infinite (for one and hence all ONB by the previous problem).

Problem 6.11. Show that, for $K \in \mathfrak{C}$, we have

$$|K| = \sum_j s_j \langle \phi_j, . \rangle \phi_j,$$

where $|K| = \sqrt{K^*K}$. Conclude that

$$\|K\|_p = (\operatorname{tr}(|K|^p))^{1/p}.$$

Problem 6.12. Suppose K, K_0 are bounded self-adjoint operators. Show that if $-K_0 \leq K \leq K_0$, then $\|K\|_p \leq \|K_0\|_p$. (Hint: Corollary 4.13.)

Problem 6.13. Suppose K is compact and let $K_{jk} = \langle \varphi_j, K\varphi_k \rangle$ be the matrix elements with respect to some ONB $\{\varphi_j\}$. Then

$$\|K\|_p \leq \left(\sum_k \left(\sum_j |K_{jk}| \right)^p \right)^{1/p}.$$

(Hint: Show that $-K_0 \leq K \leq K_0$, where K_0 is diagonal in this basis with diagonal entries given by $\sum_j |K_{jk}|$. For this, use $|K_{jk}\langle\varphi_j, .\rangle\varphi_k + K_{kj}\langle\varphi_k, .\rangle\varphi_j| = |K_{jk}|(\langle\varphi_j, .\rangle\varphi_j + \langle\varphi_k, .\rangle\varphi_k)$ by Problem 4.8.)

Problem 6.14. An operator of the form $K : \ell^2(\mathbb{N}) \to \ell^2(\mathbb{N})$, $f(n) \mapsto \sum_{j \in \mathbb{N}} k(n+j) f(j)$ is called a **Hankel operator**.

- Show that K is Hilbert–Schmidt if and only if $\sum_{j \in \mathbb{N}} j |k(j)|^2 < \infty$ and this number equals $\|K\|_2$.
- Show that K is Hilbert–Schmidt with $\|K\|_2 \leq \|c\|_1$ if $|k(n)| \leq c(n)$, where $c(n)$ is decreasing and summable.

(Hint: For the first item, use summation by parts.)

6.4. Relatively compact operators and Weyl's theorem

In the previous section, we have seen that the sum of a self-adjoint operator and a symmetric operator is again self-adjoint if the perturbing operator is *small*. In this section, we want to study the influence of perturbations on the spectrum. Our hope is that at least some parts of the spectrum remain invariant.

We introduce some notation first. The **discrete spectrum** $\sigma_d(A)$ is the set of all eigenvalues which are isolated points of the spectrum and whose corresponding eigenspace is finite dimensional. The complement of the discrete spectrum is called the **essential spectrum** $\sigma_{ess}(A) = \sigma(A)\backslash\sigma_d(A)$. Hence the essential spectrum consists of all accumulation points of the spectrum plus all isolated eigenvalues of infinite multiplicity. In particular, note that the essential spectrum is closed.

If A is self-adjoint, we might equivalently set

$$(6.29) \quad \sigma_d(A) = \{\lambda \in \sigma_p(A) | \operatorname{rank}(P_A((\lambda-\varepsilon, \lambda+\varepsilon))) < \infty \text{ for some } \varepsilon > 0\},$$

respectively,

$$(6.30) \quad \sigma_{ess}(A) = \{\lambda \in \mathbb{R} | \operatorname{rank}(P_A((\lambda-\varepsilon, \lambda+\varepsilon))) = \infty \text{ for all } \varepsilon > 0\}.$$

Hence the essential spectrum consists of the absolutely continuous spectrum, the singularly continuous spectrum, accumulation points of eigenvalues, and isolated eigenvalues of infinite multiplicity.

Example. For a self-adjoint compact operator K we have by Theorem 6.6 that

$$(6.31) \quad \sigma_{ess}(K) \subseteq \{0\},$$

where equality holds if and only if $\mathfrak{H}$ is infinite dimensional. ◇

Let A be self-adjoint. Note that if we add a multiple of the identity to A, we shift the entire spectrum. Hence, in general, we cannot expect a (relatively) bounded perturbation to leave any part of the spectrum invariant. Next, if λ_0 is in the discrete spectrum, we can easily remove this eigenvalue with a finite rank perturbation of arbitrarily small norm. In fact, consider

$$(6.32) \quad A + \varepsilon P_A(\{\lambda_0\}).$$

Our only hope is that the remainder, namely the essential spectrum, is stable under finite rank perturbations. To show this, we first need a good criterion for a point to be in the essential spectrum of A.

Lemma 6.17 (Weyl criterion). *A point λ is in the essential spectrum of a self-adjoint operator A if and only if there is a sequence $\psi_n \in \mathfrak{D}(A)$ such that $\|\psi_n\| = 1$, ψ_n converges weakly to 0, and $\|(A-\lambda)\psi_n\| \to 0$. Moreover, the*

6.4. Relatively compact operators and Weyl's theorem

sequence can be chosen orthonormal. Such a sequence is called a **singular Weyl sequence** *for λ.*

Proof. Let ψ_n be a singular Weyl sequence for the point λ. By Lemma 2.17 we have $\lambda \in \sigma(A)$ and hence it suffices to show $\lambda \notin \sigma_d(A)$. If $\lambda \in \sigma_d(A)$, we can find an $\varepsilon > 0$ such that $P_\varepsilon = P_A((\lambda - \varepsilon, \lambda + \varepsilon))$ is finite rank. Consider $\tilde{\psi}_n = P_\varepsilon \psi_n$. Clearly $\|(A - \lambda)\tilde{\psi}_n\| = \|P_\varepsilon(A - \lambda)\psi_n\| \leq \|(A - \lambda)\psi_n\| \to 0$ and Lemma 6.9 (iii) implies $\tilde{\psi}_n \to 0$. However,

$$\|\psi_n - \tilde{\psi}_n\|^2 = \int_{\mathbb{R}\setminus(\lambda-\varepsilon,\lambda+\varepsilon)} d\mu_{\psi_n} \leq \frac{1}{\varepsilon^2} \int_{\mathbb{R}\setminus(\lambda-\varepsilon,\lambda+\varepsilon)} (x-\lambda)^2 d\mu_{\psi_n}(x)$$
$$\leq \frac{1}{\varepsilon^2} \|(A-\lambda)\psi_n\|^2$$

and hence $\|\tilde{\psi}_n\| \to 1$, a contradiction.

Conversely, if $\lambda \in \sigma_{ess}(A)$ is isolated, it is an eigenvalue of infinite multiplicity and we can choose an orthogonal set of eigenfunctions. Otherwise, if $\lambda \in \sigma_{ess}(A)$ is not isolated, consider $P_n = P_A([\lambda - \frac{1}{n}, \lambda - \frac{1}{n+1}) \cup (\lambda + \frac{1}{n+1}, \lambda + \frac{1}{n}])$. Then $\text{rank}(P_{n_j}) > 0$ for an infinite subsequence n_j. Now pick $\psi_j \in \text{Ran } P_{n_j}$. □

Now let K be a self-adjoint compact operator and ψ_n a singular Weyl sequence for A. Then ψ_n converges weakly to zero and hence

(6.33) $$\|(A + K - \lambda)\psi_n\| \leq \|(A - \lambda)\psi_n\| + \|K\psi_n\| \to 0$$

since $\|(A - \lambda)\psi_n\| \to 0$ by assumption and $\|K\psi_n\| \to 0$ by Lemma 6.9 (iii). Hence $\sigma_{ess}(A) \subseteq \sigma_{ess}(A + K)$. Reversing the roles of $A + K$ and A shows $\sigma_{ess}(A + K) = \sigma_{ess}(A)$. In particular, note that A and $A + K$ have the same singular Weyl sequences.

Since we have shown that we can remove any point in the discrete spectrum by a self-adjoint finite rank operator, we obtain the following equivalent characterization of the essential spectrum.

Lemma 6.18. *The essential spectrum of a self-adjoint operator A is precisely the part which is invariant under compact perturbations. In particular,*

(6.34) $$\sigma_{ess}(A) = \bigcap_{K \in \mathfrak{C}(\mathfrak{H}), K^* = K} \sigma(A + K).$$

There is even a larger class of operators under which the essential spectrum is invariant.

Theorem 6.19 (Weyl). *Suppose A and B are self-adjoint operators. If*

(6.35) $$R_A(z) - R_B(z) \in \mathfrak{C}(\mathfrak{H})$$

for one $z \in \rho(A) \cap \rho(B)$, then

(6.36) $$\sigma_{ess}(A) = \sigma_{ess}(B).$$

Proof. In fact, suppose $\lambda \in \sigma_{ess}(A)$ and let ψ_n be a corresponding singular Weyl sequence. Then

$$(R_A(z) - \frac{1}{\lambda - z})\psi_n = \frac{R_A(z)}{z - \lambda}(A - \lambda)\psi_n$$

and thus $\|(R_A(z) - \frac{1}{\lambda - z})\psi_n\| \to 0$. Moreover, by our assumption we also have $\|(R_B(z) - \frac{1}{\lambda - z})\psi_n\| \to 0$ and thus $\|(B-\lambda)\varphi_n\| = |z - \lambda|\|(R_B(z) - \frac{1}{\lambda - z})\psi_n\| \to 0$, where $\varphi_n = R_B(z)\psi_n$. Since

$$\lim_{n \to \infty} \|\varphi_n\| = \lim_{n \to \infty} \|(\lambda - z)^{-1}\psi_n + (R_B(z) - (\lambda - z)^{-1})\psi_n\|$$
$$= |\lambda - z|^{-1} \neq 0,$$

we obtain a singular Weyl sequence $\tilde{\varphi}_n = \|\varphi_n\|^{-1}\varphi_n$ for B, showing $\lambda \in \sigma_{ess}(B)$. Now interchange the roles of A and B. $\square$

As a first consequence, note the following result:

Theorem 6.20. *Suppose A is symmetric with equal finite defect indices. Then all self-adjoint extensions have the same essential spectrum.*

Proof. By Lemma 2.30, the resolvent difference of two self-adjoint extensions is a finite rank operator if the defect indices are finite. $\square$

In addition, the following result is of interest.

Lemma 6.21. *Suppose*

(6.37) $$R_A(z) - R_B(z) \in \mathfrak{C}(\mathfrak{H})$$

for one $z \in \rho(A) \cap \rho(B)$. Then this holds for all $z \in \rho(A) \cap \rho(B)$. In addition, if A and B are self-adjoint, then

(6.38) $$f(A) - f(B) \in \mathfrak{C}(\mathfrak{H})$$

for all $f \in C_\infty(\mathbb{R})$.

Proof. If the condition holds for one z, it holds for all since we have (using both resolvent formulas)

$$R_A(z') - R_B(z')$$
$$= (1 - (z - z')R_B(z'))(R_A(z) - R_B(z))(1 - (z - z')R_A(z')).$$

Let A and B be self-adjoint. The set of all functions f for which the claim holds is a closed $*$-subalgebra of $C_\infty(\mathbb{R})$ (with sup norm). Hence the claim follows from Lemma 4.4. $\square$

6.4. Relatively compact operators and Weyl's theorem

Remember that we have called K relatively compact with respect to A if $KR_A(z)$ is compact (for one and hence for all z), and note that by the second resolvent formula the resolvent difference $R_{A+K}(z) - R_A(z)$ is compact if K is relatively compact. In particular, Theorem 6.19 applies if $B = A + K$, where K is relatively compact.

For later use, observe that the set of all operators which are relatively compact with respect to A forms a vector space (since compact operators do) and relatively compact operators have A-bound zero.

Lemma 6.22. *Let A be self-adjoint and suppose K is relatively compact with respect to A. Then the A-bound of K is zero.*

Proof. Write
$$KR_A(\lambda\mathrm{i}) = (KR_A(\mathrm{i}))((A - \mathrm{i})R_A(\lambda\mathrm{i}))$$
and observe that the first operator is compact and the second is normal and converges strongly to 0 (cf. Problem 3.7). Hence the claim follows from Lemma 6.3 and the discussion after Lemma 6.9 (since R_A is normal). □

In addition, note the following result which is a straightforward consequence of the second resolvent formula.

Lemma 6.23. *Suppose A is self-adjoint and B is symmetric with A-bound less than one. If K is relatively compact with respect to A, then it is also relatively compact with respect to $A + B$.*

Proof. Since B is A bounded with A-bound less than one, we can choose a $z \in \mathbb{C}$ such that $\|BR_A(z)\| < 1$ and hence, using the second resolvent formula,

(6.39) $$BR_{A+B}(z) = BR_A(z)(\mathbb{I} + BR_A(z))^{-1}$$

shows that B is also $A + B$ bounded and the result follows from

(6.40) $$KR_{A+B}(z) = KR_A(z)(\mathbb{I} - BR_{A+B}(z))$$

since $KR_A(z)$ is compact and $BR_{A+B}(z)$ is bounded. □

Problem 6.15. *Let A and B be self-adjoint operators. Suppose B is relatively bounded with respect to A and $A + B$ is self-adjoint. Show that if $|B|^{1/2}R_A(z)$ is Hilbert–Schmidt for one $z \in \rho(A)$, then this is true for all $z \in \rho(A)$. Moreover, $|B|^{1/2}R_{A+B}(z)$ is also Hilbert–Schmidt and $R_{A+B}(z) - R_A(z)$ is trace class.*

Problem 6.16. *Show that $A = -\frac{d^2}{dx^2} + q(x)$, $\mathfrak{D}(A) = H^2(\mathbb{R})$ is self-adjoint if $q \in L^\infty(\mathbb{R})$. Show that if $-u''(x) + q(x)u(x) = zu(x)$ has a solution for which u and u' are bounded near $+\infty$ (or $-\infty$) but u is not square integrable near $+\infty$ (or $-\infty$), then $z \in \sigma_{ess}(A)$. (Hint: Use u to construct a Weyl* sequence by restricting it to a compact set. Now modify your construction to get a singular Weyl sequence by observing that functions with disjoint support are orthogonal.)

6.5. Relatively form-bounded operators and the KLMN theorem

In Section 6.1, we have considered the case where the operators A and B have a common domain on which the operator sum is well-defined. In this section, we want to look at the case where this is no longer possible, but where it is still possible to add the corresponding quadratic forms. Under suitable conditions this form sum will give rise to an operator via Theorem 2.14.

Example. Let A be the self-adjoint operator $A = -\frac{d^2}{dx^2}$, $\mathfrak{D}(A) = \{f \in H^2[0,1] | f(0) = f(1) = 0\}$ in the Hilbert space $L^2(0,1)$. If we want to add a potential represented by a multiplication operator with a real-valued (measurable) function q, then we already have seen that q will be relatively bounded if $q \in L^2(0,1)$. Hence, if $q \notin L^2(0,1)$, we are out of luck with the theory developed so far. On the other hand, if we look at the corresponding quadratic forms, we have $\mathfrak{Q}(A) = \{f \in H^1[0,1] | f(0) = f(1) = 0\}$ and $\mathfrak{Q}(q) = \mathfrak{D}(|q|^{1/2})$. Thus we see that $\mathfrak{Q}(A) \subset \mathfrak{Q}(q)$ if $q \in L^1(0,1)$.

In summary, the operators can be added if $q \in L^2(0,1)$ while the forms can be added under the less restrictive condition $q \in L^1(0,1)$.

Finally, note that in some drastic cases, there may be no way to define the operator sum: Let x_j be an enumeration of the rational numbers in $(0,1)$ and set

$$q(x) = \sum_{j=1}^{\infty} \frac{1}{2^j \sqrt{|x - x_j|}},$$

where the sum is to be understood as a limit in $L^1(0,1)$. Then q gives rise to a self-adjoint multiplication operator in $L^2(0,1)$. However, note that $\mathfrak{D}(A) \cap \mathfrak{D}(q) = \{0\}$! In fact, let $f \in \mathfrak{D}(A) \cap \mathfrak{D}(q)$. Then f is continuous and $q(x)f(x) \in L^2(0,1)$. Now suppose $f(x_j) \neq 0$ for some rational number $x_j \in (0,1)$. Then by continuity $|f(x)| \geq \delta$ for $x \in (x_j - \varepsilon, x_j + \varepsilon)$ and $q(x)|f(x)| \geq \delta 2^{-j}|x - x_j|^{-1/2}$ for $x \in (x_j - \varepsilon, x_j + \varepsilon)$, which shows that $q(x)f(x) \notin L^2(0,1)$ and hence f must vanish at every rational point. By continuity, we conclude $f = 0$. ◇

Recall from Section 2.3 that every closed semi-bounded form $q = q_A$ corresponds to a self-adjoint operator A (Theorem 2.14).

Given a self-adjoint operator $A \geq \gamma$ and a (hermitian) form $q : \mathfrak{Q} \to \mathbb{R}$ with $\mathfrak{Q}(A) \subseteq \mathfrak{Q}$, we call q **relatively form bounded** with respect to q_A if

6.5. Relatively form-bounded operators and the KLMN theorem

there are constants $a, b \geq 0$ such that

(6.41) $\qquad |q(\psi)| \leq a\, q_{A-\gamma}(\psi) + b\|\psi\|^2, \qquad \psi \in \mathfrak{Q}(A).$

The infimum of all possible a is called the **form bound** of q with respect to q_A.

Note that we do *not* require that q be associated with some self-adjoint operator (though it will be in most cases).

Example. Let $A = -\frac{d^2}{dx^2}$, $\mathfrak{D}(A) = \{f \in H^2[0,1] | f(0) = f(1) = 0\}$. Then

$$q(f) = |f(c)|^2, \qquad f \in H^1[0,1], \quad c \in (0,1),$$

is a well-defined nonnegative form. Formally, one can interpret q as the quadratic form of the multiplication *operator* with the delta distribution at $x = c$. But for $f \in \mathfrak{Q}(A) = \{f \in H^1[0,1] | f(0) = f(1) = 0\}$, we have by Cauchy–Schwarz

$$|f(c)|^2 = 2\,\mathrm{Re}\int_0^c f(t)^* f'(t)\, dt \leq 2 \int_0^1 |f(t)^* f'(t)|\, dt \leq \varepsilon \|f'\|^2 + \frac{1}{\varepsilon}\|f\|^2.$$

Consequently, q is relatively bounded with bound 0 and hence $q_A + q$ gives rise to a well-defined operator as we will show in the next theorem. $\diamond$

The following result is the analog of the Kato–Rellich theorem and is due to Kato, Lions, Lax, Milgram, and Nelson.

Theorem 6.24 (KLMN)**.** *Suppose $q_A : \mathfrak{Q}(A) \to \mathbb{R}$ is a semi-bounded closed hermitian form and q a relatively bounded hermitian form with relative bound less than one. Then $q_A + q$ defined on $\mathfrak{Q}(A)$ is closed and hence gives rise to a semi-bounded self-adjoint operator. Explicitly we have $q_A + q \geq \gamma - b$.*

Proof. Without loss of generality, we consider only the case $\gamma = 0$. A straightforward estimate shows $q_A(\psi) + q(\psi) \geq (1-a) q_A(\psi) - b\|\psi\|^2 \geq -b\|\psi\|^2$; that is, $q_A + q$ is semi-bounded. Moreover, by

$$\|\psi\|^2_{q_A+q} = q_A(\psi) + q(\psi) + (b+1)\|\psi\|^2 \leq (1+a) q_A(\psi) + (2b+1)\|\psi\|^2$$
$$\leq (2 + a + 2b)\|\psi\|^2_{q_A}$$

and

$$\|\psi\|^2_{q_A} = q_A(\psi) + \|\psi\|^2 \leq \frac{1}{1-a}\left(q_A(\psi) + q(\psi) + b\|\psi\|^2\right) + \|\psi\|^2$$
$$\leq \frac{1}{1-a}\|\psi\|^2_{q_A+q},$$

we see that the norms $\|.\|_{q_A}$ and $\|.\|_{q_A+q}$ are equivalent. Hence $q_A + q$ is closed and the result follows from Theorem 2.14. $\square$

In the investigation of the spectrum of the operator $A + B$, a key role is played by the second resolvent formula. In our present case, we have the following analog.

Theorem 6.25. *Suppose $A - \gamma \geq 0$ is self-adjoint and let $q : \mathfrak{Q} \to \mathbb{R}$ be a hermitian form with $\mathfrak{Q}(A) \subseteq \mathfrak{Q}$. Then the hermitian form*

$$q(R_A(-\lambda)^{1/2}\psi), \qquad \psi \in \mathfrak{H}, \tag{6.42}$$

corresponds to a bounded operator $C_q(\lambda)$ with $\|C_q(\lambda)\| \leq a$ for $\lambda > \frac{b}{a} - \gamma$ if and only if q is relatively form bound with constants a and b.

In particular, the form bound is given by

$$\lim_{\lambda \to \infty} \|C_q(\lambda)\|. \tag{6.43}$$

Moreover, if $a < 1$, then

$$R_{q_A+q}(-\lambda) = R_A(-\lambda)^{1/2}(1 + C_q(\lambda))^{-1}R_A(-\lambda)^{1/2}. \tag{6.44}$$

Here $R_{q_A+q}(z)$ is the resolvent of the self-adjoint operator corresponding to $q_A + q$.

Proof. We will abbreviate $C = C_q(\lambda)$ and $R_A^{1/2} = R_A(-\lambda)^{1/2}$. If q is form bounded, we have for $\lambda > \frac{b}{a} - \gamma$ that

$$|q(R_A^{1/2}\psi)| \leq a\, q_{A-\gamma}(R_A^{1/2}\psi) + b\|R_A^{1/2}\psi\|^2$$
$$= a\langle \psi, (A - \gamma + \frac{b}{a})R_A\psi\rangle \leq a\|\psi\|^2,$$

and hence $q(R_A^{1/2}\psi)$ corresponds to a bounded operator C. The converse is similar.

If $a < 1$, then $(1 + C)^{-1}$ is a well-defined bounded operator and so is $R = R_A^{1/2}(1 + C)^{-1}R_A^{1/2}$. To see that R is the inverse of $A_1 + \lambda$, where A_1 is the operator associated with $q_A + q$, take $\varphi = R_A^{1/2}\tilde{\varphi} \in \mathfrak{Q}(A)$ and $\psi \in \mathfrak{H}$. Then

$$s_{A_1+\lambda}(\varphi, R\psi) = s_{A+\lambda}(\varphi, R\psi) + s(\varphi, R\psi)$$
$$= \langle \tilde{\varphi}, (1+C)^{-1}R_A^{1/2}\psi\rangle + \langle \tilde{\varphi}, C(1+C)^{-1}R_A^{1/2}\psi\rangle = \langle \varphi, \psi\rangle.$$

Taking $\varphi \in \mathfrak{D}(A_1) \subseteq \mathfrak{Q}(A)$, we see $\langle (A_1 + \lambda)\varphi, R\psi\rangle = \langle \varphi, \psi\rangle$ and thus $R = R_{A_1}(-\lambda)$ (Problem 6.17). □

Furthermore, we can define $C_q(\lambda)$ for all $z \in \rho(A)$, using

$$C_q(z) = ((A + \lambda)^{1/2}R_A(-z)^{1/2})^* C_q(\lambda)(A + \lambda)^{1/2}R_A(-z)^{1/2}. \tag{6.45}$$

6.5. Relatively form-bounded operators and the KLMN theorem

We will call q relatively form compact if the operator $C_q(z)$ is compact for one and hence all $z \in \rho(A)$. As in the case of relatively compact operators we have

Lemma 6.26. *Suppose $A - \gamma \geq 0$ is self-adjoint and let q be a hermitian form. If q is relatively form compact with respect to q_A, then its relative form bound is 0 and the resolvents of $q_A + q$ and q_A differ by a compact operator.*

In particular, by Weyl's theorem, the operators associated with q_A and $q_A + q$ have the same essential spectrum.

Proof. Fix $\lambda_0 > \frac{b}{a} - \gamma$ and let $\lambda \geq \lambda_0$. Consider the operator $D(\lambda) = (A+\lambda_0)^{1/2} R_A(-\lambda)^{1/2}$ and note that $D(\lambda)$ is a bounded self-adjoint operator with $\|D(\lambda)\| \leq 1$. Moreover, $D(\lambda)$ converges strongly to 0 as $\lambda \to \infty$ (cf. Problem 3.7). Hence $\|D(\lambda)C(\lambda_0)\| \to 0$ by Lemma 6.9, and the same is true for $C(\lambda) = D(\lambda)C(\lambda_0)D(\lambda)$. So the relative bound is zero by (6.43). Finally, the resolvent difference is compact by (6.44) since $(1+C)^{-1} = 1 - C(1+C)^{-1}$. $\square$

Corollary 6.27. *Suppose $A-\gamma \geq 0$ is self-adjoint and let q_1, q_2 be hermitian forms. If q_1 is relatively bounded with bound less than one and q_2 is relatively compact, then the resolvent difference of $q_A + q_1 + q_2$ and $q_A + q_1$ is compact. In particular, the operators associated with $q_A + q_1$ and $q_A + q_1 + q_2$ have the same essential spectrum.*

Proof. Just observe that $C_{q_1+q_2} = C_{q_1} + C_{q_2}$ and $(1 + C_{q_1} + C_{q_2})^{-1} = (1 + C_{q_1})^{-1} - (1 + C_{q_1})^{-1} C_{q_2} (1 + C_{q_1} + C_{q_2})^{-1}$. $\square$

Finally, we turn to the special case where $q = q_B$ for some self-adjoint operator B. In this case we have

$$C_B(z) = (|B|^{1/2} R_A(-z)^{1/2})^* \operatorname{sign}(B) |B|^{1/2} R_A(-z)^{1/2} \tag{6.46}$$

and hence

$$\|C_B(z)\| \leq \| |B|^{1/2} R_A(-z)^{1/2} \|^2 \tag{6.47}$$

with equality if $V \geq 0$. Thus the following result is not too surprising.

Lemma 6.28. *Suppose $A - \gamma \geq 0$ and B is self-adjoint. Then the following are equivalent:*

(i) *B is A form bounded.*
(ii) *$\mathfrak{Q}(A) \subseteq \mathfrak{Q}(B)$.*
(iii) *$|B|^{1/2} R_A(z)^{1/2}$ is bounded for one (and hence for all) $z \in \rho(A)$.*

Proof. (i) $\Rightarrow$ (ii) is true by definition. (ii) $\Rightarrow$ (iii) since $|B|^{1/2}R_A(z)^{1/2}$ is a closed (Problem 2.9) operator defined on all of $\mathfrak{H}$ and hence bounded by the closed graph theorem (Theorem 2.9). To see (iii) $\Rightarrow$ (i), observe $|B|^{1/2}R_A(z)^{1/2} = |B|^{1/2}R_A(z_0)^{1/2}(A-z_0)^{1/2}R_A(z)^{1/2}$, which shows that $|B|^{1/2}R_A(z)^{1/2}$ is bounded for all $z \in \rho(A)$ if it is bounded for one $z_0 \in \rho(A)$. But then (6.47) shows that (i) holds. $\square$

Clearly, $C_B(\lambda)$ will be compact if $|B|^{1/2}R_A(z)^{1/2}$ is compact. However, since $R_A^{1/2}(z)$ might be hard to compute, we provide the following more handy criterion.

Lemma 6.29. *Suppose $A - \gamma \geq 0$ and B is self-adjoint where B is relatively form bounded with bound less than one. Then the resolvent difference $R_{A+B}(z) - R_A(z)$ is compact if $|B|^{1/2}R_A(z)$ is compact and trace class if $|B|^{1/2}R_A(z)$ is Hilbert–Schmidt.*

Proof. Abbreviate $R_A = R_A(-\lambda)$, $B_1 = |B|^{1/2}$, $B_2 = \text{sign}(B)|B|^{1/2}$. Choose $\lambda > \gamma$ such that $\|C_B(\lambda)\| < 1$. Then we have

$$(1 + C_B)^{-1} = \sum_{j=0}^{\infty} (-1)^j ((B_1 R_A^{1/2})^* B_2 R_A^{1/2})^j$$

$$= 1 - (B_1 R_A^{1/2})^* \left(\sum_{j=0}^{\infty} (-1)^j (B_2 R_A^{1/2}(B_1 R_A^{1/2})^*)^j \right) B_2 R_A^{1/2}$$

$$= 1 - (B_1 R_A^{1/2})^* (1 + \tilde{C}_B)^{-1} B_2 R_A^{1/2},$$

where $\tilde{C}_B = B_2 R_A^{1/2}(B_1 R_A^{1/2})^*$. Hence by (6.44) we see $R_{A+B} - R_A = -(B_1 R_A)^*(1 + \tilde{C}_B)^{-1} B_2 R_A$ and the claim follows. $\square$

Moreover, the second resolvent formula still holds when interpreted suitably:

Lemma 6.30. *Suppose $A - \gamma \geq 0$ and B is self-adjoint. If $\mathfrak{Q}(A) \subseteq \mathfrak{Q}(B)$ and $q_A + q_B$ is a closed semi-bounded form, then*

$$R_{A+B}(z) = R_A(z) - (|B|^{1/2}R_{A+B}(z^*))^* \text{sign}(B)|B|^{1/2}R_A(z)$$

(6.48)
$$= R_A(z) - (|B|^{1/2}R_A(z^*))^* \text{sign}(B)|B|^{1/2}R_{A+B}(z)$$

for $z \in \rho(A) \cap \rho(A+B)$. Here $A + B$ is the self-adjoint operator associated with $q_A + q_B$.

Proof. Let $\varphi \in \mathfrak{D}(A+B)$ and $\psi \in \mathfrak{H}$. Denote the right-hand side in (6.48) by $R(z)$ and abbreviate $R = R(z)$, $R_A = R_A(z)$, $B_1 = |B|^{1/2}$, $B_2 =$

$\text{sign}(B)|B|^{1/2}$. Then, using $s_{A+B-z}(\varphi,\psi) = \langle (A+B+z^*)\varphi,\psi\rangle$,

$$\begin{aligned}s_{A+B-z}(\varphi, R\psi) &= s_{A+B-z}(\varphi, R_A\psi) - \langle B_1 R^*_{A+B}(A+B+z^*)\varphi, B_2 R_A\psi\rangle \\ &= s_{A+B-z}(\varphi, R_A\psi) - s_B(\varphi, R_A\psi) = s_{A-z}(\varphi, R_A\psi) \\ &= \langle\varphi,\psi\rangle.\end{aligned}$$

Thus $R = R_{A+B}(z)$ (Problem 6.17). The second equality follows after exchanging the roles of A and $A+B$. $\square$

It can be shown using abstract interpolation techniques that if B is relatively bounded with respect to A, then it is also relatively form bounded. In particular, if B is relatively bounded, then $BR_A(z)$ is bounded and it is not hard to check that (6.48) coincides with (6.5). Consequently $A+B$ defined as operator sum is the same as $A+B$ defined as form sum.

Problem 6.17. *Suppose A is closed and R is bounded. Show that $R = R_A(z)$ if and only if $\langle (A-z)^*\varphi, R\psi\rangle = \langle\varphi,\psi\rangle$ for all $\varphi \in \mathfrak{D}(A^*)$, $\psi \in \mathfrak{H}$.*

Problem 6.18. *Let q be relatively form bounded with constants a and b. Show that $C_q(\lambda)$ satisfies $\|C_q(\lambda)\| \leq \max(a, \frac{b}{\lambda+\gamma})$ for $\lambda > -\gamma$. Furthermore, show that $\|C_q(\lambda)\|$ decreases as $\lambda \to \infty$.*

6.6. Strong and norm resolvent convergence

Suppose A_n and A are self-adjoint operators. We say that A_n converges to A in the **norm**, respectively, **strong resolvent** sense, if

$$(6.49) \quad \lim_{n\to\infty} R_{A_n}(z) = R_A(z), \quad \text{respectively,} \quad \text{s-}\lim_{n\to\infty} R_{A_n}(z) = R_A(z),$$

for one $z \in \Gamma = \mathbb{C}\backslash\Sigma$, $\Sigma = \sigma(A) \cup \bigcup_n \sigma(A_n)$. In fact, in the case of strong resolvent convergence, it will be convenient to include the case if A_n is only defined on some subspace $\mathfrak{H}_n \subseteq \mathfrak{H}$, where we require $P_n \xrightarrow{s} 1$ for the orthogonal projection onto $\mathfrak{H}_n$. In this case $R_{A_n}(z)$ (respectively, any other function of A_n) has to be understood as $R_{A_n}(z)P_n$, where P_n is the orthogonal projector onto $\mathfrak{H}_n$. (This generalization will produce nothing new in the norm case, since $P_n \to 1$ implies $P_n = 1$ for sufficiently large n.)

Using the Stone–Weierstraß theorem, we obtain as a first consequence

Theorem 6.31. *Let A_n, A be self-adjoint operators and suppose A_n converges to A in the norm resolvent sense. Then $f(A_n)$ converges to $f(A)$ in norm for every bounded continuous function $f: \Sigma \to \mathbb{C}$ with $\lim_{\lambda\to-\infty} f(\lambda) = \lim_{\lambda\to\infty} f(\lambda)$.*

If A_n converges to A in the strong resolvent sense, then $f(A_n)$ converges to $f(A)$ strongly for every bounded continuous function $f: \Sigma \to \mathbb{C}$.

Proof. The set of functions for which the claim holds clearly forms a *-subalgebra (since resolvents are normal, taking adjoints is continuous even with respect to strong convergence), and since it contains $f(\lambda) = 1$ and $f(\lambda) = \frac{1}{\lambda - z_0}$, this *-subalgebra is dense by the Stone–Weierstraß theorem (cf. Problem 1.22). The usual $\frac{\varepsilon}{3}$ argument shows that this *-subalgebra is also closed.

It remains to show the strong resolvent case for arbitrary bounded continuous functions. Let χ_n be a compactly supported continuous function ($0 \leq \chi_m \leq 1$) which is one on the interval $[-m, m]$. Then $\chi_m(A_n) \xrightarrow{s} \chi_m(A)$, $f(A_n)\chi_m(A_n) \xrightarrow{s} f(A)\chi_m(A)$ by the first part and hence

$$\|(f(A_n) - f(A))\psi\| \leq \|f(A_n)\| \, \|(1 - \chi_m(A))\psi\|$$
$$+ \|f(A_n)\| \, \|(\chi_m(A) - \chi_m(A_n))\psi\|$$
$$+ \|(f(A_n)\chi_m(A_n) - f(A)\chi_m(A))\psi\|$$
$$+ \|f(A)\| \, \|(1 - \chi_m(A))\psi\|$$

can be made arbitrarily small since $\|f(.)\| \leq \|f\|_\infty$ and $\chi_m(.) \xrightarrow{s} \mathbb{I}$ by Theorem 3.1. $\square$

As a consequence, note that the point $z \in \Gamma$ is of no importance, that is,

Corollary 6.32. *Suppose A_n converges to A in the norm or strong resolvent sense for one $z_0 \in \Gamma$. Then this holds for all $z \in \Gamma$.*

Also,

Corollary 6.33. *Suppose A_n converges to A in the strong resolvent sense. Then*

(6.50) $$e^{itA_n} \xrightarrow{s} e^{itA}, \qquad t \in \mathbb{R},$$

and if all operators are semi-bounded by the same bound

(6.51) $$e^{-tA_n} \xrightarrow{s} e^{-tA}, \qquad t \geq 0.$$

Next we need some good criteria to check for norm, respectively, strong, resolvent convergence.

Lemma 6.34. *Let A_n, A be self-adjoint operators with $\mathfrak{D}(A_n) = \mathfrak{D}(A)$. Then A_n converges to A in the norm resolvent sense if there are sequences a_n and b_n converging to zero such that*

(6.52) $$\|(A_n - A)\psi\| \leq a_n \|\psi\| + b_n \|A\psi\|, \quad \psi \in \mathfrak{D}(A) = \mathfrak{D}(A_n).$$

Proof. From the second resolvent formula

$$R_{A_n}(z) - R_A(z) = R_{A_n}(z)(A - A_n)R_A(z),$$

6.6. Strong and norm resolvent convergence

we infer
$$\|(R_{A_n}(\mathrm{i}) - R_A(\mathrm{i}))\psi\| \leq \|R_{A_n}(\mathrm{i})\| \Big(a_n \|R_A(\mathrm{i})\psi\| + b_n \|AR_A(\mathrm{i})\psi\|\Big)$$
$$\leq (a_n + b_n)\|\psi\|$$
and hence $\|R_{A_n}(\mathrm{i}) - R_A(\mathrm{i})\| \leq a_n + b_n \to 0$. $\square$

In particular, norm convergence implies norm resolvent convergence:

Corollary 6.35. *Let A_n, A be bounded self-adjoint operators with $A_n \to A$. Then A_n converges to A in the norm resolvent sense.*

Similarly, if no domain problems get in the way, strong convergence implies strong resolvent convergence:

Lemma 6.36. *Let A_n, A be self-adjoint operators where A_n is defined in $\mathfrak{H}_n \subseteq \mathfrak{H}$ and P_n is the orthogonal projection onto $\mathfrak{H}_n$. Then A_n converges to A in the strong resolvent sense if there is a core $\mathfrak{D}_0$ of A such that for every $\psi \in \mathfrak{D}_0$ we have $P_n\psi \in \mathfrak{D}(A_n)$ for n sufficiently large and $A_n P_n \psi \to A\psi$.*

Proof. We begin with the case $\mathfrak{H}_n = \mathfrak{H}$. Using the second resolvent formula, we have
$$\|(R_{A_n}(\mathrm{i}) - R_A(\mathrm{i}))\psi\| \leq \|(A - A_n)R_A(\mathrm{i})\psi\| \to 0$$
for $\psi \in (A - \mathrm{i})\mathfrak{D}_0$ which is dense, since $\mathfrak{D}_0$ is a core. The rest follows from Lemma 1.14.

If $\mathfrak{H}_n \subset \mathfrak{H}$, we can consider $\tilde{A}_n = A_n \oplus 0 = A_n P_n$ and conclude $R_{\tilde{A}_n}(\mathrm{i}) \overset{s}{\to} R_A(\mathrm{i})$ from the first case. By $R_{\tilde{A}_n}(\mathrm{i}) = R_{A_n}(\mathrm{i}) \oplus \mathrm{i} = R_{A_n} P_n + \mathrm{i}(1 - P_n)$ the same is true for $R_{A_n}(\mathrm{i}) P_n$ since $1 - P_n \overset{s}{\to} 0$ by assumption. $\square$

If you wonder why we did not define weak resolvent convergence, here is the answer: it is equivalent to strong resolvent convergence.

Lemma 6.37. *Let A_n, A be self-adjoint operators. Suppose $\underset{n\to\infty}{\text{w-lim}}\, R_{A_n}(z) = R_A(z)$ for some $z \in \Gamma\backslash\mathbb{R}$. Then $\underset{n\to\infty}{\text{s-lim}}\, R_{A_n}(z) = R_A(z)$ also.*

Proof. By $R_{A_n}(z) \rightharpoonup R_A(z)$ we also have $R_{A_n}(z)^* \rightharpoonup R_A(z)^*$ and thus by the first resolvent formula
$$\|R_{A_n}(z)\psi\|^2 - \|R_A(z)\psi\|^2 = \langle \psi, R_{A_n}(z^*)R_{A_n}(z)\psi - R_A(z^*)R_A(z)\psi\rangle$$
$$= \frac{1}{z - z^*}\langle \psi, (R_{A_n}(z) - R_{A_n}(z^*) + R_A(z) - R_A(z^*))\psi\rangle \to 0.$$
Together with $R_{A_n}(z)\psi \rightharpoonup R_A(z)\psi$ we have $R_{A_n}(z)\psi \to R_A(z)\psi$ by virtue of Lemma 1.12 (iv). $\square$

Now what can we say about the spectrum?

Theorem 6.38. *Let A_n and A be self-adjoint operators. If A_n converges to A in the strong resolvent sense, we have $\sigma(A) \subseteq \lim_{n\to\infty} \sigma(A_n)$. If A_n converges to A in the norm resolvent sense, we have $\sigma(A) = \lim_{n\to\infty} \sigma(A_n)$. Here $\lim_{n\to\infty} \sigma(A_n)$ denotes the set of all λ for which there is a sequence $\lambda_n \in \sigma(A_n)$ converging to λ.*

Proof. Suppose the first claim were incorrect. Then we can find a $\lambda \in \sigma(A)$ and some $\varepsilon > 0$ such that $\sigma(A_n) \cap (\lambda - \varepsilon, \lambda + \varepsilon) = \emptyset$. Choose a bounded continuous function f which is one on $(\lambda - \frac{\varepsilon}{2}, \lambda + \frac{\varepsilon}{2})$ and which vanishes outside $(\lambda - \varepsilon, \lambda + \varepsilon)$. Then $f(A_n) = 0$ and hence $f(A)\psi = \lim f(A_n)\psi = 0$ for every ψ. On the other hand, since $\lambda \in \sigma(A)$, there is a nonzero $\psi \in \operatorname{Ran} P_A((\lambda - \frac{\varepsilon}{2}, \lambda + \frac{\varepsilon}{2}))$ implying $f(A)\psi = \psi$, a contradiction.

To see the second claim, it suffices to show that $\lambda \in \lim \sigma(A_n)$ implies $\lambda \in \sigma(A)$. To this end, recall that the norm of $R_A(z)$ is just one over the distance from the spectrum. In particular, $\lambda \notin \sigma(A)$ if and only if $\|R_A(\lambda + i)\| < 1$. So $\lambda \notin \sigma(A)$ implies $\|R_A(\lambda + i)\| < 1$, which implies $\|R_{A_n}(\lambda + i)\| < 1$ for n sufficiently large, which implies $\lambda \notin \sigma(A_n)$ for n sufficiently large. □

Example. Note that the spectrum can contract if we only have convergence in the strong resolvent sense: Let A_n be multiplication by $\frac{1}{n}x$ in $L^2(\mathbb{R})$. Then A_n converges to 0 in the strong resolvent sense, but $\sigma(A_n) = \mathbb{R}$ and $\sigma(0) = \{0\}$. ◊

Lemma 6.39. *Suppose A_n converges in the strong resolvent sense to A. If $P_A(\{\lambda\}) = 0$, then*
(6.53)
$$\operatorname*{s-lim}_{n\to\infty} P_{A_n}((-\infty, \lambda)) = \operatorname*{s-lim}_{n\to\infty} P_{A_n}((-\infty, \lambda]) = P_A((-\infty, \lambda)) = P_A((-\infty, \lambda]).$$

Proof. By Theorem 6.31, the spectral measures $\mu_{n,\psi}$ corresponding to A_n converge vaguely to those of A. Hence $\|P_{A_n}(\Omega)\psi\|^2 = \mu_{n,\psi}(\Omega)$ together with Lemma A.34 implies the claim. □

Using $P((\lambda_0, \lambda_1)) = P((-\infty, \lambda_1)) - P((-\infty, \lambda_0])$, we also obtain the following.

Corollary 6.40. *Suppose A_n converges in the strong resolvent sense to A. If $P_A(\{\lambda_0\}) = P_A(\{\lambda_1\}) = 0$, then*
(6.54)
$$\operatorname*{s-lim}_{n\to\infty} P_{A_n}((\lambda_0, \lambda_1)) = \operatorname*{s-lim}_{n\to\infty} P_{A_n}([\lambda_0, \lambda_1]) = P_A((\lambda_0, \lambda_1)) = P_A([\lambda_0, \lambda_1]).$$

Example. The following example shows that the requirement $P_A(\{\lambda\}) = 0$ is crucial, even if we have bounded operators and norm convergence. In fact,

6.6. Strong and norm resolvent convergence

let $\mathfrak{H} = \mathbb{C}^2$ and

$$(6.55) \qquad A_n = \frac{1}{n}\begin{pmatrix} 1 & 0 \\ 0 & -1 \end{pmatrix}.$$

Then $A_n \to 0$ and

$$(6.56) \qquad P_{A_n}((-\infty, 0)) = P_{A_n}((-\infty, 0]) = \begin{pmatrix} 0 & 0 \\ 0 & 1 \end{pmatrix},$$

but $P_0((-\infty, 0)) = 0$ and $P_0((-\infty, 0]) = \mathbb{I}$. $\diamond$

Problem 6.19. Suppose $A_n \to A$ in the norm resolvent sense and let A be bounded. Show that A_n are eventually bounded and $A_n \to A$ in norm. (Hint: First show $\|A_n\| \to \|A\|$ and conclude that $R_A(\mathrm{i})$ and $R_{A_n}(\mathrm{i})$ are eventually bi-Lipschitz uniformly in n. Now use the second resolvent formula.)

Problem 6.20. Show that for self-adjoint operators, strong resolvent convergence is equivalent to convergence with respect to the metric

$$(6.57) \qquad d(A, B) = \sum_{n \in \mathbb{N}} \frac{1}{2^n} \|(R_A(\mathrm{i}) - R_B(\mathrm{i}))\varphi_n\|,$$

where $\{\varphi_n\}_{n \in \mathbb{N}}$ is some (fixed) ONB.

Problem 6.21 (Weak convergence of spectral measures)**.** Suppose $A_n \to A$ in the strong resolvent sense and let $\mu_{n,\psi}$, μ_ψ be the corresponding spectral measures. Show that

$$(6.58) \qquad \int f(\lambda) d\mu_{n,\psi}(\lambda) \to \int f(\lambda) d\mu_\psi(\lambda)$$

for every bounded continuous f. Give a counterexample when f is not continuous.

Part 2

Schrödinger Operators

Chapter 7

The free Schrödinger operator

7.1. The Fourier transform

We first review some basic facts concerning the **Fourier transform** which will be needed in the following section.

Let $C^\infty(\mathbb{R}^n)$ be the set of all complex-valued functions which have partial derivatives of arbitrary order. For $f \in C^\infty(\mathbb{R}^n)$ and $\alpha \in \mathbb{N}_0^n$ we set

$$(7.1) \quad \partial_\alpha f = \frac{\partial^{|\alpha|} f}{\partial x_1^{\alpha_1} \cdots \partial x_n^{\alpha_n}}, \quad x^\alpha = x_1^{\alpha_1} \cdots x_n^{\alpha_n}, \quad |\alpha| = \alpha_1 + \cdots + \alpha_n.$$

An element $\alpha \in \mathbb{N}_0^n$ is called a **multi-index** and $|\alpha|$ is called its **order**. We will also set $(\lambda x)^\alpha = \lambda^{|\alpha|} x^\alpha$ for $\lambda \in \mathbb{R}$. Recall the Schwartz space

$$(7.2) \quad \mathcal{S}(\mathbb{R}^n) = \{f \in C^\infty(\mathbb{R}^n)| \sup_x |x^\alpha (\partial_\beta f)(x)| < \infty, \ \alpha, \beta \in \mathbb{N}_0^n\}$$

which is a subspace of $L^p(\mathbb{R}^n)$ and which is dense for $1 \leq p < \infty$ (since $C_c^\infty(\mathbb{R}^n) \subset \mathcal{S}(\mathbb{R}^n)$). Note that if $f \in \mathcal{S}(\mathbb{R}^n)$, then the same is true for $x^\alpha f(x)$ and $(\partial_\alpha f)(x)$ for every multi-index α. For $f \in \mathcal{S}(\mathbb{R}^n)$ we define its **Fourier transform** via

$$(7.3) \quad \mathcal{F}(f)(p) \equiv \hat{f}(p) = \frac{1}{(2\pi)^{n/2}} \int_{\mathbb{R}^n} e^{-ipx} f(x) d^n x.$$

Then,

Lemma 7.1. *The Fourier transform maps the Schwartz space into itself, $\mathcal{F} : \mathcal{S}(\mathbb{R}^n) \to \mathcal{S}(\mathbb{R}^n)$. Furthermore, for every multi-index $\alpha \in \mathbb{N}_0^n$ and every*

$f \in \mathcal{S}(\mathbb{R}^n)$ we have

(7.4) $\qquad (\partial_\alpha f)^\wedge(p) = (\mathrm{i}p)^\alpha \hat{f}(p), \qquad (x^\alpha f(x))^\wedge(p) = \mathrm{i}^{|\alpha|} \partial_\alpha \hat{f}(p).$

Proof. First of all, by integration by parts, we see
$$\left(\frac{\partial}{\partial x_j} f(x)\right)^\wedge(p) = \frac{1}{(2\pi)^{n/2}} \int_{\mathbb{R}^n} e^{-\mathrm{i}px} \frac{\partial}{\partial x_j} f(x) d^n x$$
$$= \frac{1}{(2\pi)^{n/2}} \int_{\mathbb{R}^n} \left(-\frac{\partial}{\partial x_j} e^{-\mathrm{i}px}\right) f(x) d^n x$$
$$= \frac{1}{(2\pi)^{n/2}} \int_{\mathbb{R}^n} \mathrm{i}p_j e^{-\mathrm{i}px} f(x) d^n x = \mathrm{i}p_j \hat{f}(p).$$

Since we can repeat this argument with an arbitrary number of derivatives, the first formula follows.

Similarly, the second formula follows from
$$(x_j f(x))^\wedge(p) = \frac{1}{(2\pi)^{n/2}} \int_{\mathbb{R}^n} x_j e^{-\mathrm{i}px} f(x) d^n x$$
$$= \frac{1}{(2\pi)^{n/2}} \int_{\mathbb{R}^n} \left(\mathrm{i}\frac{\partial}{\partial p_j} e^{-\mathrm{i}px}\right) f(x) d^n x = \mathrm{i}\frac{\partial}{\partial p_j} \hat{f}(p),$$

where interchanging the derivative and integral is permissible by Problem A.20. In particular, $\hat{f}(p)$ is differentiable.

To see that $\hat{f} \in \mathcal{S}(\mathbb{R}^n)$ if $f \in \mathcal{S}(\mathbb{R}^n)$, we begin with the observation that $\hat{f}$ is bounded; in fact, $\|\hat{f}\|_\infty \le (2\pi)^{-n/2}\|f\|_1$. But then $p^\alpha(\partial_\beta \hat{f})(p) = \mathrm{i}^{-|\alpha|-|\beta|}(\partial_\alpha x^\beta f(x))^\wedge(p)$ is bounded since $\partial_\alpha x^\beta f(x) \in \mathcal{S}(\mathbb{R}^n)$ if $f \in \mathcal{S}(\mathbb{R}^n)$.
$\square$

Hence we will sometimes write $pf(x)$ for $-\mathrm{i}\partial f(x)$, where $\partial = (\partial_1, \ldots, \partial_n)$ is the **gradient**.

Three more simple properties are left as an exercise.

Lemma 7.2. Let $f \in \mathcal{S}(\mathbb{R}^n)$. Then

(7.5) $\qquad (f(x+a))^\wedge(p) = e^{\mathrm{i}ap} \hat{f}(p), \qquad a \in \mathbb{R}^n,$

(7.6) $\qquad (e^{\mathrm{i}xa} f(x))^\wedge(p) = \hat{f}(p-a), \qquad a \in \mathbb{R}^n,$

(7.7) $\qquad (f(\lambda x))^\wedge(p) = \frac{1}{\lambda^n} \hat{f}(\frac{p}{\lambda}), \qquad \lambda > 0.$

Next, we want to compute the inverse of the Fourier transform. For this, the following lemma will be needed.

Lemma 7.3. We have $e^{-zx^2/2} \in \mathcal{S}(\mathbb{R}^n)$ for $\mathrm{Re}(z) > 0$ and

(7.8) $\qquad \mathcal{F}(e^{-zx^2/2})(p) = \frac{1}{z^{n/2}} e^{-p^2/(2z)}.$

7.1. The Fourier transform

Here $z^{n/2}$ has to be understood as $(\sqrt{z})^n$, where the branch cut of the root is chosen along the negative real axis.

Proof. Due to the product structure of the exponential, one can treat each coordinate separately, reducing the problem to the case $n = 1$.

Let $\phi_z(x) = \exp(-zx^2/2)$. Then $\phi_z'(x) + zx\phi_z(x) = 0$ and hence $i(p\hat{\phi}_z(p) + z\hat{\phi}_z'(p)) = 0$. Thus $\hat{\phi}_z(p) = c\phi_{1/z}(p)$ and (Problem A.26)

$$c = \hat{\phi}_z(0) = \frac{1}{\sqrt{2\pi}} \int_{\mathbb{R}} \exp(-zx^2/2)dx = \frac{1}{\sqrt{z}}$$

at least for $z > 0$. However, since the integral is holomorphic for $\operatorname{Re}(z) > 0$ by Problem A.22, this holds for all z with $\operatorname{Re}(z) > 0$ if we choose the branch cut of the root along the negative real axis. □

Now we can show

Theorem 7.4. *The Fourier transform $\mathcal{F}: \mathcal{S}(\mathbb{R}^n) \to \mathcal{S}(\mathbb{R}^n)$ is a bijection. Its inverse is given by*

$$(7.9) \qquad \mathcal{F}^{-1}(g)(x) \equiv \check{g}(x) = \frac{1}{(2\pi)^{n/2}} \int_{\mathbb{R}^n} e^{ipx} g(p) d^n p.$$

We have $\mathcal{F}^2(f)(x) = f(-x)$ and thus $\mathcal{F}^4 = \mathbb{I}$.

Proof. Abbreviate $\phi_\varepsilon(x) = \exp(-\varepsilon x^2/2)$. By dominated convergence we have

$$(\hat{f}(p))^\vee(x) = \frac{1}{(2\pi)^{n/2}} \int_{\mathbb{R}^n} e^{ipx} \hat{f}(p) d^n p$$

$$= \lim_{\varepsilon \to 0} \frac{1}{(2\pi)^{n/2}} \int_{\mathbb{R}^n} \phi_\varepsilon(p) e^{ipx} \hat{f}(p) d^n p$$

$$= \lim_{\varepsilon \to 0} \frac{1}{(2\pi)^n} \int_{\mathbb{R}^n} \int_{\mathbb{R}^n} \phi_\varepsilon(p) e^{ipx} f(y) e^{-ipy} d^n y d^n p,$$

and, invoking Fubini and Lemma 7.2, we further see

$$= \lim_{\varepsilon \to 0} \frac{1}{(2\pi)^{n/2}} \int_{\mathbb{R}^n} (\phi_\varepsilon(p)e^{ipx})^\wedge(y) f(y) d^n y$$

$$= \lim_{\varepsilon \to 0} \frac{1}{(2\pi)^{n/2}} \int_{\mathbb{R}^n} \frac{1}{\varepsilon^{n/2}} \phi_{1/\varepsilon}(y-x) f(y) d^n y$$

$$= \lim_{\varepsilon \to 0} \frac{1}{(2\pi)^{n/2}} \int_{\mathbb{R}^n} \phi_1(z) f(x + \sqrt{\varepsilon}z) d^n z = f(x),$$

which finishes the proof, where we used the change of coordinates $z = \frac{y-x}{\sqrt{\varepsilon}}$ and again dominated convergence in the last two steps. □

From Fubini's theorem we also obtain **Plancherel's identity**

$$\int_{\mathbb{R}^n} |\hat{f}(p)|^2 d^n p = \frac{1}{(2\pi)^{n/2}} \int_{\mathbb{R}^n} \int_{\mathbb{R}^n} f(x)^* \hat{f}(p) e^{ipx} d^n p \, d^n x$$

(7.10)
$$= \int_{\mathbb{R}^n} |f(x)|^2 d^n x$$

for $f \in \mathcal{S}(\mathbb{R}^n)$. Thus, by Theorem 0.29, we can extend $\mathcal{F}$ to all of $L^2(\mathbb{R}^n)$ by setting $\mathcal{F}(f) = \lim_{m \to \infty} \mathcal{F}(f_m)$, where f_m is an arbitrary sequence from $\mathcal{S}(\mathbb{R}^n)$ converging to f in the L^2 norm.

Theorem 7.5 (Plancherel). *The Fourier transform $\mathcal{F}$ extends to a unitary operator $\mathcal{F}: L^2(\mathbb{R}^n) \to L^2(\mathbb{R}^n)$. Its spectrum is given by*

(7.11) $$\sigma(\mathcal{F}) = \{z \in \mathbb{C} | z^4 = 1\} = \{1, -1, i, -i\}.$$

Proof. As already noted, $\mathcal{F}$ extends uniquely to a bounded operator on $L^2(\mathbb{R}^n)$. Since Parseval's identity remains valid by continuity of the norm and since its range is dense, this extension is a unitary operator. It remains to compute the spectrum. In fact, if ψ_n is a Weyl sequence, then $(\mathcal{F}^2 + z^2)(\mathcal{F} + z)(\mathcal{F} - z)\psi_n = (\mathcal{F}^4 - z^4)\psi_n = (1 - z^4)\psi_n \to 0$ implies $z^4 = 1$. Hence $\sigma(\mathcal{F}) \subseteq \{z \in \mathbb{C} | z^4 = 1\}$. We defer the proof for equality to Section 8.3, where we will explicitly compute an orthonormal basis of eigenfunctions. □

We also note that this extension is still given by (7.3) whenever the right-hand side is integrable.

Lemma 7.6. *Let $f \in L^1(\mathbb{R}^n) \cap L^2(\mathbb{R}^n)$; then (7.3) continues to hold, where $\mathcal{F}$ now denotes the extension of the Fourier transform from $\mathcal{S}(\mathbb{R}^n)$ to $L^2(\mathbb{R}^n)$.*

Proof. Fix a bounded set $X \subset \mathbb{R}^n$ and let $f \in L^2(X)$. Then we can approximate f by functions $f_n \in C_c^\infty(X)$ in the L^2 norm. Since $L^2(X)$ is continuously embedded into $L^1(X)$ (Problem 0.36), this sequence will also converge in $L^1(X)$. Extending all functions to $\mathbb{R}^n$ by setting them zero outside X, we see that the claim holds for $f \in L^2(\mathbb{R}^n)$ with compact support. Finally, for general $f \in L^1(\mathbb{R}^n) \cap L^2(\mathbb{R}^n)$, consider $f_m = f\chi_{B_m(0)}$. Then $f_m \to f$ in both $L^1(\mathbb{R}^n)$ and $L^2(\mathbb{R}^n)$ and the claim follows. □

In particular,

(7.12) $$\hat{f}(p) = \lim_{m \to \infty} \frac{1}{(2\pi)^{n/2}} \int_{|x| \leq m} e^{-ipx} f(x) d^n x,$$

where the limit has to be understood in $L^2(\mathbb{R}^n)$ and can be omitted if $f \in L^1(\mathbb{R}^n) \cap L^2(\mathbb{R}^n)$.

Clearly, we can also regard the Fourier transform as a map on $L^1(\mathbb{R}^n)$. To this end, let $C_\infty(\mathbb{R}^n)$ denote the Banach space of all continuous functions $f: \mathbb{R}^n \to \mathbb{C}$ which vanish at ∞ equipped with the sup norm (Problem 7.5).

7.1. The Fourier transform

Lemma 7.7 (Riemann-Lebesgue). *The Fourier transform as defined by (7.3) is a bounded injective map from $L^1(\mathbb{R}^n)$ into $C_\infty(\mathbb{R}^n)$ satisfying*

$$\|\hat{f}\|_\infty \leq (2\pi)^{-n/2}\|f\|_1. \tag{7.13}$$

Proof. Clearly we have $\hat{f} \in C_\infty(\mathbb{R}^n)$ if $f \in \mathcal{S}(\mathbb{R}^n)$. Moreover, since $\mathcal{S}(\mathbb{R}^n)$ is dense in $L^1(\mathbb{R}^n)$, the estimate

$$\sup_p |\hat{f}(p)| \leq \frac{1}{(2\pi)^{n/2}} \sup_p \int_{\mathbb{R}^n} |e^{-ipx} f(x)| d^n x = \frac{1}{(2\pi)^{n/2}} \int_{\mathbb{R}^n} |f(x)| d^n x$$

shows that the Fourier transform extends to a continuous map from $L^1(\mathbb{R}^n)$ into $C_\infty(\mathbb{R}^n)$.

To see that the Fourier transform is injective, suppose $\hat{f} = 0$. Then Fubini implies

$$0 = \int_{\mathbb{R}^n} \varphi(x)\hat{f}(x) d^n x = \int_{\mathbb{R}^n} \hat{\varphi}(x) f(x) d^n x$$

for every $\varphi \in \mathcal{S}(\mathbb{R}^n)$. Hence Lemma 0.41 implies $f = 0$. $\square$

Note that $\mathcal{F} : L^1(\mathbb{R}^n) \to C_\infty(\mathbb{R}^n)$ is not onto (cf. Problem 7.6). Moreover, $\mathcal{F}^{-1}\mathcal{F}f = f$ whenever $f, \mathcal{F}f \in L^1(\mathbb{R}^n)$ since this is all that was used in the proof of Theorem 7.4.

Another useful property is the convolution formula.

Lemma 7.8. *The* **convolution**

$$(f * g)(x) = \int_{\mathbb{R}^n} f(y)g(x-y) d^n y = \int_{\mathbb{R}^n} f(x-y)g(y) d^n y \tag{7.14}$$

of two functions $f, g \in L^1(\mathbb{R}^n)$ is again in $L^1(\mathbb{R}^n)$ and we have **Young's inequality**

$$\|f * g\|_1 \leq \|f\|_1 \|g\|_1. \tag{7.15}$$

Moreover, its Fourier transform is given by

$$(f * g)^\wedge(p) = (2\pi)^{n/2} \hat{f}(p)\hat{g}(p). \tag{7.16}$$

Proof. The fact that $f * g$ is in L^1 together with Young's inequality follows by applying Fubini's theorem to $h(x, y) = f(x-y)g(y)$. For the last claim we compute

$$(f * g)^\wedge(p) = \frac{1}{(2\pi)^{n/2}} \int_{\mathbb{R}^n} e^{-ipx} \int_{\mathbb{R}^n} f(y)g(x-y) d^n y \, d^n x$$

$$= \int_{\mathbb{R}^n} e^{-ipy} f(y) \frac{1}{(2\pi)^{n/2}} \int_{\mathbb{R}^n} e^{-ip(x-y)} g(x-y) d^n x \, d^n y$$

$$= \int_{\mathbb{R}^n} e^{-ipy} f(y) \hat{g}(p) d^n y = (2\pi)^{n/2} \hat{f}(p)\hat{g}(p),$$

where we have again used Fubini's theorem. □

In other words, $L^1(\mathbb{R}^n)$ together with convolution as a product is a Banach algebra (without identity). As a consequence we can also deal with the case of convolution on $\mathcal{S}(\mathbb{R}^n)$ as well as on $L^2(\mathbb{R}^n)$.

Corollary 7.9. *The convolution of two $\mathcal{S}(\mathbb{R}^n)$ functions as well as their product is in $\mathcal{S}(\mathbb{R}^n)$ and*
$$(f * g)^\wedge = (2\pi)^{n/2} \hat{f}\hat{g}, \qquad (fg)^\wedge = (2\pi)^{-n/2} \hat{f} * \hat{g}$$
in this case.

Proof. Clearly the product of two functions in $\mathcal{S}(\mathbb{R}^n)$ is again in $\mathcal{S}(\mathbb{R}^n)$ (show this!). Since $\mathcal{S}(\mathbb{R}^n) \subset L^1(\mathbb{R}^n)$, the previous lemma implies $(f * g)^\wedge = (2\pi)^{n/2} \hat{f}\hat{g} \in \mathcal{S}(\mathbb{R}^n)$. Moreover, since the Fourier transform is injective on $L^1(\mathbb{R}^n)$, we conclude $f * g = (2\pi)^{n/2}(\hat{f}\hat{g})^\vee \in \mathcal{S}(\mathbb{R}^n)$. Replacing f,g by $\check{f}, \check{g}$ in the last formula finally shows $\check{f} * \check{g} = (2\pi)^{n/2}(fg)^\vee$ and the claim follows by a simple change of variables using $\check{f}(p) = \hat{f}(-p)$. □

Corollary 7.10. *The convolution of two $L^2(\mathbb{R}^n)$ functions is in $C_\infty(\mathbb{R}^n)$ and we have $\|f * g\|_\infty \le \|f\|_2 \|g\|_2$ as well as*
$$(fg)^\wedge = (2\pi)^{-n/2} \hat{f} * \hat{g}$$
in this case.

Proof. The inequality $\|f * g\|_\infty \le \|f\|_2 \|g\|_2$ is immediate from Cauchy–Schwarz and shows that the convolution is a continuous bilinear form from $L^2(\mathbb{R}^n)$ to $L^\infty(\mathbb{R}^n)$. Now take sequences $f_n, g_n \in \mathcal{S}(\mathbb{R}^n)$ converging to $f, g \in L^2(\mathbb{R}^n)$. Then using the previous corollary together with continuity of the Fourier transform from $L^1(\mathbb{R}^n)$ to $C_\infty(\mathbb{R}^n)$ and on $L^2(\mathbb{R}^n)$, we obtain
$$(fg)^\wedge = \lim_{n\to\infty} (f_n g_n)^\wedge = (2\pi)^{-n/2} \lim_{n\to\infty} \hat{f}_n * \hat{g}_n = (2\pi)^{-n/2} \hat{f} * \hat{g}.$$
This also shows $\hat{f} * \hat{g} \in C_\infty(\mathbb{R}^n)$ by the Riemann–Lebesgue lemma. □

Finally, note that by looking at the Gaussian's $\phi_\lambda(x) = \exp(-\lambda x^2/2)$, one observes that a well-centered peak transforms into a broadly spread peak and vice versa. This turns out to be a general property of the Fourier transform known as **uncertainty principle**. One quantitative way of measuring this fact is to look at

$$(7.17) \qquad \|(x_j - x^0) f(x)\|_2^2 = \int_{\mathbb{R}^n} (x_j - x^0)^2 |f(x)|^2 d^n x$$

which will be small if f is well concentrated around x^0 in the j'th coordinate direction.

7.1. The Fourier transform

Theorem 7.11 (Heisenberg uncertainty principle). *Suppose $f \in \mathcal{S}(\mathbb{R}^n)$. Then for any $x^0, p^0 \in \mathbb{R}$, we have*

$$\|(x_j - x^0)f(x)\|_2 \|(p_j - p^0)\hat{f}(p)\|_2 \geq \frac{\|f\|_2^2}{2}. \tag{7.18}$$

Proof. Replacing $f(x)$ by $e^{ix_j p^0} f(x + x^0 e_j)$ (where e_j is the unit vector into the j'th coordinate direction) we can assume $x^0 = p^0 = 0$ by Lemma 7.2. Using integration by parts, we have

$$\|f\|_2^2 = \int_{\mathbb{R}^n} |f(x)|^2 d^n x = -\int_{\mathbb{R}^n} x_j \partial_j |f(x)|^2 d^n x = -2\,\mathrm{Re} \int_{\mathbb{R}^n} x_j f(x)^* \partial_j f(x) d^n x.$$

Hence, by Cauchy–Schwarz,

$$\|f\|_2^2 \leq 2\|x_j f(x)\|_2 \|\partial_j f(x)\|_2 = 2\|x_j f(x)\|_2 \|p_j \hat{f}(p)\|_2,$$

and the claim follows. $\square$

Recall that $|f(x)|^2$ is interpreted as the probability distribution for the position of a particle, and $|\hat{f}(x)|^2$ is interpreted as the probability distribution for its momentum. Equation (7.18) says that the variance of both distributions cannot both be small and thus one cannot simultaneously measure position and momentum of a particle with arbitrary precision. An abstract version will be given in Theorem 8.2 below.

Another version states that f and $\hat{f}$ cannot both have compact support.

Theorem 7.12. *Suppose $f \in L^2(\mathbb{R}^n)$. If both f and $\hat{f}$ have compact support, then $f = 0$.*

Proof. Let $A, B \subset \mathbb{R}^n$ be two compact sets and consider the subspace of all functions with $\mathrm{supp}(f) \subseteq A$ and $\mathrm{supp}(\hat{f}) \subseteq B$. Then

$$f(x) = \int_{\mathbb{R}^n} K(x,y) f(y) d^n y,$$

where

$$K(x,y) = \frac{1}{(2\pi)^n} \int_B e^{i(x-p)} \chi_A(y) d^n p = (2\pi)^{-n/2} \hat{\chi}_B(y-x) \chi_A(y).$$

Since $K \in L^2(\mathbb{R}^n \times \mathbb{R}^n)$, the corresponding integral operator is Hilbert–Schmidt, and thus its eigenspace corresponding to the eigenvalue 1 can be at most finite dimensional.

Now if there is a nonzero f, we can find a sequence of vectors $x^n \to 0$ such that the functions $f_n(x) = f(x - x^n)$ are linearly independent (look at their supports) and satisfy $\mathrm{supp}(f_n) \subseteq 2A$, $\mathrm{supp}(\hat{f}_n) \subseteq B$. But this a contradiction by the first part applied to the sets $2A$ and B. $\square$

Problem 7.1. Show that $\mathcal{S}(\mathbb{R}^n) \subset L^p(\mathbb{R}^n)$. (Hint: If $f \in \mathcal{S}(\mathbb{R}^n)$, then $|f(x)| \le C_m \prod_{j=1}^n (1+x_j^2)^{-m}$ for every m.)

Problem 7.2. Compute the Fourier transform of the following functions $f : \mathbb{R} \to \mathbb{C}$:

(i) $f(x) = \chi_{(-1,1)}(x)$. (ii) $f(p) = \frac{1}{p^2+k^2}$, $\operatorname{Re}(k) > 0$.

Problem 7.3. Suppose $f(x) \in L^1(\mathbb{R})$ and $g(x) = -\mathrm{i}x f(x) \in L^1(\mathbb{R})$. Then $\hat{f}$ is differentiable and $\hat{f}' = \hat{g}$.

Problem 7.4. A function $f : \mathbb{R}^n \to \mathbb{C}$ is called **spherically symmetric** if it is invariant under rotations; that is, $f(Ox) = f(x)$ for all $O \in SO(\mathbb{R}^n)$ (equivalently, f depends only on the distance to the origin $|x|$). Show that the Fourier transform of a spherically symmetric function is again spherically symmetric.

Problem 7.5. Show that $C_\infty(\mathbb{R}^n)$ is indeed a Banach space. Show that $C_c^\infty(\mathbb{R}^n)$ is dense. (Hint: Lemma 0.39.)

Problem 7.6. Show that $\mathcal{F} : L^1(\mathbb{R}^n) \to C_\infty(\mathbb{R}^n)$ is not onto as follows:

(i) The range of $\mathcal{F}$ is dense.

(ii) $\mathcal{F}$ is onto if and only if it has a bounded inverse.

(iii) $\mathcal{F}$ has no bounded inverse.

(Hint for (iii) in the case $n = 1$: Suppose $\varphi \in C_c^\infty(0,1)$ and set $f_m(x) = \sum_{k=1}^m \mathrm{e}^{\mathrm{i}kx} \varphi(x-k)$. Then $\|f_m\|_1 = m\|\varphi\|_1$ and $\|\hat{f}_m\|_\infty \le \mathrm{const}$ since $\varphi \in \mathcal{S}(\mathbb{R})$ and hence $|\varphi(p)| \le \mathrm{const}(1+|p|)^{-2}$.)

Problem 7.7 (Wiener). Suppose $f \in L^2(\mathbb{R}^n)$. Then the set $\{f(x+a)|a \in \mathbb{R}^n\}$ is total in $L^2(\mathbb{R}^n)$ if and only if $\hat{f}(p) \ne 0$ a.e. (Hint: Use Lemma 7.2 and the fact that a subspace is total if and only if its orthogonal complement is zero.)

Problem 7.8. Suppose $f(x)\mathrm{e}^{k|x|} \in L^1(\mathbb{R})$ for some $k > 0$. Then $\hat{f}(p)$ has an analytic extension to the strip $|\operatorname{Im}(p)| < k$.

7.2. Sobolev spaces

We begin by introducing the **Sobolev space**

(7.19) $$H^r(\mathbb{R}^n) = \{f \in L^2(\mathbb{R}^n) | |p|^r \hat{f}(p) \in L^2(\mathbb{R}^n)\}.$$

The most important case is when r is an integer; however, our definition makes sense for any $r \ge 0$. Moreover, note that $H^r(\mathbb{R}^n)$ becomes a Hilbert space if we introduce the scalar product

(7.20) $$\langle f, g \rangle = \int_{\mathbb{R}^n} \hat{f}(p)^* \hat{g}(p)(1+|p|^2)^r d^n p.$$

7.2. Sobolev spaces

In particular, note that by construction $\mathcal{F}$ maps $H^r(\mathbb{R}^n)$ unitarily onto $L^2(\mathbb{R}^n, (1+|p|^2)^r d^n p)$. Clearly $H^r(\mathbb{R}^n) \subset H^{r+1}(\mathbb{R}^n)$ with the embedding being continuous. Moreover, $\mathcal{S}(\mathbb{R}^n) \subset H^r(\mathbb{R}^n)$ and this subset is dense (since $\mathcal{S}(\mathbb{R}^n)$ is dense in $L^2(\mathbb{R}^n, (1+|p|^2)^r d^n p)$).

The motivation for the definition (7.19) stems from Lemma 7.1, which allows us to extend differentiation to a larger class. In fact, every function in $H^r(\mathbb{R}^n)$ has partial derivatives up to order $\lfloor r \rfloor$, which are defined via

$$(7.21) \qquad \partial_\alpha f = ((ip)^\alpha \hat{f}(p))^\vee, \qquad f \in H^r(\mathbb{R}^n), |\alpha| \leq r.$$

Example. Consider $f(x) = (1-|x|)\chi_{[-1,1]}(x)$. Then $\hat{f}(p) = \sqrt{\frac{2}{\pi}}\frac{\cos(p)-1}{p^2}$ and $f \in H^1(\mathbb{R})$. The weak derivative is $f'(x) = -\operatorname{sign}(x)\chi_{[-1,1]}(x)$. $\diamond$

By Lemma 7.1 this definition coincides with the usual one for every $f \in \mathcal{S}(\mathbb{R}^n)$ and we have

$$\int_{\mathbb{R}^n} g(x)(\partial_\alpha f)(x) d^n x = \langle g^*, (\partial_\alpha f)\rangle = \langle \hat{g}(p)^*, (ip)^\alpha \hat{f}(p)\rangle$$
$$= (-1)^{|\alpha|} \langle ((ip)^\alpha \hat{g}(p))^*, \hat{f}(p)\rangle = (-1)^{|\alpha|} \langle \partial_\alpha g^*, f\rangle$$
$$(7.22) \qquad = (-1)^{|\alpha|} \int_{\mathbb{R}^n} (\partial_\alpha g)(x) f(x) d^n x,$$

for $f, g \in H^r(\mathbb{R}^n)$. Furthermore, recall that a function $h \in L^1_{loc}(\mathbb{R}^n)$ satisfying

$$(7.23) \qquad \int_{\mathbb{R}^n} \varphi(x) h(x) d^n x = (-1)^{|\alpha|} \int_{\mathbb{R}^n} (\partial_\alpha \varphi)(x) f(x) d^n x, \qquad \varphi \in C_c^\infty(\mathbb{R}^n),$$

is also called the **weak derivative** or the derivative in the sense of distributions of f (by Lemma 0.41 such a function is unique if it exists). Hence, choosing $g = \varphi$ in (7.22), we see that $H^r(\mathbb{R}^n)$ is the set of all functions having partial derivatives (in the sense of distributions) up to order r, which are in $L^2(\mathbb{R}^n)$.

In this connection, the following norm for $H^m(\mathbb{R}^n)$ with $m \in \mathbb{N}_0$ is more common:

$$(7.24) \qquad \|f\|^2_{2,m} = \sum_{|\alpha| \leq m} \|\partial_\alpha f\|^2_2.$$

By $|p^\alpha| \leq |p|^{|\alpha|} \leq (1+|p|^2)^{m/2}$ it follows that this norm is equivalent to (7.20).

Of course, a natural question to ask is when the weak derivatives are in fact classical derivatives. To this end, observe that the Riemann–Lebesgue lemma implies that $\partial_\alpha f(x) \in C_\infty(\mathbb{R}^n)$ provided $p^\alpha \hat{f}(p) \in L^1(\mathbb{R}^n)$. Moreover, in this situation the derivatives will exist as classical derivatives:

Lemma 7.13. *Suppose $f \in L^1(\mathbb{R}^n)$ or $f \in L^2(\mathbb{R}^n)$ with $(1 + |p|^k)\hat{f}(p) \in L^1(\mathbb{R}^n)$ for some $k \in \mathbb{N}_0$. Then $f \in C_\infty^k(\mathbb{R}^n)$, the set of functions with continuous partial derivatives of order k all of which vanish at ∞. Moreover,*

$$(7.25) \qquad (\partial_\alpha f)^\wedge(p) = (ip)^\alpha \hat{f}(p), \qquad |\alpha| \leq k,$$

in this case.

Proof. We begin by observing that

$$f(x) = \frac{1}{(2\pi)^{n/2}} \int_{\mathbb{R}^n} e^{ipx} \hat{f}(p) d^n p,$$

which follows from Lemma 7.6 and the discussion after Lemma 7.7. Now the claim follows as in the proof of Lemma 7.1 by differentiating the integral using Problem A.20. □

Now we are able to prove the following embedding theorem.

Theorem 7.14 (Sobolev embedding). *Suppose $r > k + \frac{n}{2}$ for some $k \in \mathbb{N}_0$. Then $H^r(\mathbb{R}^n)$ is continuously embedded into $C_\infty^k(\mathbb{R}^n)$ with*

$$(7.26) \qquad \|\partial_\alpha f\|_\infty \leq C_{n,r} \|f\|_{2,r}, \qquad |\alpha| \leq k.$$

Proof. Abbreviate $\langle p \rangle = (1 + |p|^2)^{1/2}$. Now use $|(ip)^\alpha \hat{f}(p)| \leq \langle p \rangle^{|\alpha|} |\hat{f}(p)| = \langle p \rangle^{-s} \cdot \langle p \rangle^{|\alpha|+s} |\hat{f}(p)|$. Now $\langle p \rangle^{-s} \in L^2(\mathbb{R}^n)$ if $s > \frac{n}{2}$ (use polar coordinates to compute the norm) and $\langle p \rangle^{|\alpha|+s} |\hat{f}(p)| \in L^2(\mathbb{R}^n)$ if $s + |\alpha| \leq r$. Hence the claim follows from the previous lemma. □

In fact, we can even do a bit better.

Lemma 7.15 (Morrey inequality). *Suppose $f \in H^{n/2+\gamma}(\mathbb{R}^n)$ for some $\gamma \in (0,1)$. Then $f \in C_\infty^{0,\gamma}(\mathbb{R}^n)$, the set of functions which are Hölder continuous of exponent γ and vanish at ∞. Moreover,*

$$(7.27) \qquad |f(x) - f(y)| \leq C_{n,\gamma} \|\hat{f}(p)\|_{2,n/2+\gamma} |x-y|^\gamma$$

in this case.

Proof. We begin with

$$f(x+y) - f(x) = \frac{1}{(2\pi)^{n/2}} \int_{\mathbb{R}^n} e^{ipx}(e^{ipy} - 1)\hat{f}(p) d^n p$$

implying

$$|f(x+y) - f(x)| \leq \frac{1}{(2\pi)^{n/2}} \int_{\mathbb{R}^n} \frac{|e^{ipy} - 1|}{\langle p \rangle^{n/2+\gamma}} \langle p \rangle^{n/2+\gamma} |\hat{f}(p)| d^n p,$$

where again $\langle p \rangle = (1 + |p|^2)^{1/2}$. Hence, after applying Cauchy–Schwarz, it remains to estimate (recall (A.62))

$$\int_{\mathbb{R}^n} \frac{|e^{ipy} - 1|^2}{\langle p \rangle^{n+2\gamma}} d^n p \le S_n \int_0^{1/|y|} \frac{(|y|r)^2}{\langle r \rangle^{n+2\gamma}} r^{n-1} dr$$
$$+ S_n \int_{1/|y|}^{\infty} \frac{4}{\langle r \rangle^{n+2\gamma}} r^{n-1} dr$$
$$\le \frac{S_n}{2(1-\gamma)} |y|^{2\gamma} + \frac{S_n}{2\gamma} |y|^{2\gamma} = \frac{S_n}{2\gamma(1-\gamma)} |y|^{2\gamma},$$

where $S_n = nV_n$ is the volume of the unit sphere in $\mathbb{R}^n$. $\square$

Using this lemma we immediately obtain:

Corollary 7.16. *Suppose $r \ge k + \gamma + \frac{n}{2}$ for some $k \in \mathbb{N}_0$ and $\gamma \in (0, 1)$. Then $H^r(\mathbb{R}^n)$ is continuously embedded into $C_\infty^{k,\gamma}(\mathbb{R}^n)$, the set of functions in $C_\infty^k(\mathbb{R}^n)$ whose highest derivatives are Hölder continuous of exponent γ.*

Problem 7.9. *Suppose $f \in L^2(\mathbb{R}^n)$ show that $\varepsilon^{-1}(f(x+e_j\varepsilon) - f(x)) \to g_j(x)$ in L^2 if and only if $p_j \hat{f}(p) \in L^2$, where e_j is the unit vector into the j'th coordinate direction. Moreover, show $g_j = \partial_j f$ if $f \in H^1(\mathbb{R}^n)$.*

7.3. The free Schrödinger operator

In Section 2.1 we have seen that the Hilbert space corresponding to one particle in $\mathbb{R}^3$ is $L^2(\mathbb{R}^3)$. More generally, the Hilbert space for N particles in $\mathbb{R}^d$ is $L^2(\mathbb{R}^n)$, $n = Nd$. The corresponding nonrelativistic Hamilton operator, if the particles do not interact, is given by

(7.28) $$H_0 = -\Delta,$$

where Δ is the Laplace operator

(7.29) $$\Delta = \sum_{j=1}^n \frac{\partial^2}{\partial x_j^2}.$$

Here we have chosen units such that all relevant physical constants disappear; that is, $\hbar = 1$ and the mass of the particles is equal to $m = \frac{1}{2}$. Be aware that some authors prefer to use $m = 1$; that is, $H_0 = -\frac{1}{2}\Delta$.

Our first task is to find a good domain such that H_0 is a self-adjoint operator.

By Lemma 7.1 we have that

(7.30) $$-\Delta \psi(x) = (p^2 \hat{\psi}(p))^{\vee}(x)$$

for $\psi \in \mathcal{S}(\mathbb{R}^n)$. Moreover, if we work with weak derivatives, this even holds for $\psi \in H^2(\mathbb{R}^n)$, and the operator

(7.31) $$H_0 \psi = -\Delta \psi, \qquad \mathfrak{D}(H_0) = H^2(\mathbb{R}^n),$$

is unitarily equivalent to the maximally defined multiplication operator

(7.32) $$(\mathcal{F} H_0 \mathcal{F}^{-1}) \varphi(p) = p^2 \varphi(p), \qquad \mathfrak{D}(p^2) = \{\varphi \in L^2(\mathbb{R}^n) | p^2 \varphi(p) \in L^2(\mathbb{R}^n)\}.$$

Theorem 7.17. *The free Schrödinger operator H_0 is self-adjoint and its spectrum is characterized by*

(7.33) $$\sigma(H_0) = \sigma_{ac}(H_0) = [0, \infty), \qquad \sigma_{sc}(H_0) = \sigma_{pp}(H_0) = \emptyset.$$

Moreover, the spectral measure corresponding to ψ is purely absolutely continuous and given by

(7.34) $$d\mu_\psi(\lambda) = \frac{1}{2} \chi_{[0,\infty)}(\lambda) \lambda^{n/2-1} \left(\int_{S^{n-1}} |\hat{\psi}(\sqrt{\lambda}\omega)|^2 d^{n-1}\omega \right) d\lambda.$$

Proof. It suffices to show (7.34). First, observe that

$$\langle \psi, R_{H_0}(z)\psi \rangle = \langle \hat\psi, R_{p^2}(z)\hat\psi \rangle = \int_{\mathbb{R}^n} \frac{|\hat\psi(p)|^2}{p^2 - z} d^n p = \int_{\mathbb{R}} \frac{1}{r^2 - z} d\tilde\mu_\psi(r),$$

where

$$d\tilde\mu_\psi(r) = \chi_{[0,\infty)}(r) r^{n-1} \left(\int_{S^{n-1}} |\hat\psi(r\omega)|^2 d^{n-1}\omega \right) dr.$$

Hence, after a change of coordinates, we have

$$\langle \psi, R_{H_0}(z)\psi \rangle = \int_\mathbb{R} \frac{1}{\lambda - z} d\mu_\psi(\lambda),$$

where $d\mu_\psi$ is given by (7.34). This proves the claim. $\square$

Slightly more general, we can consider operators $p_j = \frac{1}{i} \frac{\partial}{\partial x_j}$ and define the operator $f(p)$ via

(7.35) $$f(p)\psi(x) = \mathcal{F}^{-1}(f(p)\hat\psi(p))(x)$$

for any measurable function $f : \mathbb{R}^n \to \mathbb{C}$. The corresponding operator will be self-adjoint if f is real-valued, and in the special case $f(p) = p^2$ we obtain just $H_0 = -\Delta$ and hence this provides an alternate way of defining functions of H_0. As a useful consequence of this observation we note:

Lemma 7.18. *Suppose $f \in L^2(\mathbb{R}^n)$. Then the operator $f(p)$ is an integral operator given by*

(7.36) $$f(p)\psi(x) = \frac{1}{(2\pi)^{n/2}} \int_{\mathbb{R}^n} \check{f}(x-y)\psi(y) d^n y,$$

and its range is a subset of $C_\infty(\mathbb{R}^n)$.

7.4. The time evolution in the free case

Proof. This follows from the definition since the Fourier transform maps multiplications to convolutions by Corollary 7.10. □

Recall that $f(p)$ will be bounded if and only if $f \in L^\infty(\mathbb{R}^n)$ and the operator norm is given by $\|f(p)\| = \|f\|_\infty$.

Finally, we note that the compactly supported smooth functions are a core for H_0.

Lemma 7.19. *The set $C_c^\infty(\mathbb{R}^n) = \{f \in \mathcal{S}(\mathbb{R}^n) |\, \mathrm{supp}(f)$ is compact$\}$ is a core for H_0.*

Proof. It is not hard to see that $\mathcal{S}(\mathbb{R}^n)$ is a core (Problem 7.10), and hence it suffices to show that the closure of $H_0|_{C_c^\infty(\mathbb{R}^n)}$ contains $H_0|_{\mathcal{S}(\mathbb{R}^n)}$. To see this, let $\varphi(x) \in C_c^\infty(\mathbb{R}^n)$ which is one for $|x| \leq 1$ and vanishes for $|x| \geq 2$. Set $\varphi_n(x) = \varphi(\frac{1}{n}x)$. Then $\psi_n(x) = \varphi_n(x)\psi(x)$ is in $C_c^\infty(\mathbb{R}^n)$ for every $\psi \in \mathcal{S}(\mathbb{R}^n)$ and $\psi_n \to \psi$, respectively, $\Delta\psi_n \to \Delta\psi$. □

Note also that the quadratic form of H_0 is given by

$$(7.37) \qquad q_{H_0}(\psi) = \sum_{j=1}^n \int_{\mathbb{R}^n} |\partial_j \psi(x)|^2 d^n x, \quad \psi \in \mathfrak{Q}(H_0) = H^1(\mathbb{R}^n).$$

Problem 7.10. *Show that $\mathcal{S}(\mathbb{R}^n)$ is a core for H_0. (Hint: Show that the closure of $H_0|_{\mathcal{S}(\mathbb{R}^n)}$ contains H_0.)*

Problem 7.11. *Show that $\{\psi \in \mathcal{S}(\mathbb{R}) |\, \psi(0) = 0\}$ is dense but not a core for $H_0 = -\frac{d^2}{dx^2}$.*

7.4. The time evolution in the free case

Now let us look at the time evolution. We have

$$(7.38) \qquad \mathrm{e}^{-itH_0}\psi(x) = \mathcal{F}^{-1}\mathrm{e}^{-itp^2}\hat{\psi}(p).$$

The right-hand side is a product and hence our operator should be expressible as an integral operator via the convolution formula. However, since e^{-itp^2} is not in L^2, a more careful analysis is needed.

Consider

$$(7.39) \qquad f_\varepsilon(p^2) = \mathrm{e}^{-(it+\varepsilon)p^2}, \quad \varepsilon > 0.$$

Then $f_\varepsilon(H_0)\psi \to \mathrm{e}^{-itH_0}\psi$ by Theorem 3.1. Moreover, by Lemma 7.3 and the convolution formula, we have

$$(7.40) \qquad f_\varepsilon(H_0)\psi(x) = \frac{1}{(4\pi(it+\varepsilon))^{n/2}} \int_{\mathbb{R}^n} \mathrm{e}^{-\frac{|x-y|^2}{4(it+\varepsilon)}} \psi(y) d^n y$$

and hence

(7.41) $$e^{-itH_0}\psi(x) = \frac{1}{(4\pi it)^{n/2}} \int_{\mathbb{R}^n} e^{i\frac{|x-y|^2}{4t}} \psi(y) d^n y$$

for $t \neq 0$ and $\psi \in L^1 \cap L^2$. In fact, the limit of the right-hand side exists pointwise by dominated convergence, and its pointwise limit must thus be equal to its L^2 limit. For general $\psi \in L^2$, the integral has to be understood as a limit.

Using this explicit form, it is not hard to draw some immediate consequences. For example, if $\psi \in L^2(\mathbb{R}^n) \cap L^1(\mathbb{R}^n)$, then $\psi(t) \in C(\mathbb{R}^n)$ for $t \neq 0$ (use dominated convergence and continuity of the exponential) and satisfies

(7.42) $$\|\psi(t)\|_\infty \leq \frac{1}{|4\pi t|^{n/2}} \|\psi(0)\|_1.$$

Thus we have spreading of wave functions in this case. Moreover, it is even possible to determine the asymptotic form of the wave function for large t as follows. Observe

$$e^{-itH_0}\psi(x) = \frac{e^{i\frac{x^2}{4t}}}{(4\pi it)^{n/2}} \int_{\mathbb{R}^n} e^{i\frac{y^2}{4t}} \psi(y) e^{-i\frac{xy}{2t}} d^n y$$

(7.43) $$= \left(\frac{1}{2it}\right)^{n/2} e^{i\frac{x^2}{4t}} \left(e^{i\frac{y^2}{4t}}\psi(y)\right)^{\wedge}\left(\frac{x}{2t}\right).$$

Moreover, since $\exp(i\frac{y^2}{4t})\psi(y) \to \psi(y)$ in L^2 as $|t| \to \infty$ (dominated convergence), we obtain

Lemma 7.20. *For every $\psi \in L^2(\mathbb{R}^n)$ we have*

(7.44) $$e^{-itH_0}\psi(x) - \left(\frac{1}{2it}\right)^{n/2} e^{i\frac{x^2}{4t}} \hat{\psi}\left(\frac{x}{2t}\right) \to 0$$

in L^2 as $|t| \to \infty$.

Note that this result is not too surprising from a physical point of view. In fact, if a classical particle starts at a point $x(0) = x_0$ with velocity $v = 2p$ (recall that we use units where the mass is $m = \frac{1}{2}$), then we will find it at $x = x_0 + 2pt$ at time t. Dividing by $2t$, we get $\frac{x}{2t} = p + \frac{x_0}{2t} \approx p$ for large t. Hence the probability distribution for finding a particle at a point x at time t should approach the probability distribution for the momentum at $p = \frac{x}{2t}$; that is, $|\psi(x,t)|^2 d^n x = |\hat{\psi}(\frac{x}{2t})|^2 \frac{d^n x}{(2t)^n}$. This could also be stated as follows: The probability of finding the particle in a region $\Omega \subseteq \mathbb{R}^n$ is asymptotically for $|t| \to \infty$ equal to the probability of finding the momentum of the particle in $\frac{1}{2t}\Omega$. This is sometimes known as **Dollard's theorem**.

Next, we want to apply the RAGE theorem in order to show that for every initial condition, a particle will escape to infinity.

7.5. The resolvent and Green's function

Lemma 7.21. *Let $g(x)$ be the multiplication operator by g and let $f(p)$ be the operator given by $f(p)\psi(x) = \mathcal{F}^{-1}(f(p)\hat{\psi}(p))(x)$. Denote by $L_\infty^\infty(\mathbb{R}^n)$ the bounded Borel functions which vanish at infinity. Then*

(7.45)
$$f(p)g(x) \quad \text{and} \quad g(x)f(p)$$

are compact if $f, g \in L_\infty^\infty(\mathbb{R}^n)$ and (extend to) Hilbert–Schmidt operators if $f, g \in L^2(\mathbb{R}^n)$.

Proof. By symmetry, it suffices to consider $g(x)f(p)$. Let $f, g \in L^2$. Then (Corollary 7.10)

$$g(x)f(p)\psi(x) = \frac{1}{(2\pi)^{n/2}} \int_{\mathbb{R}^n} g(x)\check{f}(x-y)\psi(y)d^ny$$

shows that $g(x)f(p)$ is Hilbert–Schmidt since $g(x)\check{f}(x-y) \in L^2(\mathbb{R}^n \times \mathbb{R}^n)$.

If f, g are bounded, then the functions $f_R(p) = \chi_{\{p\,|\,p^2 \le R\}}(p)f(p)$ and $g_R(x) = \chi_{\{x\,|\,x^2 \le R\}}(x)g(x)$ are in L^2. Thus $g_R(x)f_R(p)$ is compact and by

$$\|g(x)f(p) - g_R(x)f_R(p)\| \le \|g\|_\infty \|f - f_R\|_\infty + \|g - g_R\|_\infty \|f_R\|_\infty$$

it tends to $g(x)f(p)$ in norm since f, g vanish at infinity. $\square$

In particular, this lemma implies that

(7.46)
$$\chi_\Omega (H_0 + \mathrm{i})^{-1}$$

is compact if $\Omega \subseteq \mathbb{R}^n$ is bounded and hence

(7.47)
$$\lim_{t \to \infty} \|\chi_\Omega e^{-\mathrm{i}tH_0}\psi\|^2 = 0$$

for every $\psi \in L^2(\mathbb{R}^n)$ and every bounded subset Ω of $\mathbb{R}^n$. In other words, the particle will eventually escape to infinity since the probability of finding the particle in any bounded set tends to zero. (If $\psi \in L^1(\mathbb{R}^n)$, this of course also follows from (7.42).)

7.5. The resolvent and Green's function

Now let us compute the resolvent of H_0. We will try to use an approach similar to that for the time evolution in the previous section. Since the function $(p^2 - z)^{-1}$ is not in $L^1(\mathbb{R}^n)$ for $n > 1$, it is difficult to compute its inverse Fourier transform directly, and we would need to look at some regularization $f_\varepsilon(p) = \exp(-\varepsilon p^2)(p^2 - z)^{-1}$ as in the previous section. However, since it is highly nontrivial to compute the inverse Fourier transform of this latter function, we will use a small ruse.

Note that

(7.48)
$$R_{H_0}(z) = \int_0^\infty e^{zt}e^{-tH_0}dt, \quad \mathrm{Re}(z) < 0.$$

Indeed, by virtue of Lemma 4.1, $\lim_{r\to\infty}\int_0^r e^{zt}e^{-tH_0}dt = \lim_{r\to\infty} R_{H_0}(z)(\mathbb{I}-e^{-r(H_0-z)}) = R_{H_0}(z)$. Moreover,

$$(7.49) \qquad e^{-tH_0}\psi(x) = \frac{1}{(4\pi t)^{n/2}} \int_{\mathbb{R}^n} e^{-\frac{|x-y|^2}{4t}} \psi(y) d^n y, \quad t > 0,$$

by the same analysis as in the previous section. Hence, by Fubini, we have

$$(7.50) \qquad R_{H_0}(z)\psi(x) = \int_{\mathbb{R}^n} G_0(z,|x-y|)\psi(y) d^n y,$$

where

$$(7.51) \qquad G_0(z,r) = \int_0^\infty \frac{1}{(4\pi t)^{n/2}} e^{-\frac{r^2}{4t}+zt} dt, \quad r > 0,\ \mathrm{Re}(z) < 0.$$

The function $G_0(z,r)$ is called **Green's function** of H_0. The integral can be evaluated in terms of modified Bessel functions of the second kind as follows: First of all, it suffices to consider $z < 0$ since the remaining values will follow by analytic continuation. Then, making the substitution $t = \frac{r}{2\sqrt{-z}}e^s$, we obtain

$$\int_0^\infty \frac{1}{(4\pi t)^{n/2}} e^{-\frac{r^2}{4t}+zt} dt = \frac{1}{4\pi}\left(\frac{\sqrt{-z}}{2\pi r}\right)^{\frac{n}{2}-1} \int_{-\infty}^\infty e^{-\nu s} e^{-x\cosh(s)} ds$$

$$(7.52) \qquad = \frac{1}{2\pi}\left(\frac{\sqrt{-z}}{2\pi r}\right)^{\frac{n}{2}-1} \int_0^\infty \cosh(-\nu s) e^{-x\cosh(s)} ds,$$

where we have abbreviated $x = \sqrt{-z}r$ and $\nu = \frac{n}{2} - 1$. But the last integral is given by the modified Bessel function $K_\nu(x)$ (see [44, (10.32.9)]) and thus

$$(7.53) \qquad G_0(z,r) = \frac{1}{2\pi}\left(\frac{\sqrt{-z}}{2\pi r}\right)^{\frac{n}{2}-1} K_{\frac{n}{2}-1}(\sqrt{-z}r).$$

Note $K_\nu(x) = K_{-\nu}(x)$ and $K_\nu(x) > 0$ for $\nu, x \in \mathbb{R}$. The functions $K_\nu(x)$ satisfy the differential equation (see [44, (10.25.1)])

$$(7.54) \qquad \left(\frac{d^2}{dx^2} + \frac{1}{x}\frac{d}{dx} - 1 - \frac{\nu^2}{x^2}\right) K_\nu(x) = 0$$

and have the asymptotics (see [44, (10.30.2) and (10.30.3)])

$$(7.55) \qquad K_\nu(x) = \begin{cases} \frac{\Gamma(\nu)}{2}\left(\frac{x}{2}\right)^{-\nu} + O(x^{-\nu+2}), & \nu > 0, \\ -\log(\frac{x}{2}) + O(1), & \nu = 0, \end{cases}$$

for $|x| \to 0$ and (see [44, (10.40.2)])

$$(7.56) \qquad K_\nu(x) = \sqrt{\frac{\pi}{2x}} e^{-x}(1 + O(x^{-1}))$$

7.5. The resolvent and Green's function

for $|x| \to \infty$. For more information, see, for example, [44] or [69]. In particular, $G_0(z,r)$ has an analytic continuation for $z \in \mathbb{C}\setminus[0,\infty) = \rho(H_0)$. Hence we can define the right-hand side of (7.50) for all $z \in \rho(H_0)$ such that

$$\tag{7.57} \int_{\mathbb{R}^n}\int_{\mathbb{R}^n} \varphi(x) G_0(z, |x-y|) \psi(y) d^n y d^n x$$

is analytic for $z \in \rho(H_0)$ and $\varphi, \psi \in \mathcal{S}(\mathbb{R}^n)$ (by Morera's theorem). Since it is equal to $\langle \varphi, R_{H_0}(z)\psi \rangle$ for $\mathrm{Re}(z) < 0$, it is equal to this function for all $z \in \rho(H_0)$, since both functions are analytic in this domain. In particular, (7.50) holds for all $z \in \rho(H_0)$.

If n is odd, we have the case of spherical Bessel functions which can be expressed in terms of elementary functions. For example, we have

$$\tag{7.58} G_0(z,r) = \frac{1}{2\sqrt{-z}} e^{-\sqrt{-z}\, r}, \qquad n=1,$$

and

$$\tag{7.59} G_0(z,r) = \frac{1}{4\pi r} e^{-\sqrt{-z}\, r}, \qquad n=3.$$

Using Stone's formula we can even extend this result:

Theorem 7.22. Suppose $f(\lambda^2)\lambda^{(n-1)/2} \in L^2(\mathbb{R})$. Then

$$\tag{7.60} f(p^2)\psi(x) = \int_{\mathbb{R}^n} F(|x-y|)\psi(y) dy,$$

where

$$\tag{7.61} F(r) = (2\pi)^{-n/2} \mathcal{H}_{n/2-1}(f(s^2)(s/r)^{n/2-1})(r)$$

satisfies $F(r) r^{(n-1)/2} \in L^2(\mathbb{R})$. Here $\mathcal{H}_\nu$, $\nu \geq -\frac{1}{2}$, is the **Hankel transform** given by

$$\tag{7.62} \mathcal{H}_\nu(f)(r) = \int_0^\infty f(s) J_\nu(sr) s\, ds,$$

with $J_\nu(z)$ being the Bessel function of order ν. For $f(\lambda^2)\lambda^{(n-1)/2} \in L^2(\mathbb{R})$, the last integral has to be understood as a limit $\lim_{R\to\infty} \int_0^R$ as with the Fourier transform.

Proof. Note that Lemma 7.18 implies (7.60) with $F(|x|) = \check{g}(x)$, where $g(p) = f(p^2)$. In particular, $F(r) r^{(n-1)/2} \in L^2(\mathbb{R})$. Moreover, we can assume both f and ψ to be in some dense sets, say $f \in C_c(0,\infty)$ and $\psi \in C_c(\mathbb{R}^n)$. In order to compute F explicitly we now use Problem 4.3 which implies (using $G_0(z^*, r) = G_0(z,r)^*$)

$$f(p^2)\psi(x) = \frac{1}{\pi} \lim_{\varepsilon \downarrow 0} \int_0^\infty f(\lambda) \int_{\mathbb{R}^n} \mathrm{Im}(G_0(\lambda + i\varepsilon, |x-y|)) \psi(y) dy d\lambda,$$

where the limit has to be understood in $L^2(\mathbb{R}^n)$. Moreover, the limit will also exist pointwise for a.e. x and a suitable subsequence. Now using [44, (10.27.8) and (10.4.3)], one sees

$$\lim_{\varepsilon \downarrow 0} \frac{1}{\pi} \operatorname{Im}(G_0(\lambda + i\varepsilon, r)) = \frac{1}{4\pi} \left(\frac{\sqrt{\lambda}}{2\pi r}\right)^{n/2-1} J_{n/2-1}(r\sqrt{\lambda})$$

and using Fubini we obtain (7.60) with F given as in the theorem. □

In particular, comparing with Lemma 7.18 shows that the Fourier transform of a radial function can be expressed in terms of the Hankel transform and that the Hankel transform is unitary on $L^2([0,\infty), s\, ds)$.

Example. In the case $f(p^2) = \chi_{[0,\lambda)}(p^2)$ one can evaluate the integral using [44, (10.22.1)] to obtain

$$(7.63) \qquad P_{H_0}(\lambda)\psi(x) = \int_{\mathbb{R}^n} P_0(\lambda, |x-y|)\psi(y)dy,$$

where

$$(7.64) \qquad P_0(\lambda, r) = \left(\frac{\sqrt{\lambda}}{2\pi r}\right)^{n/2} J_{n/2}(r\sqrt{\lambda})\chi_{(0,\infty)}(\lambda).$$

In particular,

$$P_0(\lambda, r) = \chi_{(0,\infty)}(\lambda) \begin{cases} \frac{\sin(r\sqrt{\lambda})}{\pi r}, & n=1, \\ \frac{\sin(r\sqrt{\lambda}) - r\sqrt{\lambda}\cos(r\sqrt{\lambda})}{2\pi^2 r^3}, & n=3. \end{cases}$$

◇

Problem 7.12. *Verify (7.50) directly in the case $n=1$.*

Problem 7.13. *The **Bessel function** of order $\nu \in \mathbb{C}$ can be defined as*

$$J_\nu(z) = \sum_{j=0}^\infty \frac{(-1)^j}{j!\Gamma(\nu+j+1)} \left(\frac{z}{2}\right)^{2j+\nu}.$$

Show that $J_\nu(z)$ is a solution of the Bessel differential equation

$$z^2 u'' + zu' + (z^2 - \nu^2)u = 0.$$

Prove the following properties of the Bessel functions.

(i) $(z^{\pm\nu} J_\nu(z))' = \pm z^{\pm\nu} J_{\nu\mp 1}(z)$.
(ii) $J_{\nu-1}(z) + J_{\nu+1}(z) = \frac{2\nu}{z} J_\nu(z)$.
(iii) $J_{\nu-1}(z) - J_{\nu+1}(z) = 2J_\nu'(z)$.

7.5. The resolvent and Green's function

Problem 7.14. *Consider the modified Hankel transform*
$$\tilde{\mathcal{H}}_\nu(f)(r) = (2\pi)^{-\nu+1}\mathcal{H}_\nu(f(s)(s/r)^\nu)(r)$$
appearing in Theorem 7.22. Show that
$$\tilde{\mathcal{H}}_{-1/2}(f)(r) = \frac{1}{\pi}\int_0^\infty \cos(rs)f(s)\,ds = \sqrt{2\pi}\hat{f}(r)$$
if we extend f to all of $\mathbb{R}$ such that $f(r) = f(-r)$.

Moreover, show that (under suitable assumptions on f)
$$\tilde{\mathcal{H}}_{\nu+1}(f)(r) = \frac{-1}{2\pi r}\frac{d}{dr}\tilde{\mathcal{H}}_\nu(f)(r).$$

(Hint: $\frac{d}{dz}z^{-\nu}J_\nu(z) = -z^{-\nu}J_{\nu+1}(z)$ by [44, (10.6.6)].)

Chapter 8

Algebraic methods

8.1. Position and momentum

Apart from the Hamiltonian H_0, which corresponds to the kinetic energy, there are several other important observables associated with a single particle in three dimensions. Using the commutation relation between these observables, many important consequences about these observables can be derived.

First, consider the one-parameter unitary group

(8.1) $$(U_j(t)\psi)(x) = e^{-itx_j}\psi(x), \qquad 1 \leq j \leq 3.$$

For $\psi \in \mathcal{S}(\mathbb{R}^3)$, we compute

(8.2) $$\lim_{t \to 0} i\frac{e^{-itx_j}\psi(x) - \psi(x)}{t} = x_j\psi(x)$$

and hence the generator is the multiplication operator by the j'th coordinate function. By Corollary 5.4, it is essentially self-adjoint on $\psi \in \mathcal{S}(\mathbb{R}^3)$. It is customary to combine all three operators into one vector-valued operator x, which is known as the **position operator**. Moreover, it is not hard to see that the spectrum of x_j is purely absolutely continuous and given by $\sigma(x_j) = \mathbb{R}$. In fact, let $\varphi(x)$ be an orthonormal basis for $L^2(\mathbb{R})$. Then $\varphi_i(x_1)\varphi_j(x_2)\varphi_k(x_3)$ is an orthonormal basis for $L^2(\mathbb{R}^3)$, and x_1 can be written as an orthogonal sum of operators restricted to the subspaces spanned by $\varphi_j(x_2)\varphi_k(x_3)$. Each subspace is unitarily equivalent to $L^2(\mathbb{R})$, and x_1 is given by multiplication with the identity. Hence the claim follows (or use Theorem 4.17).

Next, consider the one-parameter unitary group of translations

(8.3) $$(U_j(t)\psi)(x) = \psi(x - te_j), \qquad 1 \leq j \leq 3,$$

where e_j is the unit vector in the j'th coordinate direction. For $\psi \in \mathcal{S}(\mathbb{R}^3)$, we compute

$$\text{(8.4)} \qquad \lim_{t \to 0} \mathrm{i} \frac{\psi(x - te_j) - \psi(x)}{t} = \frac{1}{\mathrm{i}} \frac{\partial}{\partial x_j} \psi(x)$$

and hence the generator is $p_j = \frac{1}{\mathrm{i}} \frac{\partial}{\partial x_j}$. Again it is essentially self-adjoint on $\psi \in \mathcal{S}(\mathbb{R}^3)$. Moreover, since it is unitarily equivalent to x_j by virtue of the Fourier transform, we conclude that the spectrum of p_j is again purely absolutely continuous and given by $\sigma(p_j) = \mathbb{R}$. The operator p is known as the **momentum operator**. Note that since

$$\text{(8.5)} \qquad [H_0, p_j]\psi(x) = 0, \qquad \psi \in \mathcal{S}(\mathbb{R}^3),$$

we have

$$\text{(8.6)} \qquad \frac{d}{dt} \langle \psi(t), p_j \psi(t) \rangle = 0, \qquad \psi(t) = \mathrm{e}^{-\mathrm{i}tH_0} \psi(0) \in \mathcal{S}(\mathbb{R}^3);$$

that is, the momentum is a conserved quantity for the free motion. More generally, we have

Theorem 8.1 (Noether)**.** *Suppose A is a self-adjoint operator which commutes with a self-adjoint operator H. Then $\mathfrak{D}(A)$ is invariant under $\mathrm{e}^{-\mathrm{i}tH}$, that is, $\mathrm{e}^{-\mathrm{i}tH} \mathfrak{D}(A) = \mathfrak{D}(A)$, and A is a conserved quantity, that is,*

$$\text{(8.7)} \qquad \langle \psi(t), A\psi(t) \rangle = \langle \psi(0), A\psi(0) \rangle, \qquad \psi(t) = \mathrm{e}^{-\mathrm{i}tH} \psi(0) \in \mathfrak{D}(A).$$

Proof. By the second part of Lemma 4.5 (with $f(\lambda) = \lambda$ and $B = \mathrm{e}^{-\mathrm{i}tH}$), we see $\mathfrak{D}(A) = \mathfrak{D}(\mathrm{e}^{-\mathrm{i}tH} A) \subseteq \mathfrak{D}(A\mathrm{e}^{-\mathrm{i}tH}) = \{\psi | \mathrm{e}^{-\mathrm{i}tH}\psi \in \mathfrak{D}(A)\}$, which implies $\mathrm{e}^{-\mathrm{i}tH} \mathfrak{D}(A) \subseteq \mathfrak{D}(A)$, and $[\mathrm{e}^{-\mathrm{i}tH}, A]\psi = 0$ for $\psi \in \mathfrak{D}(A)$. $\square$

Similarly, one has

$$\text{(8.8)} \qquad \mathrm{i}[p_j, x_k]\psi(x) = \delta_{jk}\psi(x), \qquad \psi \in \mathcal{S}(\mathbb{R}^3),$$

which is known as the **Weyl relations**. In terms of the corresponding unitary groups, they read

$$\text{(8.9)} \qquad \mathrm{e}^{-\mathrm{i}sp_j} \mathrm{e}^{-\mathrm{i}tx_k} = \mathrm{e}^{\mathrm{i}st\delta_{jk}} \mathrm{e}^{-\mathrm{i}tx_j} \mathrm{e}^{-\mathrm{i}sp_k}.$$

The Weyl relations also imply that the mean-square deviation of position and momentum cannot be made arbitrarily small simultaneously:

Theorem 8.2 (Heisenberg Uncertainty Principle)**.** *Suppose A and B are two symmetric operators. Then for every $\psi \in \mathfrak{D}(AB) \cap \mathfrak{D}(BA)$ we have*

$$\text{(8.10)} \qquad \Delta_\psi(A) \Delta_\psi(B) \geq \frac{1}{2} |\mathbb{E}_\psi([A, B])|$$

with equality if

$$\text{(8.11)} \qquad (B - \mathbb{E}_\psi(B))\psi = \mathrm{i}\lambda (A - \mathbb{E}_\psi(A))\psi, \qquad \lambda \in \mathbb{R} \setminus \{0\},$$

or if ψ is an eigenstate of A or B.

Proof. Let us fix $\psi \in \mathfrak{D}(AB) \cap \mathfrak{D}(BA)$ and abbreviate
$$\hat{A} = A - \mathbb{E}_\psi(A), \qquad \hat{B} = B - \mathbb{E}_\psi(B).$$
Then $\Delta_\psi(A) = \|\hat{A}\psi\|$, $\Delta_\psi(B) = \|\hat{B}\psi\|$ and hence by Cauchy–Schwarz
$$|\langle \hat{A}\psi, \hat{B}\psi \rangle| \leq \Delta_\psi(A)\Delta_\psi(B).$$
Now note that
$$\hat{A}\hat{B} = \frac{1}{2}\{\hat{A}, \hat{B}\} + \frac{1}{2}[A, B], \qquad \{\hat{A}, \hat{B}\} = \hat{A}\hat{B} + \hat{B}\hat{A}$$
where $\{\hat{A}, \hat{B}\}$ and $\mathrm{i}[A, B]$ are symmetric. So
$$|\langle \hat{A}\psi, \hat{B}\psi \rangle|^2 = |\langle \psi, \hat{A}\hat{B}\psi \rangle|^2 = \frac{1}{2}|\langle \psi, \{\hat{A}, \hat{B}\}\psi \rangle|^2 + \frac{1}{2}|\langle \psi, [A, B]\psi \rangle|^2,$$
which proves (8.10).

To have equality if ψ is not an eigenstate, we need $\hat{B}\psi = z\hat{A}\psi$ for equality in Cauchy–Schwarz and $\langle \psi, \{\hat{A}, \hat{B}\}\psi \rangle = 0$. Inserting the first into the second requirement gives $0 = (z - z^*)\|\hat{A}\psi\|^2$ and shows $\mathrm{Re}(z) = 0$. □

In the case of position and momentum, we have ($\|\psi\| = 1$)

(8.12) $$\Delta_\psi(p_j)\Delta_\psi(x_k) \geq \frac{\delta_{jk}}{2}$$

and the minimum is attained for the **Gaussian wave packets**

(8.13) $$\psi(x) = \left(\frac{\lambda}{\pi}\right)^{n/4} e^{-\frac{\lambda}{2}|x-x_0|^2 - \mathrm{i} p_0 x},$$

which satisfy $\mathbb{E}_\psi(x) = x_0$ and $\mathbb{E}_\psi(p) = p_0$, respectively, $\Delta_\psi(p_j)^2 = \frac{\lambda}{2}$ and $\Delta_\psi(x_k)^2 = \frac{1}{2\lambda}$.

Problem 8.1. *Check that (8.13) realizes the minimum.*

8.2. Angular momentum

Now consider the one-parameter unitary group of rotations

(8.14) $$(U_j(t)\psi)(x) = \psi(M_j(t)^{-1}x), \qquad 1 \leq j \leq 3,$$

where $M_j(t)$ is the matrix of rotation around e_j by an angle of t:

$$M_1 = \begin{pmatrix} 1 & 0 & 0 \\ 0 & \cos(t) & -\sin(t) \\ 0 & \sin(t) & \cos(t) \end{pmatrix}, \quad M_2 = \begin{pmatrix} \cos(t) & 0 & \sin(t) \\ 0 & 1 & 0 \\ -\sin(t) & 0 & \cos(t) \end{pmatrix},$$

$$(8.15) \quad M_3 = \begin{pmatrix} \cos(t) & -\sin(t) & 0 \\ \sin(t) & \cos(t) & 0 \\ 0 & 0 & 1 \end{pmatrix}.$$

For $\psi \in \mathcal{S}(\mathbb{R}^3)$, we compute

$$(8.16) \quad \lim_{t \to 0} i \frac{\psi(M_i(t)^{-1}x) - \psi(x)}{t} = \sum_{j,k=1}^{3} \varepsilon_{ijk} x_j p_k \psi(x),$$

where

$$(8.17) \quad \varepsilon_{ijk} = \begin{cases} 1 & \text{if } ijk \text{ is an even permutation of 123,} \\ -1 & \text{if } ijk \text{ is an odd permutation of 123,} \\ 0 & \text{otherwise.} \end{cases}$$

Again, one combines the three components into one vector-valued operator $L = x \wedge p$, which is known as the **angular momentum operator**. Its components are explicitly given by

$$(8.18) \quad L_i = \sum_{j,k=1}^{3} \varepsilon_{ijk} x_j p_k.$$

Since $e^{i2\pi L_j} = \mathbb{I}$, we see by Theorem 3.17 that the spectrum is a subset of $\mathbb{Z}$. In particular, the continuous spectrum is empty. We will show below that we have $\sigma(L_j) = \mathbb{Z}$. Note that since

$$(8.19) \quad [H_0, L_j]\psi(x) = 0, \quad \psi \in \mathcal{S}(\mathbb{R}^3),$$

we again have

$$(8.20) \quad \frac{d}{dt}\langle \psi(t), L_j \psi(t) \rangle = 0, \quad \psi(t) = e^{-itH_0}\psi(0) \in \mathcal{S}(\mathbb{R}^3);$$

that is, the angular momentum is a conserved quantity for the free motion as well.

Moreover, we even have

$$(8.21) \quad [L_i, K_j]\psi(x) = i \sum_{k=1}^{3} \varepsilon_{ijk} K_k \psi(x), \quad \psi \in \mathcal{S}(\mathbb{R}^3), K_j \in \{L_j, p_j, x_j\},$$

and these algebraic commutation relations are often used to derive information on the point spectra of these operators. In this respect, the domain

$$(8.22) \quad \mathfrak{D} = \operatorname{span}\{x^\alpha e^{-\frac{x^2}{2}} \,|\, \alpha \in \mathbb{N}_0^n\} \subset \mathcal{S}(\mathbb{R}^n)$$

8.2. Angular momentum

is often used. It has the nice property that the finite dimensional subspaces

(8.23) $$\mathfrak{D}_k = \text{span}\{x^\alpha e^{-\frac{x^2}{2}} | |\alpha| \leq k\}$$

are invariant under L_j (and hence they reduce L_j).

Lemma 8.3. *The subspace $\mathfrak{D} \subset L^2(\mathbb{R}^n)$ defined in (8.22) is dense.*

Proof. By Lemma 1.10, it suffices to consider the case $n = 1$. Suppose $\langle \varphi, \psi \rangle = 0$ for every $\psi \in \mathfrak{D}$. Then

$$\frac{1}{\sqrt{2\pi}} \int \overline{\varphi(x)} e^{-\frac{x^2}{2}} \sum_{j=1}^{k} \frac{(\mathrm{i}tx)^j}{j!} dx = 0$$

for every finite k and hence also in the limit $k \to \infty$ by the dominated convergence theorem. But the limit is the Fourier transform of $\overline{\varphi(x)} e^{-\frac{x^2}{2}}$, which shows that this function is zero. Hence $\varphi(x) = 0$. □

Since $\mathfrak{D}$ is invariant under the unitary groups generated by L_j, the operators L_j are essentially self-adjoint on $\mathfrak{D}$ by Corollary 5.4.

Introducing $L^2 = L_1^2 + L_2^2 + L_3^2$, it is straightforward to check

(8.24) $$[L^2, L_j]\psi(x) = 0, \qquad \psi \in \mathcal{S}(\mathbb{R}^3).$$

Moreover, $\mathfrak{D}_k$ is invariant under L^2 and L_3 and hence $\mathfrak{D}_k$ reduces L^2 and L_3. In particular, L^2 and L_3 are given by finite matrices on $\mathfrak{D}_k$. Now let $\mathfrak{H}_m = \text{Ker}(L_3 - m)$ and denote by P_k the projector onto $\mathfrak{D}_k$. Since L^2 and L_3 commute on $\mathfrak{D}_k$, the space $P_k \mathfrak{H}_m$ is invariant under L^2, which shows that we can choose an orthonormal basis consisting of eigenfunctions of L^2 for $P_k \mathfrak{H}_m$. Increasing k, we get an orthonormal set of simultaneous eigenfunctions whose span is equal to $\mathfrak{D}$. Hence there is an orthonormal basis of simultaneous eigenfunctions of L^2 and L_3.

Now let us try to draw some further consequences by using the commutation relations (8.21). (All commutation relations below hold for $\psi \in \mathcal{S}(\mathbb{R}^3)$.) Denote by $\mathfrak{H}_{l,m}$ the set of all functions in $\mathfrak{D}$ satisfying

(8.25) $$L_3 \psi = m\psi, \qquad L^2 \psi = l(l+1)\psi.$$

By $L^2 \geq 0$ and $\sigma(L_3) \subseteq \mathbb{Z}$ we can restrict our attention to the case $l \geq 0$ and $m \in \mathbb{Z}$.

First, introduce two new operators

(8.26) $$L_\pm = L_1 \pm \mathrm{i}L_2, \qquad [L_3, L_\pm] = \pm L_\pm.$$

Then, for every $\psi \in \mathfrak{H}_{l,m}$, we have

(8.27) $$L_3(L_\pm \psi) = (m \pm 1)(L_\pm \psi), \qquad L^2(L_\pm \psi) = l(l+1)(L_\pm \psi);$$

that is, $L_\pm \mathfrak{H}_{l,m} \to \mathfrak{H}_{l,m\pm 1}$. Moreover, since

(8.28) $$L^2 = L_3^2 \pm L_3 + L_\mp L_\pm,$$

we obtain

(8.29) $$\|L_\pm \psi\|^2 = \langle \psi, L_\mp L_\pm \psi \rangle = (l(l+1) - m(m \pm 1))\|\psi\|$$

for every $\psi \in \mathfrak{H}_{l,m}$. If $\psi \neq 0$, we must have $l(l+1) - m(m \pm 1) \geq 0$, which shows $\mathfrak{H}_{l,m} = \{0\}$ for $|m| > l$. Moreover, $L_\pm \mathfrak{H}_{l,m} \to \mathfrak{H}_{l,m\pm 1}$ is injective unless $|m| = l$. Hence we must have $\mathfrak{H}_{l,m} = \{0\}$ for $l \notin \mathbb{N}_0$.

Up to this point, we know $\sigma(L^2) \subseteq \{l(l+1) | l \in \mathbb{N}_0\}$, $\sigma(L_3) \subseteq \mathbb{Z}$. In order to show that equality holds in both cases, we need to show that $\mathfrak{H}_{l,m} \neq \{0\}$ for $l \in \mathbb{N}_0$, $m = -l, -l+1, \ldots, l-1, l$. First of all, we observe

(8.30) $$\psi_{0,0}(x) = \frac{1}{\pi^{3/4}} e^{-\frac{x^2}{2}} \in \mathfrak{H}_{0,0}.$$

Next, we note that (8.21) implies

$$[L_3, x_\pm] = \pm x_\pm, \qquad x_\pm = x_1 \pm ix_2,$$
$$[L_\pm, x_\pm] = 0, \qquad [L_\pm, x_\mp] = \pm 2x_3,$$
(8.31) $$[L^2, x_\pm] = 2x_\pm(1 \pm L_3) \mp 2x_3 L_\pm.$$

Hence if $\psi \in \mathfrak{H}_{l,l}$, then $(x_1 \pm ix_2)\psi \in \mathfrak{H}_{l\pm 1,l\pm 1}$. Thus

(8.32) $$\psi_{l,l}(x) = \frac{1}{\sqrt{l!}}(x_1 \pm ix_2)^l \psi_{0,0}(x) \in \mathfrak{H}_{l,l},$$

respectively,

(8.33) $$\psi_{l,m}(x) = \sqrt{\frac{(l+m)!}{(l-m)!(2l)!}} L_-^{l-m} \psi_{l,l}(x) \in \mathfrak{H}_{l,m}.$$

The constants are chosen such that $\|\psi_{l,m}\| = 1$.

In summary,

Theorem 8.4. *There exists an orthonormal basis of simultaneous eigenvectors for the operators L^2 and L_j. Moreover, their spectra are given by*

(8.34) $$\sigma(L^2) = \{l(l+1) | l \in \mathbb{N}_0\}, \qquad \sigma(L_3) = \mathbb{Z}.$$

We will give an alternate derivation of this result in Section 10.3.

8.3. The harmonic oscillator

Finally, let us consider another important model whose algebraic structure is similar to those of the angular momentum, the **harmonic oscillator**

(8.35) $$H = H_0 + \omega^2 x^2, \qquad \omega > 0.$$

8.3. The harmonic oscillator

We will choose as domain

(8.36) $$\mathfrak{D}(H) = \mathfrak{D} = \operatorname{span}\{x^\alpha e^{-\frac{x^2}{2}} \mid \alpha \in \mathbb{N}_0^3\} \subseteq L^2(\mathbb{R}^3)$$

from our previous section.

We will first consider the one-dimensional case. Introducing

(8.37) $$A_\pm = \frac{1}{\sqrt{2}}\left(\sqrt{\omega} x \mp \frac{1}{\sqrt{\omega}} \frac{d}{dx}\right), \qquad \mathfrak{D}(A_\pm) = \mathfrak{D},$$

we have

(8.38) $$[A_-, A_+] = 1$$

and

(8.39) $$H = \omega(2N+1), \quad N = A_+ A_-, \quad \mathfrak{D}(N) = \mathfrak{D},$$

for every function in $\mathfrak{D}$. In particular, note that $\mathfrak{D}$ is invariant under $A_\pm$.

Moreover, since

(8.40) $$[N, A_\pm] = \pm A_\pm,$$

we see that $N\psi = n\psi$ implies $NA_\pm \psi = (n \pm 1) A_\pm \psi$. Moreover, $\|A_+\psi\|^2 = \langle \psi, A_- A_+ \psi\rangle = (n+1)\|\psi\|^2$, respectively, $\|A_-\psi\|^2 = n\|\psi\|^2$, in this case and hence we conclude that $\sigma_p(N) \subseteq \mathbb{N}_0$.

If $N\psi_0 = 0$, then we must have $A_-\psi = 0$, and the normalized solution of this last equation is given by

(8.41) $$\psi_0(x) = \left(\frac{\omega}{\pi}\right)^{1/4} e^{-\frac{\omega x^2}{2}} \in \mathfrak{D}.$$

Hence

(8.42) $$\psi_n(x) = \frac{1}{\sqrt{n!}} A_+^n \psi_0(x)$$

is a normalized eigenfunction of N corresponding to the eigenvalue n. Moreover, since

(8.43) $$\psi_n(x) = \frac{1}{\sqrt{2^n n!}} \left(\frac{\omega}{\pi}\right)^{1/4} H_n(\sqrt{\omega} x) e^{-\frac{\omega x^2}{2}}$$

where $H_n(x)$ is a polynomial of degree n given by

(8.44) $$H_n(x) = e^{\frac{x^2}{2}} \left(x - \frac{d}{dx}\right)^n e^{-\frac{x^2}{2}} = (-1)^n e^{x^2} \frac{d^n}{dx^n} e^{-x^2},$$

we conclude $\operatorname{span}\{\psi_n\} = \mathfrak{D}$. The polynomials $H_n(x)$ are called **Hermite polynomials**.

In summary,

Theorem 8.5. *The harmonic oscillator H is essentially self-adjoint on $\mathfrak{D}$ and has an orthonormal basis of eigenfunctions*

(8.45) $$\psi_{n_1,n_2,n_3}(x) = \psi_{n_1}(x_1)\psi_{n_2}(x_2)\psi_{n_3}(x_3),$$

with $\psi_{n_j}(x_j)$ from (8.43). The spectrum is given by

(8.46) $$\sigma(H) = \{(2n+3)\omega | n \in \mathbb{N}_0\}.$$

Finally, there is also a close connection with the Fourier transformation. Without restriction we choose $\omega = 1$ and consider only one dimension. Then it easy to verify that H commutes with the Fourier transformation,

(8.47) $$\mathcal{F}H = H\mathcal{F},$$

on $\mathfrak{D}$. Moreover, by $\mathcal{F}A_\pm = \mp\mathrm{i}A_\pm\mathcal{F}$, we even infer

(8.48) $$\mathcal{F}\psi_n = \frac{1}{\sqrt{n!}}\mathcal{F}A_+^n\psi_0 = \frac{(-\mathrm{i})^n}{\sqrt{n!}}A_+^n\mathcal{F}\psi_0 = (-\mathrm{i})^n\psi_n,$$

since $\mathcal{F}\psi_0 = \psi_0$ by Lemma 7.3. In particular,

(8.49) $$\sigma(\mathcal{F}) = \{z \in \mathbb{C} | z^4 = 1\}.$$

8.4. Abstract commutation

The considerations of the previous section can be generalized as follows. First of all, the starting point was a factorization of H according to $H = A^*A$ (note that $A_\pm$ from the previous section are adjoint to each other when restricted to $\mathfrak{D}$). Then it turned out that commuting both operators just corresponds to a shift of H; that is, $AA^* = H + c$. Hence one could exploit the close spectral relation of A^*A and AA^* to compute both the eigenvalues and eigenvectors.

More generally, let A be a closed operator and recall that $H_0 = A^*A$ is a self-adjoint operator (cf. Problem 2.12) with $\mathrm{Ker}(H_0) = \mathrm{Ker}(A)$. Similarly, $H_1 = AA^*$ is a self-adjoint operator with $\mathrm{Ker}(H_1) = \mathrm{Ker}(A^*)$.

Theorem 8.6. *Let A be a densely defined closed operator and introduce $H_0 = A^*A$, $H_1 = AA^*$. Then the operators $H_0|_{\mathrm{Ker}(H_0)^\perp}$ and $H_1|_{\mathrm{Ker}(H_1)^\perp}$ are unitarily equivalent.*

If $H_0\psi_0 = E\psi_0$, $\psi_0 \in \mathfrak{D}(H_0)$, then $\psi_1 = A\psi_0 \in \mathfrak{D}(H_1)$ with $H_1\psi_1 = E\psi_1$ and $\|\psi_1\| = \sqrt{E}\|\psi_0\|$. Moreover,

(8.50) $$R_{H_1}(z) \supseteq \frac{1}{z}(AR_{H_0}(z)A^* - 1), \quad R_{H_0}(z) \supseteq \frac{1}{z}(A^*R_{H_1}(z)A - 1).$$

Proof. Introducing $|A| = H_0^{1/2}$, we have the polar decomposition (Section 4.3)
$$A = U|A|,$$

8.4. Abstract commutation

where
$$U : \text{Ker}(A)^\perp \to \text{Ker}(A^*)^\perp$$
is unitary. Taking adjoints, we have (Problem 2.3)
$$A^* = |A|U^*$$
and thus $H_1 = AA^* = U|A||A|U^* = UH_0U^*$ shows the claimed unitary equivalence.

The claims about the eigenvalues are straightforward (for the norm, note that $A\psi_0 = \sqrt{E}U\psi_0$). To see the connection between the resolvents, abbreviate $P_1 = P_{H_1}(\{0\})$. Then
$$\begin{aligned}
R_{H_1}(z) &= R_{H_1}(z)(1-P_1) + \frac{1}{z}P_1 = UR_{H_0}U^* + \frac{1}{z}P_1 \\
&\supseteq \frac{1}{z}\left(U(|H_0|^{1/2}R_{H_0}|H_0|^{1/2} - 1)U^* + P_1\right) \\
&= \frac{1}{z}\left(AR_{H_0}A^* + (1-P_1) + P_1\right) = \frac{1}{z}\left(AR_{H_0}A^* + 1\right),
\end{aligned}$$
where we have used $UU^* = 1 - P_1$. $\square$

We will use this result to compute the eigenvalues and eigenfunctions of the hydrogen atom in Section 10.4. In the physics literature, this approach is also known as **supersymmetric quantum mechanics**.

Problem 8.2. Show that $H_0 = -\frac{d^2}{dx^2} + q$ can formally (i.e., ignoring domains) be written as $H_0 = AA^*$, where $A = -\frac{d}{dx} + \phi$, if the differential equation $\psi'' + q\psi = 0$ has a positive solution. Compute $H_1 = A^*A$. (Hint: $\phi = \frac{\psi'}{\psi}$.)

Problem 8.3. Take $H_0 = -\frac{d^2}{dx^2} + \lambda$, $\lambda > 0$, and compute H_1. What about domains?

Problem 8.4. Let A be a closed operator. Show that the **supersymmetric Dirac operator**
$$D = \begin{pmatrix} 0 & A^* \\ A & 0 \end{pmatrix}, \qquad \mathfrak{D}(D) = \mathfrak{D}(A^*) \oplus \mathfrak{D}(A) \subseteq \mathfrak{H}^2$$
is self-adjoint. Compute
$$D^2 = \begin{pmatrix} A^*A & 0 \\ 0 & AA^* \end{pmatrix}, \qquad \mathfrak{D}(D^2) = \mathfrak{D}(A^*A) \oplus \mathfrak{D}(AA^*) \subseteq \mathfrak{H}^2.$$

Note that since D^2 is self-adjoint this shows that A^*A (and AA^*) is self-adjoint — see Problem 2.12.

Chapter 9

One-dimensional Schrödinger operators

9.1. Sturm–Liouville operators

In this section, we want to illustrate some of the results obtained thus far by investigating a specific example, the **Sturm–Liouville equation**

$$(9.1) \quad \tau f(x) = \frac{1}{r(x)} \left(-\frac{d}{dx} p(x) \frac{d}{dx} f(x) + q(x) f(x) \right), \quad f, pf' \in AC(a,b),$$

on an arbitrary open interval $I = (a,b) \subseteq \mathbb{R}$. Here $AC(a,b)$ denotes the set of absolutely continuous functions (cf. Section 2.7).

The case $p = r = 1$ can be viewed as the model of a particle in one dimension in the external potential q. Moreover, the case of a particle in three dimensions can in some situations be reduced to the investigation of Sturm–Liouville equations. In particular, we will see how this works when explicitly solving the hydrogen atom.

The suitable Hilbert space is

$$(9.2) \quad L^2((a,b), r(x)dx), \quad \langle f, g \rangle = \int_a^b f(x)^* g(x) r(x) dx.$$

We require

(i) $p^{-1} \in L^1_{loc}(I)$, positive,
(ii) $q \in L^1_{loc}(I)$, real-valued,
(iii) $r \in L^1_{loc}(I)$, positive.

If a is finite and if $p^{-1}, q, r \in L^1((a,c))$ ($c \in I$), then the Sturm–Liouville equation (9.1) is called **regular at** a. Similarly for b. If it is regular at both a and b, it is called **regular**.

The maximal domain of definition for τ in $L^2(I, r\, dx)$ is given by

$$(9.3) \qquad \mathfrak{D}(\tau) = \{f \in L^2(I, r\, dx) | f, pf' \in AC(I),\ \tau f \in L^2(I, r\, dx)\}.$$

It is not clear that $\mathfrak{D}(\tau)$ is dense unless (e.g.) $p \in AC(I)$, $p', q \in L^2_{loc}(I)$, $r^{-1} \in L^\infty_{loc}(I)$ since $C_0^\infty(I) \subset \mathfrak{D}(\tau)$ in this case. We will defer the general case to Lemma 9.4 below.

Since we are interested in self-adjoint operators H associated with (9.1), we perform a little calculation. Using integration by parts (twice), we obtain the **Lagrange identity** ($a < c < d < b$)

$$(9.4) \qquad \int_c^d g^*(\tau f)\, r\, dy = W_c(g^*, f) - W_d(g^*, f) + \int_c^d (\tau g)^* f\, r\, dy,$$

for $f, g, pf', pg' \in AC(I)$, where

$$(9.5) \qquad W_x(f_1, f_2) = \big(p(f_1 f_2' - f_1' f_2)\big)(x)$$

is called the **modified Wronskian**.

Equation (9.4) also shows that the Wronskian of two solutions of $\tau u = z u$ is constant:

$$(9.6) \qquad W_x(u_1, u_2) = W(u_1, u_2), \qquad \tau u_{1,2} = z u_{1,2}.$$

Moreover, it is nonzero if and only if u_1 and u_2 are linearly independent (compare Theorem 9.1 below).

If we choose $f, g \in \mathfrak{D}(\tau)$ in (9.4), then we can take the limits $c \to a$ and $d \to b$, which results in

$$(9.7) \qquad \langle g, \tau f \rangle = W_a(g^*, f) - W_b(g^*, f) + \langle \tau g, f \rangle, \qquad f, g \in \mathfrak{D}(\tau).$$

Here $W_{a,b}(g^*, f)$ has to be understood as a limit.

Finally, we recall the following well-known result from ordinary differential equations.

Theorem 9.1. *Suppose $rg \in L^1_{loc}(I)$. Then there exists a unique solution $f, pf' \in AC(I)$ of the differential equation*

$$(9.8) \qquad (\tau - z)f = g, \qquad z \in \mathbb{C},$$

satisfying the initial condition

$$(9.9) \qquad f(c) = \alpha, \quad (pf')(c) = \beta, \qquad \alpha, \beta \in \mathbb{C}, \quad c \in I.$$

In addition, f is entire with respect to z.

9.1. Sturm–Liouville operators

Proof. Introducing
$$u = \begin{pmatrix} f \\ pf' \end{pmatrix}, \quad v = \begin{pmatrix} 0 \\ rg \end{pmatrix},$$
we can rewrite (9.8) as the linear first-order system
$$u' - Au = v, \quad A(x) = \begin{pmatrix} 0 & p^{-1}(x) \\ q(x) - z\,r(x) & 0 \end{pmatrix}.$$
Integrating with respect to x, we see that this system is equivalent to the Volterra integral equation
$$u - Ku = w, \quad (Ku)(x) = \int_c^x A(y)u(y)dy, \quad w(x) = \begin{pmatrix} \alpha \\ \beta \end{pmatrix} + \int_c^x v(y)dy.$$
We will choose some $d \in (c, b)$ and consider the integral operator K in the Banach space $C([c, d])$. Then for every $h \in C([c, d])$ and $x \in [c, d]$ we have the estimate
$$|K^n(h)(x)| \le \frac{a_1(x)^n}{n!}\|h\|, \quad a_1(x) = \int_c^x a(y)dy, \quad a(x) = \|A(x)\|,$$
which follows from induction
$$|K^{n+1}(h)(x)| = \left|\int_c^x A(y)K^n(h)(y)dy\right| \le \int_c^x a(y)|K^n(h)(y)|dy$$
$$\le \|h\| \int_c^x a(y)\frac{a_1(y)^n}{n!}dy = \frac{a_1(x)^{n+1}}{(n+1)!}\|h\|.$$
Hence the unique solution of our integral equation is given by the Neumann series (show this)
$$u(x) = \sum_{n=0}^\infty K^n(w)(x).$$
To see that the solution $u(x)$ is entire with respect to z, note that the partial sums are entire (in fact polynomial) in z and hence so is the limit by uniform convergence with respect to z in compact sets. An analogous argument for $d \in (a, c)$ finishes the proof. □

Note that f, pf' can be extended continuously to a regular endpoint.

Lemma 9.2. *Suppose u_1, u_2 are two solutions of $(\tau - z)u = 0$ which satisfy $W(u_1, u_2) = 1$. Then any other solution of (9.8) can be written as $(\alpha, \beta \in \mathbb{C})$*
$$f(x) = u_1(x)\left(\alpha + \int_c^x u_2 g\,rdy\right) + u_2(x)\left(\beta - \int_c^x u_1 g\,rdy\right),$$
(9.10) $$f'(x) = u_1'(x)\left(\alpha + \int_c^x u_2 g\,rdy\right) + u_2'(x)\left(\beta - \int_c^x u_1 g\,rdy\right).$$

Note that the constants α, β coincide with those from Theorem 9.1 if $u_1(c) = (pu_2')(c) = 1$ and $(pu_1')(c) = u_2(c) = 0$.

Proof. It suffices to check $\tau f - z f = g$. Differentiating the first equation of (9.10) gives the second. Next, we compute

$$(pf')' = (pu_1')'\left(\alpha + \int u_2 g\, rdy\right) + (pu_2')'\left(\beta - \int u_1 g\, rdy\right) - W(u_1, u_2)gr$$

$$= (q - zr)u_1\left(\alpha + \int u_2 g\, rdy\right) + (q - zr)u_2\left(\beta - \int u_1 g\, rdy\right) - gr$$

$$= (q - zr)f - gr$$

which proves the claim. $\square$

Now we want to obtain a symmetric operator and hence we choose

(9.11) $\qquad A_0 f = \tau f, \qquad \mathfrak{D}(A_0) = \mathfrak{D}(\tau) \cap AC_c(I),$

where $AC_c(I)$ denotes the functions in $AC(I)$ with compact support. This definition clearly ensures that the Wronskian of two such functions vanishes on the boundary, implying that A_0 is symmetric by virtue of (9.7). Our first task is to compute the closure of A_0 and its adjoint. For this, the following elementary fact will be needed.

Lemma 9.3. *Suppose V is a vector space and $l, l_1, \ldots, l_n$ are linear functionals (defined on all of V) such that $\bigcap_{j=1}^n \mathrm{Ker}(l_j) \subseteq \mathrm{Ker}(l)$. Then $l = \sum_{j=0}^n \alpha_j l_j$ for some constants $\alpha_j \in \mathbb{C}$.*

Proof. First of all, it is no restriction to assume that the functionals l_j are linearly independent. Then the map $L : V \to \mathbb{C}^n$, $f \mapsto (l_1(f), \ldots, l_n(f))$ is surjective (since $x \in \mathrm{Ran}(L)^\perp$ implies $\sum_{j=1}^n x_j l_j(f) = 0$ for all f). Hence there are vectors $f_k \in V$ such that $l_j(f_k) = 0$ for $j \neq k$ and $l_j(f_j) = 1$. Then $f - \sum_{j=1}^n l_j(f)f_j \in \bigcap_{j=1}^n \mathrm{Ker}(l_j)$ and hence $l(f) - \sum_{j=1}^n l_j(f)l(f_j) = 0$. Thus we can choose $\alpha_j = l(f_j)$. $\square$

Now we are ready to prove

Lemma 9.4. *The operator A_0 is densely defined and its closure is given by*
(9.12)
$$\overline{A_0}f = \tau f, \quad \mathfrak{D}(\overline{A_0}) = \{f \in \mathfrak{D}(\tau) \,|\, W_a(f, g) = W_b(f, g) = 0, \, \forall g \in \mathfrak{D}(\tau)\}.$$

Its adjoint is given by

(9.13) $\qquad A_0^* f = \tau f, \qquad \mathfrak{D}(A_0^*) = \mathfrak{D}(\tau).$

Proof. We start by computing A_0^* and ignore the fact that we do not know whether $\mathfrak{D}(A_0)$ is dense for now.

By (9.7) we have $\mathfrak{D}(\tau) \subseteq \mathfrak{D}(A_0^*)$ and it remains to show $\mathfrak{D}(A_0^*) \subseteq \mathfrak{D}(\tau)$. If $h \in \mathfrak{D}(A_0^*)$, we must have

$$\langle h, A_0 f \rangle = \langle k, f \rangle, \qquad \forall f \in \mathfrak{D}(A_0),$$

9.1. Sturm–Liouville operators

for some $k \in L^2(I, r\,dx)$. Using (9.10), we can find a $\tilde{h}$ such that $\tau \tilde{h} = k$, and from integration by parts, we obtain

$$(9.14) \qquad \int_a^b (h(x) - \tilde{h}(x))^* (\tau f)(x) r(x) dx = 0, \qquad \forall f \in \mathfrak{D}(A_0).$$

Clearly we expect that $h - \tilde{h}$ will be a solution of $\tau u = 0$ and to prove this, we will invoke Lemma 9.3. Therefore, we consider the linear functionals

$$l(g) = \int_a^b (h(x) - \tilde{h}(x))^* g(x) r(x) dx, \quad l_j(g) = \int_a^b u_j(x)^* g(x) r(x) dx,$$

on $L_c^2(I, r\,dx)$, where u_j are two solutions of $\tau u = 0$ with $W(u_1, u_2) \neq 0$. Then we have $\mathrm{Ker}(l_1) \cap \mathrm{Ker}(l_2) \subseteq \mathrm{Ker}(l)$. In fact, if $g \in \mathrm{Ker}(l_1) \cap \mathrm{Ker}(l_2)$, then

$$f(x) = u_1(x) \int_a^x u_2(y) g(y) r(y) dy + u_2(x) \int_x^b u_1(y) g(y) r(y) dy$$

is in $\mathfrak{D}(A_0)$ and $g = \tau f \in \mathrm{Ker}(l)$ by (9.14). Now Lemma 9.3 implies

$$\int_a^b (h(x) - \tilde{h}(x) + \alpha_1 u_1(x) + \alpha_2 u_2(x))^* g(x) r(x) dx = 0, \quad \forall g \in L_c^2(I, r\,dx)$$

and hence $h = \tilde{h} + \alpha_1 u_1 + \alpha_2 u_2 \in \mathfrak{D}(\tau)$.

Now what if $\mathfrak{D}(A_0)$ were not dense? Then there would be some freedom in the choice of k since we could always add a component in $\mathfrak{D}(A_0)^\perp$. So suppose we have two choices $k_1 \neq k_2$. Then by the above calculation, there are corresponding functions $\tilde{h}_1$ and $\tilde{h}_2$ such that $h = \tilde{h}_1 + \alpha_{1,1} u_1 + \alpha_{1,2} u_2 = \tilde{h}_2 + \alpha_{2,1} u_1 + \alpha_{2,2} u_2$. In particular, $\tilde{h}_1 - \tilde{h}_2$ is in the kernel of τ and hence $k_1 = \tau \tilde{h}_1 = \tau \tilde{h}_2 = k_2$, a contradiction to our assumption.

Next we turn to $\overline{A_0}$. Denote the set on the right-hand side of (9.12) by $\mathfrak{D}$. Then we have $\mathfrak{D} \subseteq \mathfrak{D}(A_0^{**}) = \mathfrak{D}(\overline{A_0})$ by (9.7). Conversely, since $\overline{A_0} \subseteq A_0^*$, we can use (9.7) to conclude

$$W_a(f, h) - W_b(f, h) = 0, \quad f \in \mathfrak{D}(\overline{A_0}), h \in \mathfrak{D}(A_0^*).$$

Now replace h by a $\tilde{h} \in \mathfrak{D}(A_0^*)$ which coincides with h near a and vanishes identically near b (Problem 9.1). Then $W_a(f, h) = W_a(f, \tilde{h}) - W_b(f, \tilde{h}) = 0$. Finally, $W_b(f, h) = W_a(f, h) = 0$ shows $f \in \mathfrak{D}$. □

Example. If τ is regular at a, then $W_a(f, g) = 0$ for all $g \in \mathfrak{D}(\tau)$ if and only if $f(a) = (pf')(a) = 0$. This follows since we can prescribe the values of $g(a), (pg')(a)$ for $g \in \mathfrak{D}(\tau)$ arbitrarily. ◇

This result shows that every self-adjoint extension of A_0 must lie between $\overline{A_0}$ and A_0^*. Moreover, self-adjointness seems to be related to the Wronskian of two functions at the boundary. Hence we collect a few properties first.

Lemma 9.5. *Suppose $v \in \mathfrak{D}(\tau)$ with $W_a(v^*, v) = 0$ and suppose there is a $\hat{f} \in \mathfrak{D}(\tau)$ with $W_a(v^*, \hat{f}) \neq 0$. Then, for $f, g \in \mathfrak{D}(\tau)$, we have*

(9.15) $$W_a(v, f) = 0 \quad \Leftrightarrow \quad W_a(v, f^*) = 0$$

and

(9.16) $$W_a(v, f) = W_a(v, g) = 0 \quad \Rightarrow \quad W_a(g^*, f) = 0.$$

Proof. For all $f_1, \ldots, f_4 \in \mathfrak{D}(\tau)$, we have the **Plücker identity**
(9.17)
$$W_x(f_1, f_2)W_x(f_3, f_4) + W_x(f_1, f_3)W_x(f_4, f_2) + W_x(f_1, f_4)W_x(f_2, f_3) = 0,$$
which remains valid in the limit $x \to a$. Choosing $f_1 = v$, $f_2 = f$, $f_3 = v^*$, $f_4 = \hat{f}$, we infer (9.15). Choosing $f_1 = f$, $f_2 = g^*$, $f_3 = v$, $f_4 = \hat{f}$, we infer (9.16). $\square$

Problem 9.1. *Given $\alpha, \beta, \gamma, \delta$, show that there is a function f in $\mathfrak{D}(\tau)$ restricted to $[c, d] \subseteq (a, b)$ such that $f(c) = \alpha$, $(pf')(c) = \beta$ and $f(d) = \gamma$, $(pf')(c) = \delta$. (Hint: Lemma 9.2.)*

Problem 9.2. *Let $A_0 = -\frac{d^2}{dx^2}$, $\mathfrak{D}(A_0) = \{f \in H^2[0, 1] | f(0) = f(1) = 0\}$ and $B = q$, $\mathfrak{D}(B) = \{f \in L^2(0, 1) | qf \in L^2(0, 1)\}$. Find a $q \in L^1(0, 1)$ such that $\mathfrak{D}(A_0) \cap \mathfrak{D}(B) = \{0\}$. (Hint: Problem 0.40.)*

Problem 9.3. *Let $\phi \in L^1_{loc}(I)$. Define*

$$A_{\pm} = \pm\frac{d}{dx} + \phi, \quad \mathfrak{D}(A_{\pm}) = \{f \in L^2(I) | f \in AC(I), \pm f' + \phi f \in L^2(I)\}$$

and $A_{0,\pm} = A_{\pm}|_{AC_c(I)}$. Show $A_{0,\pm}^* = A_{\mp}$ and

$$\mathfrak{D}(\overline{A_{0,\pm}}) = \{f \in \mathfrak{D}(A_{\pm}) | \lim_{x \to a, b} f(x)g(x) = 0, \forall g \in \mathfrak{D}(A_{\mp})\}.$$

In particular, show that the limits above exist.

Problem 9.4 (Liouville normal form). *Show that every Sturm–Liouville equation can be transformed into one with $r = p = 1$ as follows: Show that the transformation $U : L^2((a, b), r\, dx) \to L^2(0, c)$, $c = \int_a^b \sqrt{\frac{r(t)}{p(t)}} dt$, defined via $u(x) \mapsto v(y)$, where*

$$y(x) = \int_a^x \sqrt{\frac{r(t)}{p(t)}} dt, \quad v(y) = \sqrt[4]{r(x(y))p(x(y))}\, u(x(y)),$$

is unitary. Moreover, if $p, r, p', r' \in AC(a, b)$, then

$$-(pu')' + qu = r\lambda u$$

transforms into

$$-v'' + Qv = \lambda v,$$

9.2. Weyl's limit circle, limit point alternative

where

$$Q = q - \frac{(pr)^{1/4}}{r}\left(p((pr)^{-1/4})'\right)'.$$

9.2. Weyl's limit circle, limit point alternative

Inspired by Lemma 9.5, we make the following definition: We call τ **limit circle** (l.c.) at a if there is a $v \in \mathfrak{D}(\tau)$ with $W_a(v^*, v) = 0$ such that $W_a(v, f) \neq 0$ for at least one $f \in \mathfrak{D}(\tau)$. Otherwise τ is called **limit point** (l.p.) at a and similarly for b.

Example. If τ is regular at a, it is limit circle at a. Since

(9.18) $$W_a(v, f) = (pf')(a)v(a) - (pv')(a)f(a),$$

any real-valued v with $(v(a), (pv')(a)) \neq (0, 0)$ works. $\diamond$

Note that if $W_a(f, v) \neq 0$, then $W_a(f, \mathrm{Re}(v)) \neq 0$ or $W_a(f, \mathrm{Im}(v)) \neq 0$. Hence it is no restriction to assume that v is real and $W_a(v^*, v) = 0$ is trivially satisfied in this case. In particular, τ is limit point if and only if $W_a(f, g) = 0$ for all $f, g \in \mathfrak{D}(\tau)$.

Theorem 9.6. *If τ is l.c. at a, then let $v \in \mathfrak{D}(\tau)$ with $W_a(v^*, v) = 0$ and $W_a(v, f) \neq 0$ for some $f \in \mathfrak{D}(\tau)$. Similarly, if τ is l.c. at b, let w be an analogous function. Then the operator*

(9.19) $$\begin{aligned} A: \mathfrak{D}(A) &\to L^2(I, r\, dx) \\ f &\mapsto \tau f \end{aligned}$$

with

(9.20) $$\mathfrak{D}(A) = \{f \in \mathfrak{D}(\tau) |\ \begin{aligned} &W_a(v, f) = 0 \text{ if l.c. at } a \\ &W_b(w, f) = 0 \text{ if l.c. at } b \end{aligned}\}$$

is self-adjoint. Moreover, the set

(9.21) $$\begin{aligned} \mathfrak{D}_1 = \{f \in \mathfrak{D}(\tau) |\ &\exists x_0 \in I : \forall x \in (a, x_0),\ W_x(v, f) = 0, \\ &\exists x_1 \in I : \forall x \in (x_1, b),\ W_x(w, f) = 0\} \end{aligned}$$

is a core for A.

Proof. By Lemma 9.5, A is symmetric and hence $A \subseteq A^* \subseteq A_0^*$. Let $g \in \mathfrak{D}(A^*)$. As in the computation of $\overline{A_0}$ we conclude $W_a(f, g) = W_b(f, g) = 0$ for all $f \in \mathfrak{D}(A)$. Moreover, we can choose f such that it coincides with v near a and hence $W_a(v, g) = 0$. Similarly, $W_b(w, g) = 0$; that is, $g \in \mathfrak{D}(A)$.

To see that $\mathfrak{D}_1$ is a core, let A_1 be the corresponding operator and observe that the argument from above, with A_1 in place of A, shows $A_1^* = A$. □

The name limit circle, respectively, limit point, stems from the original approach of Weyl, who considered the set of solutions $\tau u = zu$, $z \in \mathbb{C}\backslash\mathbb{R}$, which satisfy $W_x(u^*, u) = 0$. They can be shown to lie on a circle which converges to a circle, respectively, a point, as $x \to a$ or $x \to b$ (see Problem 9.9).

Before proceeding, let us shed some light on the number of possible boundary conditions. Suppose τ is l.c. at a and let u_1, u_2 be two real-valued solutions of $\tau u = 0$ with $W(u_1, u_2) = 1$. Abbreviate

$$(9.22) \qquad BC_x^j(f) = W_x(u_j, f), \qquad f \in \mathfrak{D}(\tau).$$

Let v be as in Theorem 9.6. Then, using Lemma 9.5, it is not hard to see that

$$(9.23) \qquad W_a(v, f) = 0 \quad \Leftrightarrow \quad \cos(\alpha)BC_a^1(f) - \sin(\alpha)BC_a^2(f) = 0,$$

where $\tan(\alpha) = \frac{BC_a^1(v)}{BC_a^2(v)}$. Hence all possible boundary conditions can be parametrized by $\alpha \in [0, \pi)$. If τ is regular at a and if we choose $u_1(a) = (pu_2')(a) = 1$ and $(pu_1')(a) = u_2(a) = 0$, then

$$(9.24) \qquad BC_a^2(f) = -f(a), \qquad BC_a^1(f) = (pf')(a),$$

and the boundary condition takes the simple form

$$(9.25) \qquad \sin(\alpha)(pf')(a) - \cos(\alpha)f(a) = 0.$$

The most common choice of $\alpha = 0$ is known as the **Dirichlet boundary condition** $f(a) = 0$. The choice $\alpha = \pi/2$ is known as the **Neumann boundary condition** $(pf')(a) = 0$.

Finally, note that if τ is l.c. at both a and b, then Theorem 9.6 does not give all possible self-adjoint extensions. For example, one could also choose

$$(9.26) \qquad BC_a^1(f) = e^{i\alpha}BC_b^1(f), \qquad BC_a^2(f) = e^{i\alpha}BC_b^2(f).$$

The case $\alpha = 0$ gives rise to **periodic boundary conditions** in the regular case.

Next we want to compute the resolvent of A.

Lemma 9.7. *Suppose $z \in \rho(A)$. Then there exists a solution $u_a(z, x)$ of $(\tau - z)u = 0$ which is in $L^2((a, c), r\, dx)$ and which satisfies the boundary condition at a if τ is l.c. at a. Similarly, there exists a solution $u_b(z, x)$ with the analogous properties near b.*

The resolvent of A is given by

$$(9.27) \qquad (A - z)^{-1}g(x) = \int_a^b G(z, x, y)g(y)r(y)dy,$$

9.2. Weyl's limit circle, limit point alternative

where

(9.28) $$G(z,x,y) = \frac{1}{W(u_b(z), u_a(z))} \begin{cases} u_b(z,x)u_a(z,y), & x \geq y, \\ u_a(z,x)u_b(z,y), & x \leq y. \end{cases}$$

Proof. Let $g \in L_c^2(I, r\,dx)$ be real-valued and consider $f = (A - z)^{-1}g \in \mathfrak{D}(A)$. Since $(\tau - z)f = 0$ near a, respectively, b, we obtain $u_a(z,x)$ by setting it equal to f near a and using the differential equation to extend it to the rest of I. Similarly, we obtain u_b. The only problem is that u_a or u_b might be identically zero. Hence we need to show that this can be avoided by choosing g properly.

Fix z and let g be supported in $(c, d) \subset I$. Since $(\tau - z)f = g$, Lemma 9.2 implies

(9.29) $$f(x) = u_1(x)\left(\alpha + \int_a^x u_2 gr\,dy\right) + u_2(x)\left(\beta + \int_x^b u_1 gr\,dy\right).$$

Near a $(x < c)$ we have $f(x) = \alpha u_1(x) + \tilde{\beta} u_2(x)$ and near b $(x > d)$ we have $f(x) = \tilde{\alpha} u_1(x) + \beta u_2(x)$, where $\tilde{\alpha} = \alpha + \int_a^b u_2 gr\,dy$ and $\tilde{\beta} = \beta + \int_a^b u_1 gr\,dy$. If f vanishes identically near both a and b, we must have $\alpha = \beta = \tilde{\alpha} = \tilde{\beta} = 0$ and thus $\alpha = \beta = 0$ and $\int_a^b u_j(y)g(y)r(y)dy = 0$, $j = 1, 2$. This case can be avoided by choosing a suitable g and hence there is at least one solution, say $u_b(z)$.

Now choose $u_1 = u_b$ and consider the behavior near b. If u_2 is not square integrable on (d, b), we must have $\beta = 0$ since $\beta u_2 = f - \tilde{\alpha} u_b$ is. If u_2 is square integrable, we can find two functions in $\mathfrak{D}(\tau)$ which coincide with u_b and u_2 near b. Since $W(u_b, u_2) = 1$, we see that τ is l.c. at b and hence $0 = W_b(u_b, f) = W_b(u_b, \tilde{\alpha} u_b + \beta u_2) = \beta$. Thus $\beta = 0$ in both cases and we have

$$f(x) = u_b(x)\left(\alpha + \int_a^x u_2 gr\,dy\right) + u_2(x)\int_x^b u_b gr\,dy.$$

Now choosing g such that $\int_a^b u_b gr\,dy \neq 0$, we infer the existence of $u_a(z)$. Choosing $u_2 = u_a$ and arguing as before, we see $\alpha = 0$ and hence

$$f(x) = u_b(x)\int_a^x u_a(y)g(y)r(y)dy + u_a(x)\int_x^b u_b(y)g(y)r(y)dy$$
$$= \int_a^b G(z,x,y)g(y)r(y)dy$$

for every $g \in L_c^2(I, r\,dx)$. Since this set is dense, the claim follows. □

Example. If τ is regular at a with a boundary condition as in the previous example, we can choose $u_a(z,x)$ to be the solution corresponding to the initial conditions $(u_a(z,a), (pu_a')(z,a)) = (\sin(\alpha), \cos(\alpha))$. In particular, $u_a(z,x)$ exists for all $z \in \mathbb{C}$.

If τ is regular at both a and b, there is a corresponding solution $u_b(z,x)$, again for all z. So the only values of z for which $(A-z)^{-1}$ does not exist must be those with $W(u_b(z), u_a(z)) = 0$. However, in this case $u_a(z,x)$ and $u_b(z,x)$ are linearly dependent and $u_a(z,x) = \gamma u_b(z,x)$ satisfies both boundary conditions. That is, z is an eigenvalue in this case.

In particular, regular operators have pure point spectrum. We will see in Theorem 9.10 below that this holds for every operator which is *l.c.* at both endpoints. $\diamond$

In the previous example, $u_a(z,x)$ is holomorphic with respect to z and satisfies $u_a(z,x)^* = u_a(z^*,x)$ (since it corresponds to real initial conditions and our differential equation has real coefficients). In general we have:

Lemma 9.8. *Suppose $z \in \rho(A)$. Then $u_a(z,x)$ from the previous lemma can be chosen locally holomorphic with respect to z such that*

(9.30) $$u_a(z,x)^* = u_a(z^*,x)$$

and similarly for $u_b(z,x)$.

Proof. Since this is a local property near a, we can assume b is regular and choose $u_b(z,x)$ such that $(u_b(z,b), (pu_b')(z,b)) = (\sin(\beta), -\cos(\beta))$ as in the example above. In addition, choose a second solution $v_b(z,x)$ such that $(v_b(z,b), (pv_b')(z,b)) = (\cos(\beta), \sin(\beta))$ and observe $W(u_b(z), v_b(z)) = 1$. If $z \in \rho(A)$, z is no eigenvalue and hence $u_a(z,x)$ cannot be a multiple of $u_b(z,x)$. Thus we can set

$$u_a(z,x) = v_b(z,x) + m(z) u_b(z,x)$$

and it remains to show that $m(z)$ is holomorphic with $m(z)^* = m(z^*)$.

Choosing h with compact support in (a,c) and g with support in (c,b), we have

$$\langle h, (A-z)^{-1} g\rangle = \langle h, u_a(z)\rangle \langle g^*, u_b(z)\rangle$$
$$= (\langle h, v_b(z)\rangle + m(z)\langle h, u_b(z)\rangle)\langle g^*, u_b(z)\rangle$$

(with a slight abuse of notation since u_b, v_b might not be square integrable). Choosing (real-valued) functions h and g such that $\langle h, u_b(z)\rangle \langle g^*, u_b(z)\rangle \neq 0$, we can solve for $m(z)$:

$$m(z) = \frac{\langle h, (A-z)^{-1} g\rangle - \langle h, v_b(z)\rangle \langle g^*, u_b(z)\rangle}{\langle h, u_b(z)\rangle \langle g^*, u_b(z)\rangle}.$$

This finishes the proof. $\square$

Example. We already know that $\tau = -\frac{d^2}{dx^2}$ on $I = (-\infty, \infty)$ gives rise to the free Schrödinger operator H_0. Furthermore,

(9.31) $$u_\pm(z,x) = e^{\mp\sqrt{-z}x}, \qquad z \in \mathbb{C},$$

9.2. Weyl's limit circle, limit point alternative

are two linearly independent solutions (for $z \neq 0$) and since $\operatorname{Re}(\sqrt{-z}) > 0$ for $z \in \mathbb{C}\backslash[0, \infty)$, there is precisely one solution (up to a constant multiple) which is square integrable near $\pm\infty$, namely $u_\pm$. In particular, the only choice for u_a is u_- and for u_b is u_+ and we get

$$(9.32) \qquad G(z, x, y) = \frac{1}{2\sqrt{-z}} e^{-\sqrt{-z}|x-y|}$$

which we already found in Section 7.5. $\diamond$

If, as in the previous example, there is only one square integrable solution, there is no choice for $G(z, x, y)$. But since different boundary conditions must give rise to different resolvents, there is no room for boundary conditions in this case. This indicates a connection between our l.c., l.p. distinction and square integrability of solutions.

Theorem 9.9 (Weyl alternative). *The operator τ is l.c. at a if and only if for one $z_0 \in \mathbb{C}$ all solutions of $(\tau - z_0)u = 0$ are square integrable near a. This then holds for all $z \in \mathbb{C}$ and similarly for b.*

Proof. If all solutions are square integrable near a, τ is l.c. at a since the Wronskian of two linearly independent solutions does not vanish.

Conversely, take two functions $v, \tilde{v} \in \mathfrak{D}(\tau)$ with $W_a(v, \tilde{v}) \neq 0$. By considering real and imaginary parts, it is no restriction to assume that v and $\tilde{v}$ are real-valued. Thus they give rise to two different self-adjoint extensions A and $\tilde{A}$ (choose any fixed w for the other endpoint). Let u_a and $\tilde{u}_a$ be the corresponding solutions from above. Then $W(u_a, \tilde{u}_a) \neq 0$ (since otherwise $A = \tilde{A}$ by Lemma 9.5) and thus there are two linearly independent solutions which are square integrable near a. Since any other solution can be written as a linear combination of those two, every solution is square integrable near a.

It remains to show that all solutions of $(\tau - z)u = 0$ for all $z \in \mathbb{C}$ are square integrable near a if τ is l.c. at a. In fact, the above argument ensures this for every $z \in \rho(A) \cap \rho(\tilde{A})$, that is, at least for all $z \in \mathbb{C}\backslash\mathbb{R}$.

Suppose there are two linearly independent solutions u_1 and u_2 of $(\tau - z_0)u = 0$. Without loss of generality, we can assume $W(u_1, u_2) = 1$. We will show that any solution u of $(\tau - z)u = 0$ is square integrable. Using $(\tau - z_0)u = (z - z_0)u$ and (9.10), we have ($a < c < x < b$)

$$u(x) = \alpha u_1(x) + \beta u_2(x) + (z - z_0) \int_c^x (u_1(x)u_2(y) - u_1(y)u_2(x))u(y)r(y)\,dy.$$

Since $u_j \in L^2((c, b), rdx)$, we can find a constant $M \geq 0$ such that

$$\int_c^b |u_{1,2}(y)|^2 r(y)\,dy \leq M.$$

Now choose c close to b such that $|z - z_0|^2 M^2 \le 1/8$. Next, estimating the integral using Cauchy–Schwarz gives

$$\left| \int_c^x (u_1(x)u_2(y) - u_1(y)u_2(x))u(y)r(y)\,dy \right|^2$$

$$\le \int_c^x |u_1(x)u_2(y) - u_1(y)u_2(x)|^2 r(y)\,dy \int_c^x |u(y)|^2 r(y)\,dy$$

$$\le M\big(|u_1(x)|^2 + |u_2(x)|^2\big) \int_c^x |u(y)|^2 r(y)\,dy$$

and hence

$$\int_c^x |u(y)|^2 r(y)\,dy \le 3(|\alpha|^2 + |\beta|^2)M + 6|z - z_0|^2 M^2 \int_c^x |u(y)|^2 r(y)\,dy$$

$$\le 3(|\alpha|^2 + |\beta|^2)M + \frac{1}{2} \int_c^x |u(y)|^2 r(y)\,dy.$$

Thus

$$\int_c^x |u(y)|^2 r(y)\,dy \le 6(|\alpha|^2 + |\beta|^2)M$$

and since $u \in AC(I)$, we have $u \in L^2((c,b), r\,dx)$ for every $c \in (a,b)$. □

Now we turn to the investigation of the spectrum of A. If τ is l.c. at both endpoints, then the spectrum of A is very simple.

Theorem 9.10. *If τ is l.c. at both endpoints, then the resolvent is a Hilbert–Schmidt operator; that is,*

(9.33) $$\int_a^b \int_a^b |G(z,x,y)|^2 r(y)dy\, r(x)dx < \infty.$$

In particular, the spectrum of every self-adjoint extension is purely discrete, and the eigenfunctions (which are simple) form an orthonormal basis.

Proof. This follows from the estimate

$$\int_a^b \left(\int_a^x |u_b(x)u_a(y)|^2 r(y)dy + \int_x^b |u_b(y)u_a(x)|^2 r(y)dy \right) r(x)dx$$

$$\le 2 \int_a^b |u_a(y)|^2 r(y)dy \int_a^b |u_b(y)|^2 r(y)dy,$$

which shows that the resolvent is Hilbert–Schmidt and hence compact. □

Note that all eigenvalues are simple. If τ is l.p. at one endpoint, this is clear, since there is at most one solution of $(\tau - \lambda)u = 0$ which is square integrable near this endpoint. If τ is l.c., this also follows since the fact that two solutions of $(\tau - \lambda)u = 0$ satisfy the same boundary condition implies that their Wronskian vanishes.

9.2. Weyl's limit circle, limit point alternative

If τ is not l.c., the situation is more complicated and we can only say something about the essential spectrum.

Theorem 9.11. *All self-adjoint extensions of A_0 have the same essential spectrum. Moreover, if A_{ac} and A_{cb} are self-adjoint extensions of τ restricted to (a, c) and (c, b) (for any $c \in I$), then*

$$\sigma_{ess}(A) = \sigma_{ess}(A_{ac}) \cup \sigma_{ess}(A_{cb}). \tag{9.34}$$

Proof. Since $(\tau - i)u = 0$ has two linearly independent solutions, the defect indices are at most two (they are zero if τ is l.p. at both endpoints, one if τ is l.c. at one and l.p. at the other endpoint, and two if τ is l.c. at both endpoints). Hence the first claim follows from Theorem 6.20.

For the second claim, restrict τ to the functions with compact support in $(a, c) \cup (c, d)$. Then, this operator is the orthogonal sum of the operators $A_{0,ac}$ and $A_{0,cb}$. Hence the same is true for the adjoints and hence the defect indices of $A_{0,ac} \oplus A_{0,cb}$ are at most four. Now note that A and $A_{ac} \oplus A_{cb}$ are both self-adjoint extensions of this operator. Thus the second claim also follows from Theorem 6.20. $\square$

In particular, this result implies that for the essential spectrum only the behaviour near the endpoints a and b is relevant.

Another useful result to determine if q is relatively compact is the following:

Lemma 9.12. *Suppose $k \in L^2_{loc}((a,b), r\,dx)$. Then $kR_A(z)$ is Hilbert–Schmidt if and only if*

$$\|kR_A(z)\|_2^2 = \frac{1}{\operatorname{Im}(z)} \int_a^b |k(x)|^2 \operatorname{Im}(G(z,x,x)) r(x) dx \tag{9.35}$$

is finite.

Proof. From the first resolvent formula, we have

$$G(z, x, y) - G(z', x, y) = (z - z') \int_a^b G(z, x, t) G(z', t, y) r(t) dt.$$

Setting $x = y$ and $z' = z^*$, we obtain

$$\operatorname{Im}(G(z, x, x)) = \operatorname{Im}(z) \int_a^b |G(z, x, t)|^2 r(t) dt. \tag{9.36}$$

Using this last formula to compute the Hilbert–Schmidt norm proves the lemma. $\square$

Problem 9.5. *Compute the spectrum and the resolvent of $\tau = -\frac{d^2}{dx^2}$, $I = (0, \infty)$ defined on $\mathfrak{D}(A) = \{f \in \mathfrak{D}(\tau) | f(0) = 0\}$.*

Problem 9.6. *Suppose τ is given on (a, ∞), where a is a regular endpoint. Suppose there are two solutions $u_\pm$ of $\tau u = zu$ satisfying $r(x)^{1/2}|u_\pm(x)| \le Ce^{\mp \alpha x}$ for some $C, \alpha > 0$. Then z is not in the essential spectrum of any self-adjoint operator corresponding to τ. (Hint: You can take any self-adjoint extension, say the one for which $u_a = u_-$ and $u_b = u_+$. Write down what you expect the resolvent to be and show that it is a bounded operator by comparison with the resolvent from the previous problem.)*

Problem 9.7. *Suppose a is regular and $\lim_{x \to b} q(x)/r(x) = \infty$. Show that $\sigma_{ess}(A) = \emptyset$ for every self-adjoint extension. (Hint: Fix some positive constant n, choose $c \in (a,b)$ such that $q(x)/r(x) \ge n$ in (c,b), and use Theorem 9.11.)*

Problem 9.8 (Approximation by regular operators). *Fix functions $v, w \in \mathfrak{D}(\tau)$ as in Theorem 9.6. Pick $I_m = (c_m, d_m)$ with $c_m \downarrow a$, $d_m \uparrow b$ and define*

$$A_m : \mathfrak{D}(A_m) \to L^2(I_m, r\, dr),$$
$$f \mapsto \tau f$$

where

$$\mathfrak{D}(A_m) = \{f \in L^2(I_m, r\, dr)|\ f, pf' \in AC(I_m),\ \tau f \in L^2(I_m, r\, dr),$$
$$W_{c_m}(v, f) = W_{d_m}(w, f) = 0\}.$$

Then A_m converges to A in the strong resolvent sense as $m \to \infty$. (Hint: Lemma 6.36.)

Problem 9.9 (Weyl circles). *Fix $z \in \mathbb{C} \backslash \mathbb{R}$ and $c \in (a,b)$. Introduce*

$$[u]_x = \frac{W(u, u^*)_x}{z - z^*} \in \mathbb{R}$$

and use (9.4) to show that

$$[u]_x = [u]_c + \int_c^x |u(y)|^2\, r(y)\, dy, \qquad (\tau - z)u = 0.$$

Hence $[u]_x$ is increasing and exists if and only if $u \in L^2((c,b), r\, dx)$.

Let $u_{1,2}$ be two solutions of $(\tau - z)u = 0$ which satisfy $[u_1]_c = [u_2]_c = 0$ and $W(u_1, u_2) = 1$. Then, all (nonzero) solutions u of $(\tau - z)u = 0$ that satisfy $[u]_b = 0$ can be written as

$$u = u_2 + m\, u_1, \qquad m \in \mathbb{C},$$

up to a complex multiple (note $[u_1]_x > 0$ for $x > c$).

Show that

$$[u_2 + m\, u_1]_x = [u_1]_x \Big(|m - M(x)|^2 - R(x)^2\Big),$$

where

$$M(x) = -\frac{W(u_2, u_1^*)_x}{W(u_1, u_1^*)_x}$$

and
$$R(x)^2 = \left(|W(u_2, u_1^*)_x|^2 + W(u_2, u_2^*)_x W(u_1, u_1^*)_x\right)\left(|z - z^*|[u_1]_x\right)^{-2}$$
$$= \left(|z - z^*|[u_1]_x\right)^{-2}.$$

Hence the numbers m for which $[u]_x = 0$ lie on a circle which either converges to a circle (if $\lim_{x \to b} R(x) > 0$) or to a point (if $\lim_{x \to b} R(x) = 0$) as $x \to b$. Show that τ is l.c. at b in the first case and l.p. in the second case.

9.3. Spectral transformations I

In this section, we want to provide some fundamental tools for investigating the spectra of Sturm–Liouville operators and, at the same time, give some nice illustrations of the spectral theorem.

Example. Consider again $\tau = -\frac{d^2}{dx^2}$ on $I = (-\infty, \infty)$. From Section 7.3 we know that the Fourier transform maps the associated operator H_0 to the multiplication operator with p^2 in $L^2(\mathbb{R})$. To get multiplication by λ, as in the spectral theorem, we set $p = \sqrt{\lambda}$ and split the Fourier integral into a positive and negative part, that is,

$$(9.37) \qquad (Uf)(\lambda) = \begin{pmatrix} \int_\mathbb{R} e^{i\sqrt{\lambda}x} f(x)\, dx \\ \int_\mathbb{R} e^{-i\sqrt{\lambda}x} f(x)\, dx \end{pmatrix}, \qquad \lambda \in \sigma(H_0) = [0, \infty).$$

Then

$$(9.38) \qquad U : L^2(\mathbb{R}) \to \bigoplus_{j=1}^2 L^2\left(\mathbb{R}, \frac{\chi_{[0,\infty)}(\lambda)}{2\sqrt{\lambda}} d\lambda\right)$$

is the spectral transformation whose existence is guaranteed by the spectral theorem (Lemma 3.4). Note, however, that the measure is *not* finite. This can be easily fixed if we replace $\exp(\pm i\sqrt{\lambda}x)$ by $\gamma(\lambda)\exp(\pm i\sqrt{\lambda}x)$. ◇

Note that in the previous example the kernel $e^{\pm i\sqrt{\lambda}x}$ of the integral transform U is just a pair of linearly independent solutions of the underlying differential equation (though no eigenfunctions, since they are not square integrable).

More generally, if

$$(9.39) \qquad U : L^2(I, r\, dx) \to L^2(\mathbb{R}, d\mu), \qquad f(x) \mapsto \int_I u(\lambda, x) f(x) r(x)\, dx$$

is an integral transformation which maps a self-adjoint Sturm–Liouville operator A to multiplication by λ, then its kernel $u(\lambda, x)$ is a solution of the underlying differential equation. This formally follows from $UAf = \lambda Uf$ which implies

$$(9.40) \quad 0 = \int_I u(\lambda, x)(\tau - \lambda)f(x)r(x)\,dx = \int_I (\tau - \lambda)u(\lambda, x)f(x)r(x)\,dx$$

and hence $(\tau - \lambda)u(\lambda, .) = 0$.

Lemma 9.13. *Suppose*

$$(9.41) \quad U : L^2(I, r\,dx) \to \bigoplus_{j=1}^{k} L^2(\mathbb{R}, d\mu_j)$$

is a spectral mapping as in Lemma 3.4. Then U is of the form

$$(9.42) \quad (Uf)(\lambda) = \int_a^b \underline{u}(\lambda, x)f(x)r(x)\,dx,$$

where $\underline{u}(\lambda, x) = (u_1(\lambda, x), \ldots, u_k(\lambda, x))$ is measurable, and for a.e. λ (with respect to μ_j), each $u_j(\lambda, .)$ is a solution of $\tau u_j = \lambda u_j$ which satisfies the boundary conditions of A (if any). Here the integral has to be understood as $\int_a^b dx = \lim_{c \downarrow a, d \uparrow b} \int_c^d dx$ with limit taken in $\bigoplus_j L^2(\mathbb{R}, d\mu_j)$.

The inverse is given by

$$(9.43) \quad (U^{-1}\underline{F})(x) = \sum_{j=1}^{k} \int_{\mathbb{R}} u_j(\lambda, x)^* F_j(\lambda)d\mu_j(\lambda).$$

Again, the integrals have to be understood as $\int_{\mathbb{R}} d\mu_j = \lim_{R \to \infty} \int_{-R}^{R} d\mu_j$ with limits taken in $L^2(I, r\,dx)$.

If the spectral measures are ordered, then the solutions $u_j(\lambda)$, $1 \le j \le l$, are linearly independent for a.e. λ with respect to μ_l. In particular, for ordered spectral measures, we always have $k \le 2$ and even $k = 1$ if τ is l.c. at one endpoint.

Proof. Using $U_j R_A(z) = \frac{1}{\lambda - z} U_j$, we have

$$U_j f(x) = (\lambda - z) U_j \int_a^b G(z, x, y)f(y)r(y)\,dy.$$

If we restrict $R_A(z)$ to a compact interval $[c, d] \subset (a, b)$, then $R_A(z)\chi_{[c,d]}$ is Hilbert–Schmidt since $G(z, x, y)\chi_{[c,d]}(y)$ is square integrable over $(a, b) \times (a, b)$. Hence $U_j \chi_{[c,d]} = (\lambda - z) U_j R_A(z) \chi_{[c,d]}$ is Hilbert–Schmidt as well and by Lemma 6.10 there is a corresponding kernel $u_j^{[c,d]}(\lambda, y)$ such that

$$(U_j \chi_{[c,d]} f)(\lambda) = \int_a^b u_j^{[c,d]}(\lambda, x)f(x)r(x)\,dx.$$

9.3. Spectral transformations I

Now take a larger compact interval $[\hat{c}, \hat{d}] \supseteq [c, d]$. Then the kernels coincide on $[c, d]$, $u_j^{[c,d]}(\lambda, .) = u_j^{[\hat{c},\hat{d}]}(\lambda, .)\chi_{[c,d]}$, since we have $U_j\chi_{[c,d]} = U_j\chi_{[\hat{c},\hat{d}]}\chi_{[c,d]}$. In particular, there is a kernel $u_j(\lambda, x)$ such that

$$U_j f(x) = \int_a^b u_j(\lambda, x) f(x) r(x) \, dx$$

for every f with compact support in (a, b). Since functions with compact support are dense and U_j is continuous, this formula holds for every f provided the integral is understood as the corresponding limit.

Using the fact that U is unitary, $\langle \underline{F}, Ug \rangle = \langle U^{-1}\underline{F}, g \rangle$, we see that

$$\sum_j \int_{\mathbb{R}} F_j(\lambda)^* \int_a^b u_j(\lambda, x) g(x) r(x) \, dx = \int_a^b (U^{-1}\underline{F})(x)^* g(x) r(x) \, dx.$$

Interchanging integrals on the right-hand side (which is permitted at least for g, $\underline{F}$ with compact support), the formula for the inverse follows.

Next, from $U_j A f = \lambda U_j f$, we have

$$\int_a^b u_j(\lambda, x)(\tau f)(x) r(x) \, dx = \lambda \int_a^b u_j(\lambda, x) f(x) r(x) \, dx$$

for a.e. λ and every $f \in \mathfrak{D}(A_0)$. Restricting everything to $[c, d] \subset (a, b)$, the above equation implies $u_j(\lambda, .)|_{[c,d]} \in \mathfrak{D}(A^*_{cd,0})$ and $A^*_{cd,0} u_j(\lambda, .)|_{[c,d]} = \lambda u_j(\lambda, .)|_{[c,d]}$. In particular, $u_j(\lambda, .)$ is a solution of $\tau u_j = \lambda u_j$. Moreover, if τ is l.c. near a, we can choose $c = a$ and allow all $f \in \mathfrak{D}(\tau)$ which satisfy the boundary condition at a and vanish identically near b.

Finally, assume the μ_j are ordered and fix $l \leq k$. Suppose

$$\sum_{j=1}^l c_j(\lambda) u_j(\lambda, x) = 0.$$

Then we have

$$\sum_{j=1}^l c_j(\lambda) F_j(\lambda) = 0, \qquad F_j = U_j f,$$

for every f. Since U is surjective, we can prescribe F_j arbitrarily on $\sigma(\mu_l)$, e.g., $F_j(\lambda) = 1$ for $j = j_0$ and $F_j(\lambda) = 0$ otherwise, which shows $c_{j_0}(\lambda) = 0$. Hence the solutions $u_j(\lambda, x)$, $1 \leq j \leq l$, are linearly independent for $\lambda \in \sigma(\mu_l)$, which shows that $k \leq 2$ since there are at most two linearly independent solutions. If τ is l.c. and $u_j(\lambda, x)$ must satisfy the boundary condition, there is only one linearly independent solution and thus $k = 1$. □

Note that since we can replace $u_j(\lambda, x)$ by $\gamma_j(\lambda) u_j(\lambda, x)$ where $|\gamma_j(\lambda)| = 1$, it is no restriction to assume that $u_j(\lambda, x)$ is real-valued.

For simplicity, we will only pursue the case where one endpoint, say a, is regular. The general case can often be reduced to this case and will be postponed until Section 9.6.

We choose a boundary condition

$$\cos(\alpha)f(a) - \sin(\alpha)p(a)f'(a) = 0 \tag{9.44}$$

and introduce two solutions $s(z, x)$ and $c(z, x)$ of $\tau u = zu$ satisfying the initial conditions

$$\begin{aligned} s(z, a) &= \sin(\alpha), & p(a)s'(z, a) &= \cos(\alpha), \\ c(z, a) &= \cos(\alpha), & p(a)c'(z, a) &= -\sin(\alpha). \end{aligned} \tag{9.45}$$

Note that $s(z, x)$ is the solution which satisfies the boundary condition at a; that is, we can choose $u_a(z, x) = s(z, x)$. In fact, if τ is not regular at a but only l.c., everything below remains valid if one chooses $s(z, x)$ to be a solution satisfying the boundary condition at a and $c(z, x)$ to be a linearly independent solution with $W(c(z), s(z)) = 1$.

Moreover, in our previous lemma we have $u_1(\lambda, x) = \gamma_a(\lambda)s(\lambda, x)$, and using the rescaling $d\mu(\lambda) = |\gamma_a(\lambda)|^2 d\mu_a(\lambda)$ and $(U_1 f)(\lambda) = \gamma_a(\lambda)(Uf)(\lambda)$, we obtain a unitary map

$$U : L^2(I, r\, dx) \to L^2(\mathbb{R}, d\mu), \qquad (Uf)(\lambda) = \int_a^b s(\lambda, x)f(x)r(x)dx \tag{9.46}$$

with inverse

$$(U^{-1}F)(x) = \int_\mathbb{R} s(\lambda, x)F(\lambda)d\mu(\lambda). \tag{9.47}$$

Note, however, that while this rescaling gets rid of the unknown factor $\gamma_a(\lambda)$, it destroys the normalization of the measure μ. For μ_1 we know $\mu_1(\mathbb{R})$ (if the corresponding vector is normalized), but μ might not even be bounded! In fact, it turns out that μ is indeed unbounded.

So up to this point we have our spectral transformation U which maps A to multiplication by λ, but we know nothing about the measure μ. Furthermore, the measure μ is the object of desire since it contains all the spectral information of A. So our next aim must be to compute μ. If A has only pure point spectrum (i.e., only eigenvalues), this is straightforward as the following example shows.

Example. Suppose $E \in \sigma_p(A)$ is an eigenvalue. Then $s(E, x)$ is the corresponding eigenfunction and the same is true for $S_E(\lambda) = (Us(E))(\lambda)$. In particular, $\chi_{\{E\}}(A)s(E, x) = s(E, x)$ shows $S_E(\lambda) = (U\chi_{\{E\}}(A)s(E))(\lambda) = \chi_{\{E\}}(\lambda)S_E(\lambda)$; that is,

$$S_E(\lambda) = \begin{cases} \|s(E)\|^2, & \lambda = E, \\ 0, & \lambda \neq E. \end{cases} \tag{9.48}$$

9.3. Spectral transformations I

Moreover, since U is unitary, we have

$$(9.49) \quad \|s(E)\|^2 = \int_a^b s(E,x)^2 r(x)dx = \int_\mathbb{R} S_E(\lambda)^2 d\mu(\lambda) = \|s(E)\|^4 \mu(\{E\});$$

that is, $\mu(\{E\}) = \|s(E)\|^{-2}$. In particular, if A has pure point spectrum (e.g., if τ is limit circle at both endpoints), we have

$$(9.50) \quad d\mu(\lambda) = \sum_{j=1}^\infty \frac{1}{\|s(E_j)\|^2} d\Theta(\lambda - E_j), \quad \sigma_p(A) = \{E_j\}_{j=1}^\infty,$$

where $d\Theta$ is the Dirac measure centered at 0. For arbitrary A, the above formula holds at least for the pure point part μ_{pp}. ◇

In the general case, we have to work a bit harder. Since $c(z,x)$ and $s(z,x)$ are linearly independent solutions,

$$(9.51) \quad W(c(z), s(z)) = 1,$$

we can write $u_b(z, x) = \gamma_b(z)(c(z,x) + m_b(z)s(z,x))$, where

$$(9.52) \quad m_b(z) = -\frac{W(c(z), u_b(z))}{W(s(z), u_b(z))} = \frac{\cos(\alpha)p(a)u_b'(z,a) + \sin(\alpha)u_b(z,a)}{\cos(\alpha)u_b(z,a) - \sin(\alpha)p(a)u_b'(z,a)}, \quad z \in \rho(A),$$

is known as the **Weyl–Titchmarsh m-function**. Note that $m_b(z)$ is holomorphic in $\rho(A)$ and that

$$(9.53) \quad m_b(z)^* = m_b(z^*)$$

since the same is true for $u_b(z, x)$ (the denominator in (9.52) only vanishes if $u_b(z, x)$ satisfies the boundary condition at a, that is, if z is an eigenvalue). Moreover, the constant $\gamma_b(z)$ is of no importance and can be chosen equal to one,

$$(9.54) \quad u_b(z,x) = c(z,x) + m_b(z)s(z,x).$$

Lemma 9.14. *The Weyl m-function satisfies*

$$(9.55) \quad \frac{m_b(z) - m_b(\hat{z})}{z - \hat{z}} = \int_a^b u_b(z,y)u_b(\hat{z}, y)r(y)\,dy,$$

where $u_b(z,x)$ is normalized as in (9.54). In particular, for $\hat{z} = z^*$ we see that it is a Herglotz–Nevanlinna function satisfying

$$(9.56) \quad \mathrm{Im}(m_b(z)) = \mathrm{Im}(z)\int_a^b |u_b(z,x)|^2 r(x)\,dx.$$

Proof. Given two solutions $u(x)$, $v(x)$ of $\tau u = zu$, $\tau v = \hat{z}v$, respectively, it follows from the Lagrange identity (9.4) that

$$(z - \hat{z})\int_a^x u(y)v(y)r(y)\,dy = W_x(u,v) - W_a(u,v).$$

Now choose $u(x) = u_b(z, x)$ and $v(x) = u_b(\hat{z}, x)$,

$$(z - \hat{z}) \int_a^x u_b(z, y) u_b(\hat{z}, y) r(y) \, dy = W_x(u_b(z), u_b(\hat{z})) + (m_b(z) - m_b(\hat{z})),$$

and observe that $W_x(u_b(z), u_b(\hat{z}))$ vanishes as $x \uparrow b$, since both $u_b(z)$ and $u_b(\hat{z})$ are in $\mathfrak{D}(\tau)$ near b. $\square$

Lemma 9.15. *Let*

(9.57) $$G(z, x, y) = \begin{cases} s(z, x) u_b(z, y), & y \geq x, \\ s(z, y) u_b(z, x), & y \leq x, \end{cases}$$

be the Green function of A. Then
(9.58)
$$(UG(z, x, .))(\lambda) = \frac{s(\lambda, x)}{\lambda - z} \quad \text{and} \quad (Up(x)\partial_x G(z, x, .))(\lambda) = \frac{p(x) s'(\lambda, x)}{\lambda - z}$$

for every $x \in (a, b)$ and every $z \in \rho(A)$.

Proof. First of all, note that $G(z, x, .) \in L^2((a, b), r \, dx)$ for every $x \in (a, b)$ and $z \subset \rho(A)$. Moreover, from $R_A(z) f = U^{-1} \frac{1}{\lambda - z} Uf$, we have

(9.59) $$\int_a^b G(z, x, y) f(y) r(y) \, dy = \int_{\mathbb{R}} \frac{s(\lambda, x) F(\lambda)}{\lambda - z} d\mu(\lambda),$$

where $F = Uf$. Here equality is to be understood in L^2, that is, for a.e. x. However, the left-hand side is continuous with respect to x and so is the right-hand side, at least if F has compact support. Since this set is dense, the first equality follows. Similarly, the second follows after differentiating (9.59) with respect to x. $\square$

Corollary 9.16. *We have*

(9.60) $$(Uu_b(z))(\lambda) = \frac{1}{\lambda - z},$$

where $u_b(z, x)$ is normalized as in (9.54).

Proof. Choosing $x = a$ in the lemma, we obtain the claim from the first identity if $\sin(\alpha) \neq 0$ and from the second if $\cos(\alpha) \neq 0$. $\square$

Now combining Lemma 9.14 and Corollary 9.16, we infer from unitarity of U that

(9.61) $$\operatorname{Im}(m_b(z)) = \operatorname{Im}(z) \int_a^b |u_b(z, x)|^2 r(x) \, dx = \operatorname{Im}(z) \int_{\mathbb{R}} \frac{1}{|\lambda - z|^2} d\mu(\lambda)$$

and since a holomorphic function is determined up to a real constant by its imaginary part, we obtain

9.3. Spectral transformations I

Theorem 9.17. *The Weyl m-function is given by*

$$(9.62) \qquad m_b(z) = d + \int_{\mathbb{R}} \left(\frac{1}{\lambda - z} - \frac{\lambda}{1 + \lambda^2} \right) d\mu(\lambda), \quad d \in \mathbb{R},$$

and

$$(9.63) \qquad d = \operatorname{Re}(m_b(i)), \qquad \int_{\mathbb{R}} \frac{1}{1 + \lambda^2} d\mu(\lambda) = \operatorname{Im}(m_b(i)) < \infty.$$

Moreover, μ is given by the Stieltjes inversion formula

$$(9.64) \qquad \mu(\lambda) = \lim_{\delta \downarrow 0} \lim_{\varepsilon \downarrow 0} \frac{1}{\pi} \int_{\delta}^{\lambda + \delta} \operatorname{Im}(m_b(t + i\varepsilon)) dt,$$

where

$$(9.65) \qquad \operatorname{Im}(m_b(\lambda + i\varepsilon)) = \varepsilon \int_{a}^{b} |u_b(\lambda + i\varepsilon, x)|^2 r(x) \, dx.$$

Proof. Choosing $z = i$ in (9.61) shows (9.63) and hence the right-hand side of (9.62) is a well-defined holomorphic function in $\mathbb{C}\backslash\mathbb{R}$. By

$$\operatorname{Im}(\frac{1}{\lambda - z} - \frac{\lambda}{1 + \lambda^2}) = \frac{\operatorname{Im}(z)}{|\lambda - z|^2},$$

its imaginary part coincides with that of $m_b(z)$ and hence equality follows. The Stieltjes inversion formula follows from Theorem 3.23. $\square$

Example. Consider $\tau = -\frac{d^2}{dx^2}$ on $I = (0, \infty)$. Then

$$(9.66) \qquad c(z, x) = \cos(\alpha) \cos(\sqrt{z}x) - \frac{\sin(\alpha)}{\sqrt{z}} \sin(\sqrt{z}x)$$

and

$$(9.67) \qquad s(z, x) = \sin(\alpha) \cos(\sqrt{z}x) + \frac{\cos(\alpha)}{\sqrt{z}} \sin(\sqrt{z}x).$$

Moreover,

$$(9.68) \qquad u_b(z, x) = u_b(z, 0) e^{-\sqrt{-z}x}$$

and thus

$$(9.69) \qquad m_b(z) = \frac{\sin(\alpha) - \sqrt{-z} \cos(\alpha)}{\cos(\alpha) + \sqrt{-z} \sin(\alpha)},$$

respectively,

$$(9.70) \qquad d\mu(\lambda) = \frac{\sqrt{\lambda}}{\pi(\cos(\alpha)^2 + \lambda \sin(\alpha)^2)} d\lambda.$$

Note that if $\alpha \ne 0$, we even have $\int \frac{1}{|\lambda-z|} d\mu(\lambda) < 0$ in the previous example and hence

$$(9.71) \qquad m_b(z) = -\cot(\alpha) + \int_{\mathbb{R}} \frac{1}{\lambda - z} d\mu(\lambda)$$

in this case (the factor $-\cot(\alpha)$ follows by considering the limit $|z| \to \infty$ of both sides). Formally, this even follows in the general case by choosing $x = a$ in $u_b(z,x) = (U^{-1}\frac{1}{\lambda-z})(x)$; however, since we know equality only for a.e. x, a more careful analysis is needed. We will address this problem in the next section.

Problem 9.10. *Show*

$$(9.72) \qquad m_{b,\alpha}(z) = \frac{\cos(\alpha-\beta)m_{b,\beta}(z) + \sin(\alpha-\beta)}{\cos(\alpha-\beta) - \sin(\alpha-\beta)m_{b,\beta}(z)}.$$

(Hint: The case $\beta = 0$ is (9.52).)

Problem 9.11. *Suppose τ is l.c. at a. Let $\phi_0(x)$, $\theta_0(x)$ be two real-valued solutions of $\tau u = \lambda_0 u$ for some fixed $\lambda_0 \in \mathbb{R}$ such that $W(\theta_0, \phi_0) = 1$. Show that the limits*

$$(9.73) \qquad \lim_{x \to a} W_x(\phi_0, u(z)), \qquad \lim_{x \to a} W_x(\theta_0, u(z))$$

exist for every solution $u(z)$ of $\tau u = zu$.

Introduce

$$(9.74) \qquad \begin{aligned} \phi(z,x) &= W_a(c(z), \phi_0) s(z,x) - W_a(s(z), \phi_0) c(z,x), \\ \theta(z,x) &= W_a(c(z), \theta_0) s(z,x) - W_a(s(z), \theta_0) c(z,x), \end{aligned}$$

where $c(z,x)$ and $s(z,x)$ are chosen with respect to some base point $c \in (a,b)$ and a singular Weyl m-function $M_b(z)$ such that

$$(9.75) \qquad \psi(z,x) = \theta(z,x) + M_b(z)\phi(z,x) \in L^2(c,b).$$

Show that all claims from this section still hold true in this case for the operator associated with the boundary condition $W_a(\phi_0, f) = 0$. (Hint: Plücker's identity.)

9.4. Inverse spectral theory

In this section, we want to show that the Weyl m-function (respectively, the corresponding spectral measure) uniquely determines the operator. For simplicity, we only consider the case $p = r \equiv 1$.

We begin with some asymptotics for large z away from the spectrum. We recall that $\sqrt{z}$ always denotes the branch with $\arg(z) \in (-\pi, \pi]$. In particular, $\operatorname{Re}(\sqrt{-z}) > 0$ for all $z \in \mathbb{C}\backslash[0, \infty)$ and $\operatorname{Re}(\sqrt{-z}) = 0$ for all

9.4. Inverse spectral theory

$z \in [0, \infty)$. We will write $c(z,x) = c_\alpha(z,x)$ and $s(z,x) = s_\alpha(z,x)$ to display the dependence on α whenever necessary. Clearly

$$s_\alpha(z,x) = \sin(\alpha)c_0(z,x) + \cos(\alpha)s_0(z,x),$$
(9.76)
$$c_\alpha(z,x) = \cos(\alpha)c_0(z,x) - \sin(\alpha)s_0(z,x).$$

We first observe (Problem 9.12)

Lemma 9.18. *For $\alpha = 0$, we have*

(9.77)
$$c_0(z,x) = \cosh(\sqrt{-z}(x-a)) + O(\frac{1}{\sqrt{-z}}e^{\sqrt{-z}(x-a)}),$$
$$s_0(z,x) = \frac{1}{\sqrt{-z}}\sinh(\sqrt{-z}(x-a)) + O(\frac{1}{z}e^{\sqrt{-z}(x-a)}),$$

uniformly for $x \in (a,c)$ as $|z| \to \infty$.

Note that for $z \in \mathbb{C}\backslash[0,\infty)$, this can be written as

(9.78)
$$c_0(z,x) = \frac{1}{2}e^{\sqrt{-z}(x-a)}\left(1 + O(\frac{1}{\sqrt{-z}})\right),$$
$$s_0(z,x) = \frac{1}{2\sqrt{-z}}e^{\sqrt{-z}(x-a)}\left(1 + O(\frac{1}{\sqrt{-z}})\right),$$

for $\text{Im}(z) \to \infty$, and for $z = \lambda \in [0,\infty)$, we have

(9.79)
$$c_0(\lambda, x) = \cos(\sqrt{\lambda}(x-a)) + O(\frac{1}{\sqrt{\lambda}}),$$
$$s_0(\lambda, x) = \frac{1}{\sqrt{\lambda}}\sin(\sqrt{\lambda}(x-a)) + O(\frac{1}{\lambda}),$$

as $\lambda \to \infty$.

From this lemma we obtain

Lemma 9.19. *The Weyl m-function satisfies*

(9.80)
$$m_b(z) = \begin{cases} -\cot(\alpha) + \frac{1}{\sin(\alpha)^2\sqrt{-z}} + O(\frac{1}{z}), & \alpha \neq 0, \\ -\sqrt{-z} + O(1), & \alpha = 0, \end{cases}$$

as $z \to \infty$ in any sector $\varepsilon|\text{Re}(z)| \leq |\text{Im}(z)|$, $\varepsilon > 0$.

Proof. Evaluating (9.36) using Lemma 9.15 (as in (9.61)), we conclude as in the proof of Theorem 9.17 that

$$G(z,x,x) = d(x) + \int_\mathbb{R} \left(\frac{1}{\lambda - z} - \frac{\lambda}{1+\lambda^2}\right) s(\lambda, x)^2 d\mu(\lambda).$$

Hence, Problem 3.24 implies $G(z,x,x) = o(z)$ as $z \to \infty$ in any sector $\varepsilon|\text{Re}(z)| \leq |\text{Im}(z)|$. Now solving $G(z,x,x) = s(z,x)u_b(z,x)$ for $m_b(z)$ and

using the asymptotic expansions from Lemma 9.18, we see
$$m_b(z) = -\frac{c(z,x)}{s(z,x)} + o(z^2 e^{-2\sqrt{-z}(x-a)})$$
from which the case $\alpha = 0$ and the leading term in the case $\alpha \neq 0$ follows. The next term in the case $\alpha \neq 0$ follows using (9.72) with $\beta = 0$. $\square$

Note that assuming $q \in C^k([a,b))$, one can obtain further asymptotic terms in Lemma 9.18 and hence also in the expansion of $m_b(z)$.

Corollary 9.20. *The following asymptotics are valid:*

(9.81) $$G(z,x,y) = \frac{e^{-\sqrt{-z}|y-x|}}{2\sqrt{-z}}\left(1 + O(z^{-1/2})\right),$$

(9.82) $$u_b(z,x) = e^{-\sqrt{-z}(x-a)}\begin{cases}\cos(\alpha) + O(\frac{1}{\sqrt{-z}}), & \alpha \neq \frac{\pi}{2},\\ \frac{1}{\sqrt{-z}} + O(\frac{1}{z}), & \alpha = \frac{\pi}{2},\end{cases}$$

(9.83) $$m_b(z) = -\frac{c(z,x)}{s(z,x)} + O(z^{-1/2}e^{-2\sqrt{-z}(x-a)}),$$

as $z \to \infty$ in any sector $\varepsilon|\mathrm{Re}(z)| \leq |\mathrm{Im}(z)|$, $\varepsilon > 0$.

Proof. First of all, note that we can write
$$G(z,x,x) = \frac{u_a(z,x)u_b(z,x)}{W(u_a(z), u_b(z))} = \frac{-1}{m_a(z,x) + m_b(z,x)},$$
where
$$m_a(z,x) = -\frac{u_a'(z,x)}{u_a(z,x)} \quad \text{and} \quad m_b(z,x) = \frac{u_b'(z,x)}{u_b(z,x)}$$
are the Weyl m-functions associated with H restricted to (a,x) and (x,b) with a Dirichlet boundary condition at x, respectively. Hence (9.80) implies (9.81) in the case $y = x$. Furthermore, the asymptotics for $u_b(z,x)$ follow from $u_b(z,x) = G(z,x,x)/s(z,x)$ which in turn implies the $x \neq y$ case since $G(z,x,y) = G(z,x,y)u_b(z,y)/u_b(z,x)$, $y \geq x$. Finally, (9.83) follows after solving (9.81) for $m_b(z)$. $\square$

Now we come to our main result of this section:

Theorem 9.21. *Suppose τ_j, $j = 0,1$, are given on (a,b) and both are regular at a. Moreover, A_j are some self-adjoint operators associated with τ_j and the same boundary condition at a.*

Let $c \in (a,b)$. Then $q_0(x) = q_1(x)$ for $x \in (a,c)$ if and only if $m_{1,b}(z) - m_{0,b}(z) = o(e^{-2(c-a)\,\mathrm{Re}(\sqrt{-z})})$ as $z \to \infty$ along some nonreal ray.

9.4. Inverse spectral theory

Proof. The direct claim is immediate from (9.83). To see the converse, first note that by (9.77) we have $s_1(z,x)/s_0(z,x) \to 1$ as $z \to \infty$ along any nonreal ray. Moreover, (9.81) in the case $y = x$ shows $s_0(z,x)u_{1,b}(z,x) \to 0$ and $s_1(z,x)u_{0,b}(z,x) \to 0$ as well. Hence the entire function

$$s_1(z,x)c_0(z,x) - s_0(z,x)c_1(z,x) = s_1(z,x)u_{0,b}(z,x) - s_0(z,x)u_{1,b}(z,x)$$
$$+ (m_{1,b}(z) - m_{0,b}(z))s_0(z,x)s_1(z,x)$$

vanishes as $z \to \infty$ along any nonreal ray for fixed $x \in (a,c)$ by the assumption on $m_{1,b}(z) - m_{0,b}(z)$. Moreover, by (9.77), this function has an order of growth $\le 1/2$ and thus by the Phragmén–Lindelöf theorem (e.g., [62, Thm. 4.3.4]) is bounded on all of $\mathbb{C}$. By Liouville's theorem it must be constant and since it vanishes along rays, it must be zero; that is, $s_1(z,x)c_0(z,x) = s_0(z,x)c_1(z,x)$ for all $z \in \mathbb{C}$ and $x \in (a,c)$. Differentiating this identity with respect to x and using $W(c_j(z), s_j(z)) = 1$ shows $s_1(z,x)^2 = s_0(z,x)^2$. Taking the logarithmic derivative further gives $s_1'(z,x)/s_1(z,x) = s_0'(z,x)/s_0(z,x)$ and differentiating once more shows $s_1''(z,x)/s_1(z,x) = s_0''(z,x)/s_0(z,x)$. This finishes the proof since $q_j(x) = z + s_j''(z,x)/s_j(z,x)$. □

By virtue of Lemma 3.30, the asymptotics of $m_b(z)$ in turn tell us more about $L^2(\mathbb{R}, d\mu)$. For example, using (9.80) and (9.79), we obtain

$$(9.84) \qquad \int s(\lambda, x)^2 (1+|\lambda|)^{-\gamma} d\mu(\lambda) < \infty \quad \text{for} \quad \gamma > \frac{1}{2}.$$

As a consequence we obtain:

Lemma 9.22. *We have*

$$(9.85) \qquad G(z,x,y) = \int_{\mathbb{R}} \frac{s(\lambda, x)s(\lambda, y)}{\lambda - z} d\mu(\lambda),$$

where the integrand is integrable.

Proof. By Lemma 9.15, this formula holds for given fixed x in the sense of $L^2((a,b), r\, dx)$ with respect to y. Since our above observation implies integrability for all y (uniformly for $x, y \in (a,c)$), both sides are continuous and the claim follows. □

Problem 9.12. *Prove Lemma 9.18. (Hint: Without loss, set $a = 0$. Now use that*

$$c(z,x) = \cos(\alpha)\cosh(\sqrt{-z}x) - \frac{\sin(\alpha)}{\sqrt{-z}}\sinh(\sqrt{-z}x)$$
$$+ \frac{1}{\sqrt{-z}}\int_0^x \sinh(\sqrt{-z}(x-y))q(y)c(z,y)dy$$

by Lemma 9.2 and consider $\tilde{c}(z,x) = e^{-\sqrt{-z}x}c(z,x)$.)

9.5. Absolutely continuous spectrum

In this section, we will show how to locate the absolutely continuous spectrum. We will again assume that a is a regular endpoint. Moreover, we assume that b is l.p. since otherwise the spectrum is discrete and there will be no absolutely continuous spectrum.

In this case, we have seen in Section 9.3 that A is unitarily equivalent to multiplication by λ in the space $L^2(\mathbb{R}, d\mu)$, where μ is the measure associated to the Weyl m-function. Hence by Theorem 3.27 we conclude that the set

$$(9.86) \qquad M_s = \{\lambda | \limsup_{\varepsilon \downarrow 0} \operatorname{Im}(m_b(\lambda + i\varepsilon)) = \infty\}$$

is a support for the singularly continuous part and

$$(9.87) \qquad M_{ac} = \{\lambda | 0 < \limsup_{\varepsilon \downarrow 0} \operatorname{Im}(m_b(\lambda + i\varepsilon)) < \infty\}$$

is a minimal support for the absolutely continuous part. Moreover, $\sigma_{ac}(A)$ can be recovered from the essential closure of M_{ac}; that is,

$$(9.88) \qquad \sigma_{ac}(A) = \overline{M_{ac}}^{ess}.$$

Compare also Section 3.2.

We now begin our investigation with a crucial estimate on $\operatorname{Im}(m_b(\lambda+i\varepsilon))$. Set

$$(9.89) \qquad \|f\|_{(a,x)} = \sqrt{\int_a^x |f(y)|^2 r(y) dy}, \qquad x \in (a,b).$$

Lemma 9.23. *Let*

$$(9.90) \qquad \varepsilon = (2\|s(\lambda)\|_{(a,x)} \|c(\lambda)\|_{(a,x)})^{-1}$$

and note that since b is l.p., there is a one-to-one correspondence between $\varepsilon \in (0, \infty)$ and $x \in (a, b)$. Then

$$(9.91) \qquad 5 - \sqrt{24} \leq |m_b(\lambda+i\varepsilon)| \frac{\|s(\lambda)\|_{(a,x)}}{\|c(\lambda)\|_{(a,x)}} \leq 5 + \sqrt{24}.$$

Proof. Let $x > a$. Then by Lemma 9.2,

$$u_b(\lambda + i\varepsilon, x) = c(\lambda, x) - m_b(\lambda + i\varepsilon)s(\lambda, x)$$
$$- i\varepsilon \int_a^x \big(c(\lambda, x)s(\lambda, y) - c(\lambda, y)s(\lambda, x)\big) u_b(\lambda + i\varepsilon, y) r(y) dy.$$

Hence one obtains after a little calculation (as in the proof of Theorem 9.9)

$$\|c(\lambda) - m_b(\lambda + i\varepsilon)s(\lambda)\|_{(a,x)} \leq \|u_b(\lambda + i\varepsilon)\|_{(a,x)}$$
$$+ 2\varepsilon \|s(\lambda)\|_{(a,x)} \|c(\lambda)\|_{(a,x)} \|u_b(\lambda + i\varepsilon)\|_{(a,x)}.$$

9.5. Absolutely continuous spectrum

Using the definition of ε and (9.56), we obtain

$$\|c(\lambda) - m_b(\lambda + \mathrm{i}\varepsilon)s(\lambda)\|_{(a,x)}^2 \leq 4\|u_b(\lambda + \mathrm{i}\varepsilon)\|_{(a,x)}^2$$
$$\leq 4\|u_b(\lambda + \mathrm{i}\varepsilon)\|_{(a,b)}^2 = \frac{4}{\varepsilon}\mathrm{Im}(m_b(\lambda + \mathrm{i}\varepsilon))$$
$$\leq 8\|s(\lambda)\|_{(a,x)}\|c(\lambda)\|_{(a,x)}\mathrm{Im}(m_b(\lambda + \mathrm{i}\varepsilon)).$$

Combining this estimate with

$$\|c(\lambda) - m_b(\lambda + \mathrm{i}\varepsilon)s(\lambda)\|_{(a,x)}^2 \geq \left(\|c(\lambda)\|_{(a,x)} - |m_b(\lambda + \mathrm{i}\varepsilon)|\|s(\lambda)\|_{(a,x)}\right)^2$$

shows $(1-t)^2 \leq 8t$, where $t = |m_b(\lambda + \mathrm{i}\varepsilon)|\|s(\lambda)\|_{(a,x)}\|c(\lambda)\|_{(a,x)}^{-1}$. □

We now introduce the concept of **subordinacy**. A nonzero solution u of $\tau u = zu$ is called **sequentially subordinate** at b with respect to another solution v if

$$(9.92) \qquad \liminf_{x \to b} \frac{\|u\|_{(a,x)}}{\|v\|_{(a,x)}} = 0.$$

If the lim inf can be replaced by a lim, the solution is called **subordinate**. Both concepts will eventually lead to the same results (cf. Remark 9.26 below). We will work with (9.92) since this will simplify proofs later on and hence we will drop the additional *sequentially*.

It is easy to see that if u is subordinate with respect to v, then it is subordinate with respect to any linearly independent solution. In particular, a subordinate solution is unique up to a constant. Moreover, if a solution u of $\tau u = \lambda u$, $\lambda \in \mathbb{R}$, is subordinate, then it is real up to a constant, since both the real and the imaginary parts are subordinate. For $z \in \mathbb{C}\backslash\mathbb{R}$ we know that there is always a subordinate solution near b, namely $u_b(z,x)$. The following result considers the case $z \in \mathbb{R}$.

Lemma 9.24. *Let $\lambda \in \mathbb{R}$. There is a subordinate solution $u(\lambda)$ at b if and only if there is a sequence $\varepsilon_n \downarrow 0$ such that $m_b(\lambda + \mathrm{i}\varepsilon_n)$ converges to a limit in $\mathbb{R} \cup \{\infty\}$ as $n \to \infty$. Moreover,*

$$(9.93) \qquad \lim_{n \to \infty} m_b(\lambda + \mathrm{i}\varepsilon_n) = \frac{\cos(\alpha)p(a)u'(\lambda,a) + \sin(\alpha)u(\lambda,a)}{\cos(\alpha)u(\lambda,a) - \sin(\alpha)p(a)u'(\lambda,a)}$$

in this case (compare (9.52)).

Proof. We will consider the number α fixing the boundary condition as a parameter and write $s_\alpha(z,x)$, $c_\alpha(z,x)$, $m_{b,\alpha}$, etc., to emphasize the dependence on α.

Every solution can (up to a constant) be written as $s_\beta(\lambda, x)$ for some $\beta \in [0, \pi)$. But by Lemma 9.23, $s_\beta(\lambda, x)$ is subordinate if and only there is a sequence $\varepsilon_n \downarrow 0$ such that $\lim_{n\to\infty} m_{b,\beta}(\lambda + i\varepsilon_n) = \infty$, and by (9.72), this is the case if and only if

$$\lim_{n\to\infty} m_{b,\alpha}(\lambda + i\varepsilon_n) = \lim_{n\to\infty} \frac{\cos(\alpha - \beta)m_{b,\beta}(\lambda + i\varepsilon_n) + \sin(\alpha - \beta)}{\cos(\alpha - \beta) - \sin(\alpha - \beta)m_{b,\beta}(\lambda + i\varepsilon_n)} = \cot(\alpha - \beta)$$

is a number in $\mathbb{R} \cup \{\infty\}$. $\square$

We are interested in $N(\tau)$, the set of all $\lambda \in \mathbb{R}$ for which no subordinate solution exists, that is,

(9.94) $\quad N(\tau) = \{\lambda \in \mathbb{R} | \text{No solution of } \tau u = \lambda u \text{ is subordinate at } b\}$

and the set

(9.95) $\quad S_\alpha(\tau) = \{\lambda | s_\alpha(\lambda, x) \text{ is subordinate at } b\}.$

From the previous lemma we obtain

Corollary 9.25. *We have $\lambda \in N(\tau)$ if and only if*

$$\liminf_{\varepsilon \downarrow 0} \operatorname{Im}(m_b(\lambda + i\varepsilon)) > 0 \quad \text{and} \quad \limsup_{\varepsilon \downarrow 0} |m_b(\lambda + i\varepsilon)| < \infty.$$

Similarly, $\lambda \in S_\alpha(\tau)$ if and only if $\limsup_{\varepsilon \downarrow 0} |m_b(\lambda + i\varepsilon)| = \infty$.

Remark 9.26. *Since the set, for which the limit $\lim_{\varepsilon \downarrow 0} m_b(\lambda + i\varepsilon)$ does not exist, is of zero spectral and Lebesgue measure (Corollary 3.29), changing the $\liminf$ in (9.92) to a $\lim$ will affect $N(\tau)$ only on such a set (which is irrelevant for our purpose). Moreover, by (9.72), the set where the limit exists (finitely or infinitely) is independent of the boundary condition α.*

Then, as a consequence of the previous corollary, we have

Theorem 9.27. *The set $N(\tau) \subseteq M_{ac}$ is a minimal support for the absolutely continuous spectrum of H. In particular,*

(9.96) $\quad \sigma_{ac}(H) = \overline{N(\tau)}^{ess}.$

Moreover, the set $S_\alpha(\tau) \supseteq M_s$ is a minimal support for the singular spectrum of H.

Proof. By our corollary we have $N(\tau) \subseteq M_{ac}$. Moreover, if $\lambda \in M_{ac} \backslash N(\tau)$, then either $0 = \liminf \operatorname{Im}(m_b) < \limsup \operatorname{Im}(m_b)$ or $\limsup \operatorname{Re}(m_b) = \infty$. The first case can only happen on a set of Lebesgue measure zero by Theorem 3.27, and the same is true for the second by Corollary 3.29.

Similarly, by our corollary we also have $S_\alpha(\tau) \supseteq M_s$, and $\lambda \in S_\alpha(\tau) \backslash M_s$ happens precisely when $\limsup \operatorname{Re}(m_b) = \infty$, which can only happen on a set of Lebesgue measure zero by Corollary 3.29. $\square$

Note that if $(\lambda_1, \lambda_2) \subseteq N(\tau)$, then the spectrum of every self-adjoint extension H of τ is purely absolutely continuous in the interval (λ_1, λ_2).

Example. Consider $H_0 = -\frac{d^2}{dx^2}$ on $(0, \infty)$ with a Dirichlet boundary condition at $x = 0$. Then it is easy to check $H_0 \geq 0$ and $N(\tau_0) = (0, \infty)$. Hence $\sigma_{ac}(H_0) = [0, \infty)$. Moreover, since the singular spectrum is supported on $[0, \infty) \backslash N(\tau_0) = \{0\}$, we see $\sigma_{sc}(H_0) = \emptyset$ (since the singular continuous spectrum cannot be supported on a finite set) and $\sigma_{pp}(H_0) \subseteq \{0\}$. Since 0 is no eigenvalue, we have $\sigma_{pp}(H_0) = \emptyset$. $\diamond$

Problem 9.13. *Determine the spectrum of $H_0 = -\frac{d^2}{dx^2}$ on $(0, \infty)$ with a general boundary condition (9.44) at $a = 0$.*

9.6. Spectral transformations II

In Section 9.3, we have looked at the case of one regular endpoint. In this section, we want to remove this restriction. In the case of a regular endpoint (or more generally an *l.c.* endpoint), the choice of $u(\lambda, x)$ in Lemma 9.13 was dictated by the fact that $u(\lambda, x)$ is required to satisfy the boundary condition at the regular (*l.c.*) endpoint. We begin by showing that in the general case we can choose any pair of linearly independent solutions. We will choose some arbitrary point $c \in I$ and two linearly independent solutions according to the initial conditions

(9.97) $\quad c(z, c) = 1, \quad p(c)c'(z, c) = 0, \quad\quad s(z, c) = 0, \quad p(c)s'(z, c) = 1.$

We will abbreviate

(9.98) $$\underline{s}(z, x) = \begin{pmatrix} c(z, x) \\ s(z, x) \end{pmatrix}.$$

Lemma 9.28. *There is a measure $d\mu(\lambda)$ and a nonnegative matrix $R(\lambda)$ with trace one such that*

(9.99) $\quad\begin{aligned} U: \quad L^2(I, r\,dx) &\to L^2(\mathbb{R}, R\,d\mu) \\ f(x) &\mapsto \int_a^b \underline{s}(\lambda, x) f(x) r(x)\, dx \end{aligned}$

is a spectral mapping as in Lemma 9.13. As before, the integral has to be understood as $\int_a^b dx = \lim_{c \downarrow a, d \uparrow b} \int_c^d dx$ with limit taken in $L^2(\mathbb{R}, R\,d\mu)$, where $L^2(\mathbb{R}, R\,d\mu)$ is the Hilbert space of all $\mathbb{C}^2$-valued measurable functions with scalar product

(9.100) $$\langle \underline{f}, \underline{g} \rangle = \int_\mathbb{R} \underline{f}^* R \underline{g}\, d\mu.$$

The inverse is given by

(9.101) $$(U^{-1} \underline{F})(x) = \int_\mathbb{R} \underline{s}(\lambda, x) R(\lambda) \underline{F}(\lambda) d\mu(\lambda).$$

Proof. Let U_0 be a spectral transformation as in Lemma 9.13 with corresponding real solutions $u_j(\lambda, x)$ and measures $d\mu_j(x)$, $1 \le j \le k$. Without loss of generality, we can assume $k = 2$ since we can always choose $d\mu_2 = 0$ and $u_2(\lambda, x)$ such that u_1 and u_2 are linearly independent.

Now define the 2×2 matrix $C(\lambda)$ via

$$\begin{pmatrix} u_1(\lambda, x) \\ u_2(\lambda, x) \end{pmatrix} = C(\lambda) \begin{pmatrix} c(\lambda, x) \\ s(\lambda, x) \end{pmatrix}$$

and note that $C(\lambda)$ is nonsingular since u_1, u_2 as well as s, c are linearly independent.

Set $d\tilde{\mu} = d\mu_1 + d\mu_2$. Then $d\mu_j = r_j d\tilde{\mu}$ and we can introduce $\tilde{R} = C^* \begin{pmatrix} r_1 & 0 \\ 0 & r_2 \end{pmatrix} C$. By construction, $\tilde{R}$ is a (symmetric) nonnegative matrix. Moreover, since $C(\lambda)$ is nonsingular, $\operatorname{tr}(\tilde{R})$ is positive a.e. with respect to $\tilde{\mu}$. Thus we can set $R = \operatorname{tr}(\tilde{R})^{-1} \tilde{R}$ and $d\mu = \operatorname{tr}(\tilde{R})^{-1} d\tilde{\mu}$.

This matrix gives rise to an operator

$$C : L^2(\mathbb{R}, R\, d\mu) \to \bigoplus_j L^2(\mathbb{R}, d\mu_j), \qquad F(\lambda) \mapsto C(\lambda) F(\lambda),$$

which, by our choice of $R\, d\mu$, is norm preserving. By $CU = U_0$ it is onto and hence it is unitary (this also shows that $L^2(\mathbb{R}, R\, d\mu)$ is a Hilbert space, i.e., complete).

It is left as an exercise to check that C maps multiplication by λ in $L^2(\mathbb{R}, R\, d\mu)$ to multiplication by λ in $\bigoplus_j L^2(\mathbb{R}, d\mu_j)$ and the formula for U^{-1}. □

Clearly the matrix-valued measure $R\, d\mu$ contains all the spectral information of A. Hence it remains to relate it to the resolvent of A as in Section 9.3.

For our base point $x = c$ there are corresponding Weyl m-functions $m_a(z)$ and $m_b(z)$ such that

(9.102) $\quad u_a(z) = c(z, x) - m_a(z) s(z, x), \qquad u_b(z) = c(z, x) + m_b(z) s(z, x).$

The different sign in front of $m_a(z)$ is introduced such that $m_a(z)$ will again be a Herglotz–Nevanlinna function. In fact, this follows using reflection at c, $x - c \mapsto -(x - c)$, which will interchange the roles of $m_a(z)$ and $m_b(z)$. In particular, all considerations from Section 9.3 hold for $m_a(z)$ as well.

Furthermore, we will introduce the **Weyl M-matrix**
(9.103)
$$M(z) = \frac{1}{m_a(z) + m_b(z)} \begin{pmatrix} -1 & (m_a(z) - m_b(z))/2 \\ (m_a(z) - m_b(z))/2 & m_a(z) m_b(z) \end{pmatrix}.$$

9.6. Spectral transformations II

Note $\det(M(z)) = -\frac{1}{4}$. Since by virtue of (9.52)

(9.104) $\qquad m_a(z) = -\frac{p(c)u'_a(z,c)}{u_a(z,c)} \quad \text{and} \quad m_b(z) = \frac{p(c)u'_b(z,c)}{u_b(z,c)},$

it follows that $W(u_a(z), u_b(z)) = m_a(z) + m_b(z)$ and

$M(z) =$

(9.105)
$$\begin{pmatrix} G(z,x,x) & (p(x)\partial_x + p(y)\partial_y)G(z,x,y)/2 \\ (p(x)\partial_x + p(y)\partial_y)G(z,x,y)/2 & p(x)\partial_x p(y)\partial_y G(z,x,y) \end{pmatrix}\bigg|_{x=y=c},$$

where $G(z,x,y)$ is the Green function of A. Note that while $p(x)\partial_x G(z,x,y)$ has different limits as $y \to x$ from $y > x$, respectively, $y < x$, the above off-diagonal elements are continuous.

We begin by showing

Lemma 9.29. *Let U be the spectral mapping from the previous lemma. Then*

(9.106)
$$(UG(z,x,.))(\lambda) = \frac{1}{\lambda - z}\underline{s}(\lambda, x),$$
$$(Up(x)\partial_x G(z,x,.))(\lambda) = \frac{1}{\lambda - z}p(x)\underline{s}'(\lambda, x)$$

for every $x \in (a,b)$ and every $z \in \rho(A)$.

Proof. First of all, note that $G(z,x,.) \in L^2((a,b), r\,dx)$ for every $x \in (a,b)$ and $z \in \rho(A)$. Moreover, from $R_A(z)f = U^{-1}\frac{1}{\lambda - z}Uf$, we have

$$\int_a^b G(z,x,y)f(y)r(y)dy = \int_{\mathbb{R}} \frac{1}{\lambda - z}\underline{s}(\lambda, x)R(\lambda)\underline{F}(\lambda)d\mu(\lambda)$$

where $\underline{F} = Uf$. Now proceed as in the proof of Lemma 9.15. $\square$

With the aid of this lemma, we can now show

Theorem 9.30. *The Weyl M-matrix is given by*

(9.107) $\qquad M(z) = D + \int_{\mathbb{R}} \left(\frac{1}{\lambda - z} - \frac{\lambda}{1 + \lambda^2}\right)R(\lambda)d\mu(\lambda), \quad D_{jk} \in \mathbb{R},$

and

(9.108) $\qquad D = \operatorname{Re}(M(\mathrm{i})), \quad \int_{\mathbb{R}} \frac{1}{1+\lambda^2}R(\lambda)d\mu(\lambda) = \operatorname{Im}(M(\mathrm{i})),$

where

(9.109) $\operatorname{Re}(M(z)) = \frac{1}{2}(M(z) + M^*(z)), \quad \operatorname{Im}(M(z)) = \frac{1}{2\mathrm{i}}(M(z) - M^*(z)).$

Proof. By the previous lemma, we have
$$\int_a^b |G(z,c,y)|^2 r(y) dy = \int_{\mathbb{R}} \frac{1}{|z-\lambda|^2} R_{11}(\lambda) d\mu(\lambda).$$
Moreover, by (9.28), (9.56), and (9.102), we infer
$$\int_a^b |G(z,c,y)|^2 r(y) dy = \frac{1}{|W(u_a,u_b)|^2}\left(|u_b(z,c)|^2 \int_a^c |u_a(z,y)|^2 r(y) dy \right.$$
$$\left. + |u_a(z,c)|^2 \int_c^b |u_b(z,y)|^2 r(y) dy\right) = \frac{\mathrm{Im}(M_{11}(z))}{\mathrm{Im}(z)}.$$
Similarly, we obtain
$$\int_{\mathbb{R}} \frac{1}{|z-\lambda|^2} R_{22}(\lambda) d\mu(\lambda) = \frac{\mathrm{Im}(M_{22}(z))}{\mathrm{Im}(z)}$$
and
$$\int_{\mathbb{R}} \frac{1}{|z-\lambda|^2} R_{12}(\lambda) d\mu(\lambda) = \frac{\mathrm{Im}(M_{12}(z))}{\mathrm{Im}(z)}.$$
Hence the result follows as in the proof of Theorem 9.17. □

Example. Consider $\tau = -\frac{d^2}{dx^2}$ on $I = (-\infty, \infty)$. Then we already know from the example in Section 9.3 that $m_b(z)$ is given by (9.69). Moreover, by symmetry, we have $m_a(z) = m_b(z)$, implying

(9.110) $$M(z) = \frac{1}{2}\begin{pmatrix} \frac{1}{\sqrt{-z}} & 0 \\ 0 & -\sqrt{-z} \end{pmatrix}$$

and

(9.111) $$R(\lambda) d\mu(\lambda) = \frac{1}{2\pi}\begin{pmatrix} \frac{1}{\sqrt{\lambda}} & 0 \\ 0 & \sqrt{\lambda} \end{pmatrix} d\lambda.$$

◇

Now we are also able to extend Theorem 9.27. Note that by

(9.112) $$\mathrm{tr}(M(z)) = M_{11}(z) + M_{22}(z) = d + \int_{\mathbb{R}}\left(\frac{1}{\lambda - z} - \frac{\lambda}{1+\lambda^2}\right) d\mu(\lambda)$$

(with $d = \mathrm{tr}(D) \in \mathbb{R}$), we have that the set

(9.113) $$M_s = \{\lambda | \limsup_{\varepsilon \downarrow 0} \mathrm{Im}(\mathrm{tr}(M(\lambda + i\varepsilon))) = \infty\}$$

is a support for the singularly continuous part and

(9.114) $$M_{ac} = \{\lambda | 0 < \limsup_{\varepsilon \downarrow 0} \mathrm{Im}(\mathrm{tr}(M(\lambda + i\varepsilon))) < \infty\}$$

is a minimal support for the absolutely continuous part.

9.6. Spectral transformations II

Theorem 9.31. *The set $N_a(\tau) \cup N_b(\tau) \subseteq M_{ac}$ is a minimal support for the absolutely continuous spectrum of A. In particular,*

$$\sigma_{ac}(A) = \overline{N_a(\tau) \cup N_b(\tau)}^{ess}. \tag{9.115}$$

Moreover, the set

$$\bigcup_{\alpha \in [0,\pi)} S_{a,\alpha}(\tau) \cap S_{b,\alpha}(\tau) \supseteq M_s \tag{9.116}$$

is a support for the singular spectrum of A.

Proof. By Corollary 9.25 we have $0 < \liminf \operatorname{Im}(m_a)$ and $\limsup |m_a| < \infty$ if and only if $\lambda \in N_a(\tau)$ and similarly for m_b.

Now suppose $\lambda \in N_a(\tau)$. Then $\limsup |M_{11}| < \infty$ since $\limsup |M_{11}| = \infty$ is impossible by $0 = \liminf |M_{11}^{-1}| = \liminf |m_a + m_b| \geq \liminf \operatorname{Im}(m_a) > 0$. Similarly, $\limsup |M_{22}| < \infty$. Moreover, if $\limsup |m_b| < \infty$, we also have

$$\liminf \operatorname{Im}(M_{11}) = \liminf \frac{\operatorname{Im}(m_a + m_b)}{|m_a + m_b|^2} \geq \frac{\liminf \operatorname{Im}(m_a)}{\limsup |m_a|^2 + \limsup |m_b|^2} > 0$$

and if $\limsup |m_b| = \infty$, we have

$$\liminf \operatorname{Im}(M_{22}) = \liminf \operatorname{Im}\left(\frac{m_a}{1 + \frac{m_a}{m_b}}\right) \geq \liminf \operatorname{Im}(m_a) > 0.$$

Thus $N_a(\tau) \subseteq M_{ac}$ and similarly $N_b(\tau) \subseteq M_{ac}$.

Conversely, let $\lambda \in M_{ac}$. By Corollary 3.29, we can assume that the limits $\lim m_a$ and $\lim m_b$ both exist and are finite after disregarding a set of Lebesgue measure zero. For such λ, $\lim \operatorname{Im}(M_{11})$ and $\lim \operatorname{Im}(M_{22})$ both exist and are finite. Moreover, either $\lim \operatorname{Im}(M_{11}) > 0$, in which case $\lim \operatorname{Im}(m_a + m_b) > 0$, or $\lim \operatorname{Im}(M_{11}) = 0$, in which case

$$0 < \lim \operatorname{Im}(M_{22}) = \lim \frac{|m_a|^2 \operatorname{Im}(m_b) + |m_b|^2 \operatorname{Im}(m_a)}{|m_a|^2 + |m_b|^2} = 0$$

yields a contradiction. Thus $\lambda \in N_a(\tau) \cup N_b(\tau)$ and the first part is proven.

To prove the second part, let $\lambda \in M_s$. If $\limsup \operatorname{Im}(M_{11}) = \infty$, we have $\limsup |M_{11}| = \infty$ and thus $\liminf |m_a + m_b| = 0$. But this implies that there is some subsequence such that $\lim m_b = -\lim m_a = \cot(\alpha) \in \mathbb{R} \cup \{\infty\}$. Similarly, if $\limsup \operatorname{Im}(M_{22}) = \infty$, we have $\liminf |m_a^{-1} + m_b^{-1}| = 0$, and there is some subsequence such that $\lim m_b^{-1} = -\lim m_a^{-1} = \tan(\alpha) \in \mathbb{R} \cup \{\infty\}$. This shows $M_s \subseteq \bigcup_\alpha S_{a,\alpha}(\tau) \cap S_{b,\alpha}(\tau)$. □

Problem 9.14. *Show*

$$R(\lambda) d\mu_{ac}(\lambda) = \begin{pmatrix} \frac{\operatorname{Im}(m_a(\lambda) + m_b(\lambda))}{|m_a(\lambda)|^2 + |m_b(\lambda)|^2} & \frac{\operatorname{Im}(m_a(\lambda) m_b^*(\lambda))}{|m_a(\lambda)|^2 + |m_b(\lambda)|^2} \\ \frac{\operatorname{Im}(m_a(\lambda) m_b^*(\lambda))}{|m_a(\lambda)|^2 + |m_b(\lambda)|^2} & \frac{|m_a(\lambda)|^2 \operatorname{Im}(m_b(\lambda)) + |m_b(\lambda)|^2 \operatorname{Im}(m_a(\lambda))}{|m_a(\lambda)|^2 + |m_b(\lambda)|^2} \end{pmatrix} \frac{d\lambda}{\pi},$$

where $m_a(\lambda) = \lim_{\varepsilon \downarrow 0} m_a(\lambda + i\varepsilon)$ and similarly for $m_b(\lambda)$.

Moreover, show that the choice of solutions
$$\begin{pmatrix} u_b(\lambda, x) \\ u_a(\lambda, x) \end{pmatrix} = V(\lambda) \begin{pmatrix} c(\lambda, x) \\ s(\lambda, x) \end{pmatrix},$$
where
$$V(\lambda) = \frac{1}{m_a(\lambda) + m_b(\lambda)} \begin{pmatrix} 1 & m_b(\lambda) \\ 1 & -m_a(\lambda) \end{pmatrix},$$
diagonalizes the absolutely continuous part,
$$V^{-1}(\lambda)^* R(\lambda) V(\lambda)^{-1} d\mu_{ac}(\lambda) = \frac{1}{\pi} \begin{pmatrix} \mathrm{Im}(m_a(\lambda)) & 0 \\ 0 & \mathrm{Im}(m_b(\lambda)) \end{pmatrix} d\lambda.$$

Problem 9.15. *Show that at an eigenvalue E, we have*
$$R(E)\mu(\{E\}) = \frac{1}{\int_a^b |u(E,x)|^2 dx} \begin{pmatrix} |u(E,0)|^2 & u(E,0)u'(E,0)^* \\ u(E,0)^* u'(E,0) & |u'(E,0)|^2 \end{pmatrix},$$
where $u(E, x)$ is a corresponding eigenfunction.

9.7. The spectra of one-dimensional Schrödinger operators

In this section, we want to look at the case of one-dimensional Schrödinger operators; that is, $r = p = 1$ on $(a, b) = (0, \infty)$.

Recall that
$$(9.117) \qquad H_0 = -\frac{d^2}{dx^2}, \qquad \mathfrak{D}(H_0) = H^2(\mathbb{R}),$$
is self-adjoint and
$$(9.118) \qquad q_{H_0}(f) = \|f'\|^2, \qquad \mathfrak{Q}(H_0) = H^1(\mathbb{R}).$$
Hence we can try to apply the results from Chapter 6. We begin with a simple estimate:

Lemma 9.32. *Suppose $f \in H^1(0, 1)$. Then*
$$(9.119) \qquad \sup_{x \in [0,1]} |f(x)|^2 \leq \varepsilon \int_0^1 |f'(x)|^2 dx + \left(1 + \frac{1}{\varepsilon}\right) \int_0^1 |f(x)|^2 dx$$
for every $\varepsilon > 0$.

Proof. First, note that
$$|f(x)|^2 = |f(c)|^2 + 2\int_c^x \mathrm{Re}(f(t)^* f'(t)) dt \leq |f(c)|^2 + 2\int_0^1 \sqrt{\varepsilon} |f(t)| \frac{|f'(t)|}{\sqrt{\varepsilon}} dt$$
$$\leq |f(c)|^2 + \int_0^1 \left(\varepsilon |f'(t)|^2 + \frac{1}{\varepsilon} |f(t)|^2\right) dt$$

9.7. The spectra of one-dimensional Schrödinger operators

for any $c \in [0, 1]$. But by the mean value theorem there is a $c \in (0, 1)$ such that $|f(c)|^2 = \int_0^1 |f(t)|^2 dt$. $\square$

As a consequence we obtain

Lemma 9.33. *Suppose $q \in L^2_{loc}(\mathbb{R})$ and*

$$(9.120) \qquad \sup_{n \in \mathbb{Z}} \int_n^{n+1} |q(x)|^2 dx < \infty.$$

Then q is relatively bounded with respect to H_0 with bound zero.

Similarly, if $q \in L^1_{loc}(\mathbb{R})$ and

$$(9.121) \qquad \sup_{n \in \mathbb{Z}} \int_n^{n+1} |q(x)| dx < \infty,$$

then q is relatively form bounded with respect to H_0 with bound zero.

Proof. Let Q be in $L^2_{loc}(\mathbb{R})$ and abbreviate $M = \sup_{n \in \mathbb{Z}} \int_n^{n+1} |Q(x)|^2 dx$. Using the previous lemma, we have for $f \in H^1(\mathbb{R})$ that

$$\|Qf\|^2 \leq \sum_{n \in \mathbb{Z}} \int_n^{n+1} |Q(x)f(x)|^2 dx \leq M \sum_{n \in \mathbb{Z}} \sup_{x \in [n, n+1]} |f(x)|^2$$

$$\leq M \sum_{n \in \mathbb{Z}} \left(\varepsilon \int_n^{n+1} |f'(x)|^2 dx + \left(1 + \frac{1}{\varepsilon}\right) \int_n^{n+1} |f(x)|^2 dx \right)$$

$$= M \left(\varepsilon \|f'\|^2 + \left(1 + \frac{1}{\varepsilon}\right) \|f\|^2 \right).$$

Choosing $Q = |q|^{1/2}$, this already proves the form case since $\|f'\|^2 = q_{H_0}(f)$. Choosing $Q = q$ and observing $q_{H_0}(f) = \langle f, H_0 f \rangle \leq \|H_0 f\| \|f\| \leq \frac{1}{2}(\|H_0 f\|^2 + \|f\|^2)$ for $f \in H^2(\mathbb{R})$ shows the operator case. $\square$

Hence, in both cases, $H_0 + q$ is a well-defined (semi-bounded) operator defined as operator sum on $\mathfrak{D}(H_0 + q) = \mathfrak{D}(H_0) = H^2(\mathbb{R})$ in the first case and as form sum on $\mathfrak{Q}(H_0 + q) = \mathfrak{Q}(H_0) = H^1(\mathbb{R})$ in the second case. Note also that the first case implies the second one since by Cauchy–Schwarz we have

$$(9.122) \qquad \int_n^{n+1} |q(x)| dx \leq \left(\int_n^{n+1} |q(x)|^2 dx \right)^{1/2}.$$

This is not too surprising since we already know how to turn $H_0 + q$ into a self-adjoint operator without imposing any conditions on q (except for $L^1_{loc}(\mathbb{R})$) at all. However, we get at least a simple description of the (form) domains, and by requiring a bit more, we can even compute the essential spectrum of the perturbed operator.

Lemma 9.34. *Suppose $q \in L^1(\mathbb{R})$. Then the resolvent difference of H_0 and $H_0 + q$ is trace class.*

Proof. Using $G_0(z,x,x) = 1/(2\sqrt{-z})$, Lemma 9.12 implies that $|q|^{1/2} R_{H_0}(z)$ is Hilbert–Schmidt and hence the result follows from Lemma 6.29. $\square$

Lemma 9.35. *Suppose $q \in L^1_{loc}(\mathbb{R})$ and*

(9.123) $$\lim_{|n|\to\infty} \int_n^{n+1} |q(x)| dx = 0.$$

Then $R_{H_0+q}(z) - R_{H_0}(z)$ is compact and hence $\sigma_{ess}(H_0 + q) = \sigma_{ess}(H_0) = [0, \infty)$.

Proof. By Weyl's theorem it suffices to show that the resolvent difference is compact. Let $q_n(x) = q(x)\chi_{\mathbb{R}\setminus[-n,n]}(x)$. Then $R_{H_0+q}(z) - R_{H_0+q_n}(z)$ is trace class, which can be shown as in the previous theorem since $q - q_n$ has compact support (no information on the corresponding diagonal Green's function is needed since by continuity it is bounded on every compact set). Moreover, by the proof of Lemma 9.33, q_n is form bounded with respect to H_0 with constants $a = M_n$ and $b = 2M_n$, where $M_n = \sup_{|m|\geq n} \int_m^{m+1} |q(x)|^2 dx$. Hence by Theorem 6.25 we see

$$R_{H_0+q_n}(-\lambda) = R_{H_0}(-\lambda)^{1/2} (1 - C_{q_n}(\lambda))^{-1} R_{H_0}(-\lambda)^{1/2}, \qquad \lambda > 2,$$

with $\|C_{q_n}(\lambda)\| \leq M_n$. So we conclude

$$R_{H_0+q_n}(-\lambda) - R_{H_0}(-\lambda) = -R_{H_0}(-\lambda)^{1/2} C_{q_n}(\lambda)(1 - C_{q_n}(\lambda))^{-1} R_{H_0}(-\lambda)^{1/2},$$

$\lambda > 2$, which implies that the sequence of compact operators $R_{H_0+q}(-\lambda) - R_{H_0+q_n}(-\lambda)$ converges to $R_{H_0+q}(-\lambda) - R_{H_0}(-\lambda)$ in norm, which implies that the limit is also compact and finishes the proof. $\square$

Using Lemma 6.23, respectively, Corollary 6.27, we even obtain

Corollary 9.36. *Let $q = q_1 + q_2$ where q_1 and q_2 satisfy the assumptions of Lemma 9.33 and Lemma 9.35, respectively. Then $H_0 + q_1 + q_2$ is self-adjoint and $\sigma_{ess}(H_0 + q_1 + q_2) = \sigma_{ess}(H_0 + q_1)$.*

This result applies, for example, in the case where q_2 is a decaying perturbation of a periodic potential q_1.

Finally we turn to the absolutely continuous spectrum.

Lemma 9.37. *Suppose $q = q_1 + q_2$, where $q_1 \in L^1(0, \infty)$ and $q_2 \in AC[0, \infty)$ with $q_2' \in L^1(0, \infty)$ and $\lim_{x\to\infty} q_2(x) = 0$. Then there are two solutions $u_\pm(\lambda, x)$ of $\tau u = \lambda u$, $\lambda > 0$, of the form*

(9.124) $\quad u_\pm(\lambda, x) = (1 + o(1))u_{0,\pm}(\lambda, x), \quad u'_\pm(\lambda, x) = (1 + o(1))u'_{0,\pm}(\lambda, x)$

9.7. The spectra of one-dimensional Schrödinger operators

as $x \to \infty$, where

$$(9.125) \qquad u_{0,\pm}(\lambda, x) = \exp\left(\pm i \int_0^x \sqrt{\lambda - q_2(y)} dy\right).$$

Proof. We will omit the dependence on λ for notational simplicity. Moreover, we will choose x so large that $W_x(u_{0,-}, u_{0,+}) = 2i\sqrt{\lambda - q_2(x)} \neq 0$. Write

$$\underline{u}(x) = U_0(x)\underline{a}(x), \quad U_0(x) = \begin{pmatrix} u_{0,+}(x) & u_{0,-}(x) \\ u'_{0,+}(x) & u'_{0,-}(x) \end{pmatrix}, \quad \underline{a}(x) = \begin{pmatrix} a_+(x) \\ a_-(x) \end{pmatrix}.$$

Then

$$\underline{u}'(x) = \begin{pmatrix} 0 & 1 \\ q(x) - \lambda & 0 \end{pmatrix} \underline{u}(x)$$

$$- \begin{pmatrix} 0 & 0 \\ q_+(x)u_{0,+}(x) & q_-(x)u_{0,-}(x) \end{pmatrix} \underline{a}(x) + U_0(x)\underline{a}'(x),$$

where

$$q_\pm(x) = q_1(x) \pm i \frac{q'_2(x)}{2\sqrt{\lambda - q_2(x)}}.$$

Hence, $u(x)$ will solve $\tau u = \lambda u$ if

$$\underline{a}'(x) = \frac{1}{W_x(u_{0,-}, u_{0,+})} \begin{pmatrix} -q_+(x) & -q_-(x)u_{0,-}(x)^2 \\ q_+(x)u_{0,+}(x)^2 & q_-(x) \end{pmatrix} \underline{a}(x).$$

Since the coefficient matrix of this linear system is integrable, the claim follows by a simple application of Gronwall's inequality. $\square$

Theorem 9.38 (Weidmann)**.** *Let q_1 and q_2 be as in the previous lemma and suppose $q = q_1 + q_2$ satisfies the assumptions of Lemma 9.35. Let $H = H_0 + q_1 + q_2$. Then $\sigma_{ac}(H) = [0, \infty)$, $\sigma_{sc}(H) = \emptyset$, and $\sigma_p(H) \subseteq (-\infty, 0]$.*

Proof. By the previous lemma, there is no subordinate solution for $\lambda > 0$ on $(0, \infty)$ and hence $0 < \text{Im}(m_b(\lambda + i0)) < \infty$. Similarly, there is no subordinate solution $(-\infty, 0)$ and hence $0 < \text{Im}(m_a(\lambda + i0)) < \infty$. Thus the same is true for the diagonal entries $M_{jj}(z)$ of the Weyl M-matrix, $0 < \text{Im}(M_{jj}(\lambda + i0)) < \infty$, and hence $d\mu$ is purely absolutely continuous on $(0, \infty)$. Since $\sigma_{ess}(H) = [0, \infty)$, we conclude $\sigma_{ac}(H) = [0, \infty)$ and $\sigma_{sc}(H) \subseteq \{0\}$. Since the singular continuous part cannot live on a single point, we are done. $\square$

Note that the same results hold for operators on $[0, \infty)$ rather than $\mathbb{R}$. Moreover, observe that the conditions from Lemma 9.37 are only imposed near $+\infty$ but not near $-\infty$. The conditions from Lemma 9.35 are only used to ensure that there is no essential spectrum in $(-\infty, 0)$.

Having dealt with the essential spectrum, let us next look at the discrete spectrum. In the case of decaying potentials, as in the previous theorem, one key question is whether the number of eigenvalues below the essential spectrum is finite.

As preparation, we shall prove Sturm's comparison theorem:

Theorem 9.39 (Sturm)**.** *Let τ_0, τ_1 be associated with $q_0 \geq q_1$ on (a,b), respectively. Let $(c,d) \subseteq (a,b)$ and $\tau_0 u = 0$, $\tau_1 v = 0$. Suppose at each end of (c,d) either $W_x(u,v) = 0$ or, if $c, d \in (a,b)$, $u = 0$. Then v is either a multiple of u in (c,d) or v must vanish at some point in (c,d).*

Proof. By decreasing d to the first zero of u in $(c,d]$ (and perhaps flipping signs), we can suppose $u > 0$ on (c,d). If v has no zeros in (c,d), we can suppose $v > 0$ on (c,d) again by perhaps flipping signs. At each endpoint, $W(u,v)$ vanishes or else $u = 0$, $v > 0$, and $u'(c) > 0$ (or $u'(d) < 0$). Thus, $W_c(u,v) \leq 0$, $W_d(u,v) \geq 0$. But this is inconsistent with

$$(9.126) \qquad W_d(u,v) - W_c(u,v) = \int_c^d (q_0(t) - q_1(t))u(t)v(t)\, dt,$$

unless both sides vanish. $\square$

In particular, choosing $q_0 = q - \lambda_0$ and $q_1 = q - \lambda_1$, this result holds for solutions of $\tau u = \lambda_0 u$ and $\tau v = \lambda_1 v$.

Now we can prove

Theorem 9.40. *Suppose q satisfies (9.121) such that H is semi-bounded and $\mathfrak{Q}(H) = H^1(\mathbb{R})$. Let $\lambda_0 < \cdots < \lambda_n < \cdots$ be its eigenvalues below the essential spectrum and $\psi_0, \ldots, \psi_n, \ldots$ the corresponding eigenfunctions. Then ψ_n has n zeros.*

Proof. We first prove that ψ_n has at least n zeros and then that if ψ_n has m zeros, then $(-\infty, \lambda_n]$ has at least $(m+1)$ eigenvalues. If ψ_n has m zeros at $x_1, x_2, \ldots, x_m$ and we let $x_0 = a$, $x_{m+1} = b$, then by Theorem 9.39, ψ_{n+1} must have at least one zero in each of $(x_0, x_1), (x_1, x_2), \ldots, (x_m, x_{m+1})$; that is, ψ_{n+1} has at least $m+1$ zeros. It follows by induction that ψ_n has at least n zeros.

On the other hand, if ψ_n has m zeros $x_1, \ldots, x_m$, define

$$(9.127) \qquad \eta_j(x) = \begin{cases} \psi_n(x), & x_j \leq x \leq x_{j+1}, \\ 0 & \text{otherwise}, \end{cases} \quad j = 0, \ldots, m,$$

where we set $x_0 = -\infty$ and $x_{m+1} = \infty$. Then η_j is in the form domain of H and satisfies $\langle \eta_j, H\eta_j \rangle = \lambda_n \|\eta_j\|^2$. Hence if $\eta = \sum_{j=0}^m c_j \eta_j$, then $\langle \eta, H\eta \rangle = \lambda_n \|\eta\|^2$ and it follows by Theorem 4.15 (i) that there are at least $m+1$ eigenvalues in $(-\infty, \lambda_n]$. $\square$

9.7. The spectra of one-dimensional Schrödinger operators

Note that by Theorem 9.39, the zeros of ψ_n interlace the zeros of ψ_n. The second part of the proof also shows

Corollary 9.41. *Let H be as in the previous theorem. If the Weyl solution $u_\pm(\lambda,.)$ has m zeros, then $\dim \operatorname{Ran}_{(-\infty,\lambda)}(H) \geq m$. In particular, λ below the spectrum of H implies that $u_\pm(\lambda,.)$ has no zeros.*

The equation $(\tau - \lambda)u$ is called **oscillating** if one solution has an infinite number of zeros. Theorem 9.39 implies that this is then true for all solutions. By our previous considerations this is the case if and only if $\sigma(H)$ has infinitely many points below λ. Hence it remains to find a good oscillation criterion.

Theorem 9.42 (Kneser). *Consider q on $(0, \infty)$. Then*

$$(9.128) \qquad \liminf_{x \to \infty} (x^2 q(x)) > -\frac{1}{4} \text{ implies nonoscillation of } \tau \text{ near } \infty$$

and

$$(9.129) \qquad \limsup_{x \to \infty} (x^2 q(x)) < -\frac{1}{4} \text{ implies oscillation of } \tau \text{ near } \infty.$$

Proof. The key idea is that the equation

$$\tau_0 = -\frac{d^2}{dx^2} + \frac{\mu}{x^2}$$

is of Euler type. Hence it is explicitly solvable with a fundamental system given by

$$x^{\frac{1}{2} \pm \sqrt{\mu + \frac{1}{4}}}.$$

There are two cases to distinguish. If $\mu \geq -1/4$, all solutions are nonoscillatory. If $\mu < -1/4$, one has to take real/imaginary parts and all solutions are oscillatory. Hence a straightforward application of Sturm's comparison theorem between τ_0 and τ yields the result. $\square$

Corollary 9.43. *Suppose q satisfies (9.121). Then H has finitely many eigenvalues below the infimum of the essential spectrum 0 if*

$$(9.130) \qquad \liminf_{|x| \to \infty} (x^2 q(x)) > -\frac{1}{4}$$

and infinitely many if

$$(9.131) \qquad \limsup_{|x| \to \infty} (x^2 q(x)) < -\frac{1}{4}.$$

Problem 9.16. *Suppose $q(x)$ is symmetric, $q(-x) = q(x)$. Show that all eigenvectors are either symmetric or anti-symmetric. Show that under the assumptions of Theorem 9.40 the lowest eigenvalue is symmetric.*

Problem 9.17. Show that if q is relatively bounded with respect to H_0, then necessarily $q \in L^2_{loc}(\mathbb{R})$ and (9.120) holds. Similarly, if q is relatively form bounded with respect to H_0, then necessarily $q \in L^1_{loc}(\mathbb{R})$ and (9.121) holds.

Problem 9.18. Suppose $q \in L^1(\mathbb{R})$ and consider $H = -\frac{d^2}{dx^2} + q$. Show that $\inf \sigma(H) \le \int_\mathbb{R} q(x) dx$. In particular, there is at least one eigenvalue below the essential spectrum if $\int_\mathbb{R} q(x) dx < 0$. (Hint: Let $\varphi \in C_c^\infty(\mathbb{R})$ with $\varphi(x) = 1$ for $|x| \le 1$ and investigate $q_H(\varphi_n)$, where $\varphi_n(x) = \varphi(x/n)$.)

Chapter 10

One-particle Schrödinger operators

10.1. Self-adjointness and spectrum

Our next goal is to apply these results to Schrödinger operators. The Hamiltonian of one particle in d dimensions is given by

(10.1) $$H = H_0 + V,$$

where $V : \mathbb{R}^d \to \mathbb{R}$ is the potential energy of the particle. We are mainly interested in the case $1 \leq d \leq 3$ and want to find classes of potentials which are relatively bounded, respectively, relatively compact. To do this, we need a better understanding of the functions in the domain of H_0.

Lemma 10.1. *Suppose $n \leq 3$ and $\psi \in H^2(\mathbb{R}^n)$. Then $\psi \in C_\infty(\mathbb{R}^n)$ and for every $a > 0$ there is a $b > 0$ such that*

(10.2) $$\|\psi\|_\infty \leq a\|H_0\psi\| + b\|\psi\|.$$

Proof. The important observation is that $(p^2 + \gamma^2)^{-1} \in L^2(\mathbb{R}^n)$ if $n \leq 3$. Hence, since $(p^2 + \gamma^2)\hat\psi \in L^2(\mathbb{R}^n)$, the Cauchy–Schwarz inequality

$$\|\hat\psi\|_1 = \|(p^2 + \gamma^2)^{-1}(p^2 + \gamma^2)\hat\psi(p)\|_1$$
$$\leq \|(p^2 + \gamma^2)^{-1}\|\,\|(p^2 + \gamma^2)\hat\psi(p)\|$$

shows $\hat\psi \in L^1(\mathbb{R}^n)$. But now everything follows from the Riemann-Lebesgue lemma, that is,

$$\|\psi\|_\infty \leq (2\pi)^{-n/2}\|(p^2 + \gamma^2)^{-1}\|(\|p^2\hat\psi(p)\| + \gamma^2\|\hat\psi(p)\|)$$
$$= (\gamma/2\pi)^{n/2}\|(p^2 + 1)^{-1}\|(\gamma^{-2}\|H_0\psi\| + \|\psi\|),$$

which finishes the proof. □

Now we come to our first result.

Theorem 10.2. *Let V be real-valued and $V \in L^\infty_\infty(\mathbb{R}^n)$ if $n > 3$ and $V \in L^\infty_\infty(\mathbb{R}^n) + L^2(\mathbb{R}^n)$ if $n \leq 3$. Then V is relatively compact with respect to H_0. In particular,*

(10.3) $$H = H_0 + V, \qquad \mathfrak{D}(H) = H^2(\mathbb{R}^n),$$

is self-adjoint, bounded from below and

(10.4) $$\sigma_{ess}(H) = [0, \infty).$$

Moreover, $C_c^\infty(\mathbb{R}^n)$ is a core for H.

Proof. Write $V = V_1 + V_2$ with $V_1 \in L^\infty_\infty(\mathbb{R}^n)$ and $V_2 \in L^2(\mathbb{R}^n)$ if $n \leq 3$ and $V_2 = 0$ otherwise. Clearly $\mathfrak{D}(H_0) \subseteq \mathfrak{D}(V_1)$ and our previous lemma shows that $\mathfrak{D}(H_0) \subseteq \mathfrak{D}(V_2)$ as well. Moreover, invoking Lemma 7.21 with $f(p) = (p^2 - z)^{-1}$, $z \in \rho(H_0)$, and $g(x) = V_j(x)$ (note that $f \in L^\infty_\infty(\mathbb{R}^n) \cap L^2(\mathbb{R}^n)$ for $n \leq 3$) shows that both V_1 and V_2 are relatively compact. Hence $V = V_1 + V_2$ is relatively compact. Since $C_c^\infty(\mathbb{R}^n)$ is a core for H_0 by Lemma 7.19, the same is true for H by the Kato–Rellich theorem. $\square$

Observe that since $C_c^\infty(\mathbb{R}^n) \subset \mathfrak{D}(H_0)$, we must have $V \in L^2_{loc}(\mathbb{R}^n)$ if $\mathfrak{D}(H_0) \subseteq \mathfrak{D}(V)$.

Corollary 10.3. *Let V be as in the previous theorem. Then χ_Ω, with $\Omega \subseteq \mathbb{R}^n$ bounded, is relatively compact with respect to $H = H_0 + V$. In particular, the operators $K_n = \chi_{B_n(0)}$ satisfy the assumptions of the RAGE theorem.*

Proof. This follows from Lemma 6.23 and the discussion after Lemma 7.21. $\square$

10.2. The hydrogen atom

We begin with the simple model of a single electron in $\mathbb{R}^3$ moving in the external potential V generated by a nucleus (which is assumed to be fixed at the origin). If one takes only the electrostatic force into account, then V is given by the Coulomb potential and the corresponding Hamiltonian is given by

(10.5) $$H^{(1)} = -\Delta - \frac{\gamma}{|x|}, \qquad \mathfrak{D}(H^{(1)}) = H^2(\mathbb{R}^3).$$

If the potential is attracting, that is, if $\gamma > 0$, then it describes the hydrogen atom and is probably the most famous model in quantum mechanics.

We have chosen as domain $\mathfrak{D}(H^{(1)}) = \mathfrak{D}(H_0) \cap \mathfrak{D}(\frac{1}{|x|}) = \mathfrak{D}(H_0)$, and by Theorem 10.2, we conclude that $H^{(1)}$ is self-adjoint. Moreover, Theorem 10.2

10.2. The hydrogen atom

also tells us

(10.6) $$\sigma_{ess}(H^{(1)}) = [0, \infty)$$

and that $H^{(1)}$ is bounded from below,

(10.7) $$E_0 = \inf \sigma(H^{(1)}) > -\infty.$$

If $\gamma \leq 0$, we have $H^{(1)} \geq 0$ and hence $E_0 = 0$, but if $\gamma > 0$, we might have $E_0 < 0$ and there might be some discrete eigenvalues below the essential spectrum.

In order to say more about the eigenvalues of $H^{(1)}$, we will use the fact that both H_0 and $V^{(1)} = -\gamma/|x|$ have a simple behavior with respect to scaling. Consider the **dilation group**

(10.8) $$U(s)\psi(x) = e^{-ns/2}\psi(e^{-s}x), \qquad s \in \mathbb{R},$$

which is a strongly continuous one-parameter unitary group. The generator can be easily computed:

(10.9) $$D\psi(x) = \frac{1}{2}(xp + px)\psi(x) = (xp - \frac{in}{2})\psi(x), \qquad \psi \in \mathcal{S}(\mathbb{R}^n).$$

Now let us investigate the action of $U(s)$ on $H^{(1)}$:
(10.10)
$$H^{(1)}(s) = U(-s)H^{(1)}U(s) = e^{-2s}H_0 + e^{-s}V^{(1)}, \quad \mathfrak{D}(H^{(1)}(s)) = \mathfrak{D}(H^{(1)}).$$

Now suppose $H^{(1)}\psi = \lambda\psi$. Then

(10.11) $$\langle \psi, [U(s), H^{(1)}]\psi \rangle = \langle U(-s)\psi, \lambda\psi \rangle - \langle \lambda\psi, U(s)\psi \rangle = 0$$

and hence

$$0 = \lim_{s \to 0} \frac{1}{s}\langle \psi, [U(s), H^{(1)}]\psi \rangle = \lim_{s \to 0}\langle U(-s)\psi, \frac{H^{(1)} - H^{(1)}(s)}{s}\psi \rangle$$
(10.12) $$= -\langle \psi, (2H_0 + V^{(1)})\psi \rangle.$$

Thus we have proven the **virial theorem**.

Theorem 10.4. *Suppose $H = H_0 + V$ with V symmetric, relatively bounded, and $U(-s)VU(s) = e^{-s}V$. Then every normalized eigenfunction ψ corresponding to an eigenvalue λ satisfies*

(10.13) $$\lambda = -\langle \psi, H_0\psi \rangle = \frac{1}{2}\langle \psi, V\psi \rangle.$$

In particular, all eigenvalues must be negative.

This result even has some further consequences for the point spectrum of $H^{(1)}$.

Corollary 10.5. *Suppose $\gamma > 0$. Then*
$$(10.14) \quad \sigma_p(H^{(1)}) = \sigma_d(H^{(1)}) = \{E_j\}_{j \in \mathbb{N}_0}, \quad E_0 < E_j < E_{j+1} < 0,$$
with $\lim_{j \to \infty} E_j = 0$.

Proof. Choose $\psi \in C_c^\infty(\mathbb{R} \backslash \{0\})$ and set $\psi(s) = U(-s)\psi$. Then
$$\langle \psi(s), H^{(1)} \psi(s) \rangle = e^{-2s} \langle \psi, H_0 \psi \rangle + e^{-s} \langle \psi, V^{(1)} \psi \rangle$$
which is negative for s large. Now choose a sequence $s_n \to \infty$ such that we have $\mathrm{supp}(\psi(s_n)) \cap \mathrm{supp}(\psi(s_m)) = \emptyset$ for $n \neq m$. Then Theorem 4.15 (i) shows that $\mathrm{rank}(P_{H^{(1)}}((-\infty, 0))) = \infty$. Since each eigenvalue E_j has finite multiplicity (it lies in the discrete spectrum), there must be an infinite number of eigenvalues which accumulate at 0. $\square$

If $\gamma \leq 0$, we have $\sigma_d(H^{(1)}) = \emptyset$ since $H^{(1)} \geq 0$ in this case.

Hence we have obtained quite a complete picture of the spectrum of $H^{(1)}$. Next, we could try to compute the eigenvalues of $H^{(1)}$ (in the case $\gamma > 0$) by solving the corresponding eigenvalue equation, which is given by the partial differential equation
$$(10.15) \quad -\Delta \psi(x) - \frac{\gamma}{|x|} \psi(x) = \lambda \psi(x).$$

For a general potential this is hopeless, but in our case we can use the rotational symmetry of our operator to reduce our partial differential equation to ordinary ones.

First of all, it suggests a switch from Cartesian coordinates $x = (x_1, x_2, x_3)$ to **spherical coordinates** (r, θ, φ) defined by
$$(10.16) \quad x_1 = r \sin(\theta) \cos(\varphi), \quad x_2 = r \sin(\theta) \sin(\varphi), \quad x_3 = r \cos(\theta),$$
where $r \in [0, \infty)$, $\theta \in [0, \pi]$, and $\varphi \in (-\pi, \pi]$. This change of coordinates corresponds to a unitary transform
$$(10.17) \quad L^2(\mathbb{R}^3) \to L^2((0, \infty), r^2 dr) \otimes L^2((0, \pi), \sin(\theta) d\theta) \otimes L^2((0, 2\pi), d\varphi).$$

In these new coordinates (r, θ, φ), our operator reads
$$(10.18) \quad H^{(1)} = -\frac{1}{r^2} \frac{\partial}{\partial r} r^2 \frac{\partial}{\partial r} + \frac{1}{r^2} L^2 + V(r), \quad V(r) = -\frac{\gamma}{r},$$
where
$$(10.19) \quad L^2 = L_1^2 + L_2^2 + L_3^2 = -\frac{1}{\sin(\theta)} \frac{\partial}{\partial \theta} \sin(\theta) \frac{\partial}{\partial \theta} - \frac{1}{\sin(\theta)^2} \frac{\partial^2}{\partial \varphi^2}.$$
(Recall the angular momentum operators L_j from Section 8.2.)

Making the product ansatz (separation of variables)
$$(10.20) \quad \psi(r, \theta, \varphi) = R(r) \Theta(\theta) \Phi(\varphi),$$

we obtain the three Sturm–Liouville equations

$$\left(-\frac{1}{r^2}\frac{d}{dr}r^2\frac{d}{dr} + \frac{l(l+1)}{r^2} + V(r)\right)R(r) = \lambda R(r),$$

$$\frac{1}{\sin(\theta)}\left(-\frac{d}{d\theta}\sin(\theta)\frac{d}{d\theta} + \frac{m^2}{\sin(\theta)}\right)\Theta(\theta) = l(l+1)\Theta(\theta),$$

(10.21)
$$-\frac{d^2}{d\varphi^2}\Phi(\varphi) = m^2\Phi(\varphi).$$

The form chosen for the constants $l(l+1)$ and m^2 is for convenience later on. These equations will be investigated in the following sections.

Problem 10.1. *Generalize the virial theorem to the case $U(-s)VU(s) = e^{-\alpha s}V$, $\alpha \in \mathbb{R}\backslash\{0\}$. What about Corollary 10.5?*

10.3. Angular momentum

We start by investigating the equation for $\Phi(\varphi)$ which is associated with the Sturm–Liouville equation

(10.22) $$\tau\Phi = -\Phi'', \quad I = (0, 2\pi).$$

Since we want ψ defined via (10.20) to be in the domain of H_0 (in particular continuous), we choose periodic boundary conditions the Sturm–Liouville equation

(10.23)
$$A\Phi = \tau\Phi, \quad \mathfrak{D}(A) = \{\Phi \in L^2(0, 2\pi) | \ \Phi \in AC^1[0, 2\pi],$$
$$\Phi(0) = \Phi(2\pi), \Phi'(0) = \Phi'(2\pi)\}.$$

From our analysis in Section 9.1, we immediately obtain

Theorem 10.6. *The operator A defined via (10.22) is self-adjoint. Its spectrum is purely discrete, that is,*

(10.24) $$\sigma(A) = \sigma_d(A) = \{m^2 | m \in \mathbb{Z}\},$$

and the corresponding eigenfunctions

(10.25) $$\Phi_m(\varphi) = \frac{1}{\sqrt{2\pi}}e^{im\varphi}, \quad m \in \mathbb{Z},$$

form an orthonormal basis for $L^2(0, 2\pi)$.

Note that except for the lowest eigenvalue, all eigenvalues are twice degenerate.

We note that this operator is essentially the square of the angular momentum in the third coordinate direction, since in polar coordinates

(10.26) $$L_3 = \frac{1}{i}\frac{\partial}{\partial\varphi}.$$

Now we turn to the equation for $\Theta(\theta)$:
$$\tau_m \Theta(\theta) = \frac{1}{\sin(\theta)} \left(-\frac{d}{d\theta} \sin(\theta) \frac{d}{d\theta} + \frac{m^2}{\sin(\theta)} \right) \Theta(\theta), \quad I = (0, \pi), m \in \mathbb{N}_0. \tag{10.27}$$

For the investigation of the corresponding operator we use the unitary transform
$$L^2((0, \pi), \sin(\theta) d\theta) \to L^2((-1, 1), dx), \quad \Theta(\theta) \mapsto f(x) = \Theta(\arccos(x)). \tag{10.28}$$

The operator τ transforms to the somewhat simpler form
$$\tau_m = -\frac{d}{dx}(1 - x^2) \frac{d}{dx} - \frac{m^2}{1 - x^2}. \tag{10.29}$$

The corresponding eigenvalue equation
$$\tau_m u = l(l+1) u \tag{10.30}$$

is the **associated Legendre equation**. For $l \in \mathbb{N}_0$ it is solved by the **associated Legendre functions** [44, (14.6.1)]
$$P_l^m(x) = (-1)^m (1 - x^2)^{m/2} \frac{d^m}{dx^m} P_l(x), \quad |m| \leq l, \tag{10.31}$$

where the
$$P_l(x) = \frac{1}{2^l l!} \frac{d^l}{dx^l} (x^2 - 1)^l, \quad l \in \mathbb{N}_0, \tag{10.32}$$

are the **Legendre polynomials** [44, (14.7.13)] (Problem 10.2). Moreover, note that the $P_l(x)$ are (nonzero) polynomials of degree l and since τ_m depends only on m^2, there must be a relation between $P_l^m(x)$ and $P_l^{-m}(x)$. In fact (Problem 10.3),
$$P_l^{-m}(x) = (-1)^m \frac{(l+m)!}{(l-m)!} P_l^m. \tag{10.33}$$

A second, linearly independent, solution is given by
$$Q_l^m(x) = P_l^m(x) \int_0^x \frac{dt}{(1 - t^2) P_l^m(t)^2}. \tag{10.34}$$

In fact, for every Sturm–Liouville equation, $v(x) = u(x) \int^x \frac{dt}{p(t) u(t)^2}$ satisfies $\tau v = 0$ whenever $\tau u = 0$. Now fix $l = 0$ and note $P_0(x) = 1$. For $m = 0$ we have $Q_0^0 = \operatorname{arctanh}(x) \in L^2$ and so τ_0 is l.c. at both endpoints. For $m > 0$ we have $Q_0^m = (x \pm 1)^{-m/2}(C + O(x \pm 1))$ which shows that it is not square integrable. Thus τ_m is l.c. for $m = 0$ and l.p. for $m > 0$ at both endpoints. In order to make sure that the eigenfunctions for $m = 0$ are continuous (such

10.3. Angular momentum

that ψ defined via (10.20) is continuous), we choose the boundary condition generated by $P_0(x) = 1$ in this case:

$$A_m f = \tau_m f,$$

(10.35) $\mathfrak{D}(A_m) = \{f \in L^2(-1,1) |\ f \in AC^1(-1,1), \tau_m f \in L^2(-1,1),$
$$\lim_{x \to \pm 1}(1-x^2)f'(x) = 0 \text{ if } m = 0\}.$$

Theorem 10.7. *The operator A_m, $m \in \mathbb{N}_0$, defined via (10.35) is self-adjoint. Its spectrum is purely discrete, that is,*

(10.36) $$\sigma(A_m) = \sigma_d(A_m) = \{l(l+1)| l \in \mathbb{N}_0, l \geq m\},$$

and the corresponding eigenfunctions

(10.37) $$u_{l,m}(x) = \sqrt{\frac{2l+1}{2}\frac{(l-m)!}{(l+m)!}} P_l^m(x), \qquad l \in \mathbb{N}_0, l \geq m,$$

form an orthonormal basis for $L^2(-1,1)$.

Proof. By Theorem 9.6, A_m is self-adjoint. Moreover, P_l^m is an eigenfunction corresponding to the eigenvalue $l(l+1)$, and it suffices to show that the P_l^m form a basis. To prove this, it suffices to show that the functions $P_l^m(x)$ are dense. Since $(1-x^2) > 0$ for $x \in (-1,1)$, it suffices to show that the functions $(1-x^2)^{-m/2}P_l^m(x)$ are dense. But the span of these functions contains every polynomial. Every continuous function can be approximated by polynomials (in the sup norm, Theorem 0.20, and hence in the L^2 norm), and since the continuous functions are dense, so are the polynomials.

For the normalization of the eigenfunctions, see Problem 10.7, respectively, [44, (14.17.6)]. $\square$

Returning to our original setting, we conclude that the

(10.38) $$\Theta_l^m(\theta) = \sqrt{\frac{2l+1}{2}\frac{(l+m)!}{(l-m)!}} P_l^m(\cos(\theta)), \quad |m| \leq l,$$

form an orthonormal basis for $L^2((0,\pi), \sin(\theta)d\theta)$ for every fixed $m \in \mathbb{N}_0$.

Theorem 10.8. *The operator L^2 on $L^2((0,\pi), \sin(\theta)d\theta) \otimes L^2((0,2\pi))$ has a purely discrete spectrum given*

(10.39) $$\sigma(L^2) = \{l(l+1)| l \in \mathbb{N}_0\}.$$

The **spherical harmonics**
(10.40)
$$Y_l^m(\theta,\varphi) = \Theta_l^m(\theta)\Phi_m(\varphi) = \sqrt{\frac{2l+1}{4\pi}\frac{(l-m)!}{(l+m)!}} P_l^m(\cos(\theta))e^{im\varphi}, \quad |m| \leq l,$$

form an orthonormal basis and satisfy $L^2 Y_l^m = l(l+1) Y_l^m$ and $L_3 Y_l^m = m Y_l^m$.

Proof. Everything follows from our construction, if we can show that the Y_l^m form a basis. But this follows as in the proof of Lemma 1.10. □

Note that transforming the Y_l^m back to Cartesian coordinates gives
(10.41)
$$Y_l^{\pm m}(x) = (-1)^m \sqrt{\frac{2l+1}{4\pi} \frac{(l-|m|)!}{(l+|m|)!}} \tilde{P}_l^m(\frac{x_3}{r}) \left(\frac{x_1 \pm ix_2}{r}\right)^m, \quad r = |x|,$$

where $\tilde{P}_l^m$ is a polynomial of degree $l - m$ given by

(10.42)
$$\tilde{P}_l^m(x) = (1-x^2)^{-m/2} P_l^m(x) = \frac{d^{l+m}}{dx^{l+m}}(1-x^2)^l.$$

In particular, the Y_l^m are smooth away from the origin, and by construction, they satisfy

(10.43)
$$-\Delta Y_l^m = \frac{l(l+1)}{r^2} Y_l^m.$$

Problem 10.2. Show that the associated Legendre functions satisfy the differential equation (10.30). (Hint: Start with the Legendre polynomials (10.32) which correspond to $m = 0$. Set $v(x) = (x^2 - 1)^l$ and observe $(x^2 - 1) v'(x) = 2lx\, v(x)$. Then differentiate this identity $l + 1$ times using Leibniz's rule. For the case of the associated Legendre functions, substitute $v(x) = (1 - x^2)^{m/2} u(x)$ in (10.30) and differentiate the resulting equation once.)

Problem 10.3. Show (10.33). (Hint: Write $(x^2 - 1)^l = (x-1)^l (x+1)^l$ and use Leibniz's rule.)

Problem 10.4 (Orthogonal polynomials)**.** Suppose the monic polynomials $P_j(x) = x^j + \beta_j x^{j-1} + \ldots$ are orthogonal with respect to the weight function $w(x)$:
$$\int_a^b P_i(x) P_j(x) w(x) dx = \begin{cases} \alpha_j^2, & i = j, \\ 0, & \text{otherwise.} \end{cases}$$

Note that they are uniquely determined by the Gram–Schmidt procedure. Let $\bar{P}_j(x) = \alpha_j^{-1} P(x)$ and show that they satisfy the three term recurrence relation
$$a_j \bar{P}_{j+1}(x) + b_j \bar{P}_j(x) + a_{j-1} \bar{P}_{j-1}(x) = x \bar{P}_j(x),$$

where
$$a_j = \int_a^b x \bar{P}_{j+1}(x) \bar{P}_j(x) w(x) dx, \quad b_j = \int_a^b x \bar{P}_j(x)^2 w(x) dx.$$

Moreover, show
$$a_j = \frac{\alpha_{j+1}}{\alpha_j}, \qquad b_j = \beta_j - \beta_{j+1}.$$
(Note that $w(x)dx$ could be replaced by a measure $d\mu(x)$.)

Problem 10.5. Consider the orthogonal polynomials with respect to the weight function $w(x)$ as in the previous problem. Suppose $|w(x)| \le Ce^{-k|x|}$ for some $C, k > 0$. Show that the orthogonal polynomials are dense in $L^2(\mathbb{R}, w(x)dx)$. (Hint: It suffices to show that $\int f(x)x^j w(x)dx = 0$ for all $j \in \mathbb{N}_0$ implies $f = 0$. Consider the Fourier transform of $f(x)w(x)$ and note that it has an analytic extension by Problem 7.8. Hence this Fourier transform will be zero if, e.g., all derivatives at $p = 0$ are zero (cf. Problem 7.3).)

Problem 10.6. Show
$$P_l(x) = \sum_{k=0}^{\lfloor l/2 \rfloor} \frac{(-1)^k (2l-2k)!}{2^l k! (l-k)! (l-2k)!} x^{l-2k}.$$
Moreover, by Problem 10.4, there is a recurrence relation of the form $P_{l+1}(x) = (\tilde{a}_l + \tilde{b}_l x) P_l(x) + \tilde{c}_l P_{l-1}(x)$. Find the coefficients by comparing the highest powers in x and conclude
$$(l+1)P_{l+1}(x) = (2l+1)xP_l(x) - lP_{l-1}.$$
Use this to prove
$$\int_{-1}^{1} P_l(x)^2 dx = \frac{2}{2l+1}.$$

Problem 10.7. Prove
$$\int_{-1}^{1} P_l^m(x)^2 dx = \frac{2}{2l+1} \frac{(l+m)!}{(l-m)!}.$$
(Hint: Use (10.33) to compute $\int_{-1}^{1} P_l^m(x) P_l^{-m}(x) dx$ by integrating by parts until you can use the case $m = 0$ from the previous problem.)

10.4. The eigenvalues of the hydrogen atom

Now we want to use the considerations from the previous section to decompose the Hamiltonian of the hydrogen atom. In fact, we can even admit any spherically symmetric potential $V(x) = V(|x|)$ with

(10.44) $\qquad V(r) \in L^\infty_\infty((0, \infty)) + L^2((0, \infty), r^2 dr)$

such that Theorem 10.2 holds.

The important observation is that the spaces

(10.45) $\qquad \mathfrak{H}_{l,m} = \{\psi(x) = R(r)Y_l^m(\theta, \varphi) | R(r) \in L^2((0, \infty), r^2 dr)\}$

with corresponding projectors
(10.46)
$$P_l^m \psi(r, \theta, \varphi) = \left(\int_0^{2\pi} \int_0^\pi \psi(r, \theta', \varphi') Y_l^m(\theta', \varphi') \sin(\theta') d\theta' d\varphi' \right) Y_l^m(\theta, \varphi)$$

reduce our operator $H = H_0 + V$. By Lemma 2.25 it suffices to check this for H restricted to $C_c^\infty(\mathbb{R}^3)$, which is straightforward. Hence, again by Lemma 2.25,

(10.47)
$$H = H_0 + V = \bigoplus_{l,m} \tilde{H}_l,$$

where

$$\tilde{H}_l R(r) = \tilde{\tau}_l R(r), \quad \tilde{\tau}_l = -\frac{1}{r^2} \frac{d}{dr} r^2 \frac{d}{dr} + \frac{l(l+1)}{r^2} + V(r),$$
(10.48)
$$\mathfrak{D}(H_l) \subseteq L^2((0,\infty), r^2 dr).$$

Using the unitary transformation

(10.49) $\quad L^2((0,\infty), r^2 dr) \to L^2((0,\infty)), \quad R(r) \mapsto u(r) = r R(r),$

our operator transforms to

$$A_l f = \tau_l f, \quad \tau_l = -\frac{d^2}{dr^2} + \frac{l(l+1)}{r^2} + V(r),$$
(10.50)
$$\mathfrak{D}(A_l) = P_l^m \mathfrak{D}(H) \subseteq L^2((0,\infty)).$$

It remains to investigate this operator (that its domain is indeed independent of m follows from the next theorem).

Theorem 10.9. *Suppose* (10.44). *The domain of the operator A_l is given by*

(10.51) $\quad \mathfrak{D}(A_l) = \{f \in L^2((0,\infty)) |\ f, f' \in AC((0,\infty)), \tau f \in L^2((0,\infty)),$
$$\lim_{r \to 0}(f(r) - r f'(r)) = 0 \text{ if } l = 0\},$$

where $I = (0,\infty)$. Moreover,

(10.52) $\quad \sigma_{ess}(A_l) = [0, \infty).$

If, in addition, $L_\infty^\infty((0,\infty)) \cap AC([0,\infty]) + L^2((0,\infty), r^2 dr)$, then

(10.53) $\quad \sigma_{ess}(A_l) = \sigma_{ac}(A_l) = [0, \infty), \quad \sigma_{sc}(A_l) = \emptyset, \quad \sigma_p \subset (-\infty, 0].$

Proof. By construction of A_l, we know that it is self-adjoint and satisfies $\sigma_{ess}(A_l) \subseteq [0, \infty)$ (Problem 10.8). So it remains to compute the domain. We know at least $\mathfrak{D}(A_l) \subseteq \mathfrak{D}(\tau)$ and since $\mathfrak{D}(H) = \mathfrak{D}(H_0)$, it suffices to consider the case $V = 0$. In this case, the solutions of $-u''(r) + \frac{l(l+1)}{r^2} u(r) = 0$ are given by $u(r) = \alpha r^{l+1} + \beta r^{-l}$. Thus we are in the l.p. case at ∞ for every $l \in \mathbb{N}_0$. However, at 0 we are in the l.p. case only if $l > 0$; that is, we need an additional boundary condition at 0 if $l = 0$. Since we need $R(r) = \frac{u(r)}{r}$ to

10.4. The eigenvalues of the hydrogen atom

be bounded (such that (10.20) is in the domain of H_0, that is, continuous), we have to take the boundary condition generated by $u(r) = r$.

Concerning the last claim, note that under this additional assumption we can apply Lemma 9.37. Hence we have $(0, \infty) \subseteq N_\infty(\tau_l)$ and Theorem 9.31 implies $\sigma_{ac}(A_l) = [0, \infty)$, $\sigma_{sc}(A_l) = \emptyset$, and $\sigma_p \subset (-\infty, 0]$. □

Finally, let us turn to some special choices for V, where the corresponding differential equation can be explicitly solved. The simplest case is $V = 0$. In this case, the solutions of

$$-u''(r) + \frac{l(l+1)}{r^2} u(r) = z u(r) \tag{10.54}$$

are given by

$$u(r) = \alpha \, z^{-l/2} r \, j_l(\sqrt{z} r) + \beta \, z^{(l+1)/2} r \, y_l(\sqrt{z} r), \tag{10.55}$$

where $j_l(r)$ and $y_l(r)$ are the **spherical Bessel**, respectively, **spherical Neumann**, functions

$$j_l(r) = \sqrt{\frac{\pi}{2r}} J_{l+1/2}(r) = (-r)^l \left(\frac{1}{r}\frac{d}{dr}\right)^l \frac{\sin(r)}{r},$$

$$y_l(r) = \sqrt{\frac{\pi}{2r}} Y_{l+1/2}(r) = -(-r)^l \left(\frac{1}{r}\frac{d}{dr}\right)^l \frac{\cos(r)}{r}. \tag{10.56}$$

Note that $z^{-l/2} r \, j_l(\sqrt{z} r)$ and $z^{(l+1)/2} r \, y_l(\sqrt{z} r)$ are entire as functions of z and their Wronskian is given by $W(z^{-l/2} r \, j_l(\sqrt{z} r), z^{(l+1)/2} r \, y_l(\sqrt{z} r)) = 1$. See [44, Chapter 10]. In particular,

$$u_a(z, r) = \frac{r}{z^{l/2}} j_l(\sqrt{z} r) = \frac{2^l l!}{(2l+1)!} r^{l+1} (1 + O(r^2)),$$

$$u_b(z, r) = \sqrt{-z} r \left(j_l(i\sqrt{-z} r) + i y_l(i\sqrt{-z} r) \right) = e^{-\sqrt{-z} r + i l \pi/2} (1 + O(\tfrac{1}{r})) \tag{10.57}$$

are the functions which are square integrable and satisfy the boundary condition (if any) near $a = 0$ and $b = \infty$, respectively.

The second case is that of our Coulomb potential

$$V(r) = -\frac{\gamma}{r}, \qquad \gamma > 0, \tag{10.58}$$

where we will try to compute the eigenvalues plus corresponding eigenfunctions. It turns out that they can be expressed in terms of the **Laguerre polynomials** ([44, (18.5.5)])

$$L_j(r) = \frac{e^r}{j!} \frac{d^j}{dr^j} e^{-r} r^j \tag{10.59}$$

and the **generalized Laguerre polynomials** ([44, (18.5.5)])

(10.60) $$L_j^{(k)}(r) = (-1)^k \frac{d^k}{dr^k} L_{j+k}(r).$$

Note that the $L_j^{(k)}(r)$ are polynomials of degree $j - k$ which are explicitly given by

(10.61) $$L_j^{(k)}(r) = \sum_{i=0}^{j} (-1)^i \binom{j+k}{j-i} \frac{r^i}{i!}$$

and satisfy the differential equation (Problem 10.9)

(10.62) $$r\, y''(r) + (k+1-r) y'(r) + j\, y(r) = 0.$$

Moreover, they are orthogonal in the Hilbert space $L^2((0,\infty), r^k e^{-r} dr)$ (Problem 10.11):

(10.63) $$\int_0^\infty L_i^{(k)}(r) L_j^{(k)}(r) r^k e^{-r} dr = \begin{cases} \frac{(j+k)!}{j!}, & i = j, \\ 0, & \text{otherwise.} \end{cases}$$

Theorem 10.10. *The eigenvalues of $H^{(1)}$ are explicitly given by*

(10.64) $$E_n' = -\left(\frac{\gamma}{2(n+1)}\right)^2, \qquad n \in \mathbb{N}_0.$$

An orthonormal basis for the corresponding eigenspace is given by the $(n+1)^2$ functions

(10.65) $$\psi_{n,l,m}(x) = R_{n,l}(r) Y_l^m(x), \qquad |m| \le l \le n,$$

where
(10.66)
$$R_{n,l}(r) = \sqrt{\frac{\gamma^3 (n-l)!}{2(n+1)^4 (n+l+1)!}} \left(\frac{\gamma r}{n+1}\right)^l e^{-\frac{\gamma r}{2(n+1)}} L_{n-l}^{(2l+1)}\!\left(\frac{\gamma r}{n+1}\right).$$

In particular, the lowest eigenvalue $E_0 = -\frac{\gamma^2}{4}$ is simple and the corresponding eigenfunction $\psi_{000}(x) = \sqrt{\frac{\gamma^3}{2}} e^{-\gamma r/2}$ is positive.

Proof. Since all eigenvalues are negative, we need to look at the equation

$$-u''(r) + \left(\frac{l(l+1)}{r^2} - \frac{\gamma}{r}\right) u(r) = \lambda u(r)$$

for $\lambda < 0$. Introducing new variables $x = 2\sqrt{-\lambda}\, r$ and $v(x) = \frac{e^{x/2}}{x^{l+1}} u(\frac{x}{2\sqrt{-\lambda}})$, this equation transforms into Kummer's equation

$$x v''(x) + (k+1-x) v'(x) + j\, v(x) = 0, \quad k = 2l+1,\ j = \frac{\gamma}{2\sqrt{-\lambda}} - (l+1).$$

10.4. The eigenvalues of the hydrogen atom

Now let us search for a solution which can be expanded into a convergent power series

(10.67) $$v(x) = \sum_{i=0}^{\infty} v_i x^i, \quad v_0 = 1.$$

The corresponding $u(r)$ is square integrable near 0 and satisfies the boundary condition (if any). Thus we need to find those values of λ for which it is square integrable near $+\infty$.

Substituting the ansatz (10.67) into our differential equation and comparing powers of x gives the following recursion for the coefficients:

$$v_{i+1} = \frac{(i-j)}{(i+1)(i+k+1)} v_i$$

and thus

$$v_i = \frac{1}{i!} \prod_{\ell=0}^{i-1} \frac{\ell - j}{\ell + k + 1}.$$

Now there are two cases to distinguish. If $j \in \mathbb{N}_0$, then $v_i = 0$ for $i > j$ and $v(x)$ is a polynomial; namely

$$v(x) = \binom{j+k}{j}^{-1} L_j^{(k)}(x).$$

In this case, $u(r)$ is square integrable and hence an eigenfunction corresponding to the eigenvalue $\lambda_j = -(\frac{\gamma}{2(n+1)})^2$, $n = j + l$. This proves the formula for $R_{n,l}(r)$ except for the normalization which follows from (Problem 10.11)

(10.68) $$\int_0^\infty L_j^{(k)}(r)^2 r^{k+1} e^{-r} dr = \frac{(j+k)!}{j!}(2j+k+1).$$

It remains to show that we have found all eigenfunctions, that is, that there are no other square integrable solutions. Otherwise, if $j \notin \mathbb{N}$, we have $\frac{v_{i+1}}{v_i} \geq \frac{(1-\varepsilon)}{i+1}$ for i sufficiently large. Hence by adding a polynomial to $v(x)$ (and perhaps flipping its sign), we can get a function $\tilde{v}(x)$ such that $\tilde{v}_i \geq \frac{(1-\varepsilon)^i}{i!}$ for all i. But then $\tilde{v}(x) \geq \exp((1-\varepsilon)x)$ and thus the corresponding $u(r)$ is not square integrable near $+\infty$. □

Finally, let us also look at an alternative algebraic approach for computing the eigenvalues and eigenfunctions of A_l based on the commutation methods from Section 8.4. We begin by introducing

$$Q_l f = -\frac{d}{dr} + \frac{l+1}{r} - \frac{\gamma}{2(l+1)},$$

(10.69) $$\mathfrak{D}(Q_l) = \{f \in L^2((0,\infty)) | f \in AC((0,\infty)), Q_l f \in L^2((0,\infty))\}.$$

Then (Problem 9.3) Q_l is closed, and its adjoint is given by

$$Q_l^* f = \frac{d}{dr} + \frac{l+1}{r} - \frac{\gamma}{2(l+1)},$$

(10.70) $\mathfrak{D}(Q_l^*) = \{f \in L^2((0,\infty)) |\ f \in AC((0,\infty)), Q_l^* f \in L^2((0,\infty)),$
$$\lim_{x \to 0, \infty} f(x)g(x) = 0,\ \forall g \in \mathfrak{D}(Q_l)\}.$$

It is straightforward to check

(10.71) $\qquad \operatorname{Ker}(Q_l) = \operatorname{span}\{u_{l,0}\}, \qquad \operatorname{Ker}(Q_l^*) = \{0\},$

where

(10.72) $\qquad u_{l,0}(r) = \dfrac{1}{\sqrt{(2l+2)!}} \left(\dfrac{\gamma}{l+1}\right)^{(l+1)+1/2} r^{l+1} e^{-\frac{\gamma}{2(l+1)} r}$

is normalized.

Theorem 10.11. *The radial Schrödinger operator A_l with Coulomb potential (10.58) satisfies*

(10.73) $\qquad A_l = Q_l^* Q_l - c_l^2, \qquad A_{l+1} = Q_l Q_l^* - c_l^2,$

where

(10.74) $\qquad c_l = \dfrac{\gamma}{2(l+1)}.$

Proof. Equality is easy to check for $f \in AC^2$ with compact support. Hence $Q_l^* Q_l - c_l^2$ is a self-adjoint extension of τ_l restricted to this set. If $l > 0$, there is only one self-adjoint extension and equality follows. If $l = 0$, we know $u_{0,0} \in \mathfrak{D}(Q_l^* Q_l)$, and since A_l is the only self-adjoint extension with $u_{0,0} \in \mathfrak{D}(A_l)$, equality follows in this case as well. □

Hence, as a consequence of Theorem 8.6 we see $\sigma(A_l) = \sigma(A_{l+1}) \cup \{-c_l^2\}$, or, equivalently,

(10.75) $\qquad \sigma_p(A_l) = \{-c_j^2 | j \geq l\}$

if we use that $\sigma_p(A_l) \subset (-\infty, 0)$, which already follows from the virial theorem. Moreover, using Q_l, we can turn any eigenfunction of H_l into one of H_{l+1}. However, we only know the lowest eigenfunction $u_{l,0}$, which is mapped to 0 by Q_l. On the other hand, we can also use Q_l^* to turn an eigenfunction of H_{l+1} into one of H_l. Hence $Q_l^* u_{l+1,0}$ will give the second eigenfunction of H_l. Proceeding inductively, the normalized eigenfunction of H_l corresponding to the eigenvalue $-c_{l+j}^2$ is given by

(10.76) $\qquad u_{l,j} = \left(\displaystyle\prod_{k=0}^{j-1}(c_{l+j} - c_{l+k})\right)^{-1} Q_l^* Q_{l+1}^* \cdots Q_{l+j-1}^* u_{l+j,0}.$

10.4. The eigenvalues of the hydrogen atom

The connection with Theorem 10.10 is given by

(10.77) $$R_{n,l}(r) = \frac{1}{r}u_{l,n-l}(r).$$

Problem 10.8. Let $A = \bigoplus_n A_n$. Then $\overline{\bigcup_n \sigma_{ess}(A_n)} \subseteq \sigma_{ess}(A)$. Give an example where equality does not hold.

Problem 10.9. Show that the generalized Laguerre polynomials satisfy the differential equation (10.62). (Hint: Start with the Laguerre polynomials (10.59) which correspond to $k = 0$. Set $v(r) = r^j e^{-r}$ and observe $r v'(r) = (j - r)v(r)$. Then differentiate this identity $j + 1$ times using Leibniz's rule. For the case of the generalized Laguerre polynomials, start with the differential equation for $L_{j+k}(r)$ and differentiate k times.)

Problem 10.10. Show that the differential equation (10.59) can be rewritten in Sturm–Liouville form as

$$-r^{-k}e^r \frac{d}{dr} r^{k+1} e^{-r} \frac{d}{dr} u = ju.$$

We have found one entire solution in the proof of Theorem 10.10. Show that any linearly independent solution behaves like $\log(r)$ if $k = 0$, respectively, like r^{-k} otherwise. Show that it is l.c. at the endpoint $r = 0$ if $k = 0$ and l.p. otherwise.

Let $\mathfrak{H} = L^2((0, \infty), r^k e^{-r} dr)$. The operator

$$A_k f = \tau f = -r^{-k} e^r \frac{d}{dr} r^{k+1} e^{-r} \frac{d}{dr} f,$$

$$\mathfrak{D}(A_k) = \{f \in \mathfrak{H} | \; f \in AC^1(0, \infty), \tau_k f \in \mathfrak{H},$$
$$\lim_{r \to 0} rf'(r) = 0 \text{ if } k = 0\}$$

for $k \in \mathbb{N}_0$ is self-adjoint. Its spectrum is purely discrete, that is,

(10.78) $$\sigma(A_k) = \sigma_d(A_k) = \mathbb{N}_0,$$

and the corresponding eigenfunctions

(10.79) $$L_j^{(k)}(r), \quad j \in \mathbb{N}_0,$$

form an orthogonal base for $\mathfrak{H}$. (Hint: Compare the argument for the associated Legendre equation and Problem 10.5.)

Problem 10.11. By Problem 10.4 there is a recurrence relation of the form $L_{j+1}^{(k)}(r) = (\tilde{a}_j + \tilde{b}_j r) L_j^{(k)}(r) + \tilde{c}_j L_{j-1}^{(k)}(r)$. Find the coefficients by comparing the highest powers in r and conclude

$$L_{j+1}^{(k)}(r) = \frac{1}{1+j}\left((2j + k + 1 - r)L_j^{(k)}(r) - (j + k)L_{j-1}^{(k)}(r)\right).$$

Use this to prove (10.63) and (10.68). (Hint: Orthogonality follows from the previous problem, and to see the normalization in (10.63), use orthogonality and the above recursion (twice) to relate the integral for j to the one for $j-1$.)

10.5. Nondegeneracy of the ground state

The lowest eigenvalue (below the essential spectrum) of a Schrödinger operator is called the **ground state**. Since the laws of physics state that a quantum system will transfer energy to its surroundings (e.g., an atom emits radiation) until it eventually reaches its ground state, this state is in some sense the most important state. We have seen that the hydrogen atom has a nondegenerate (simple) ground state with a corresponding positive eigenfunction. In particular, the hydrogen atom is stable in the sense that there is a lowest possible energy. This is quite surprising since the corresponding classical mechanical system is not — the electron could fall into the nucleus!

Our aim in this section is to show that the ground state is simple with a corresponding positive eigenfunction. Note that it suffices to show that any ground state eigenfunction is positive since nondegeneracy then follows for free: two positive functions cannot be orthogonal.

To set the stage, let us introduce some notation. Let $\mathfrak{H} = L^2(\mathbb{R}^n)$. We call $f \in L^2(\mathbb{R}^n)$ **positive** if $f \geq 0$ a.c. and $f \not\equiv 0$. We call f **strictly positive** if $f > 0$ a.e. A bounded operator A is called **positivity preserving** if $f \geq 0$ implies $Af \geq 0$ and **positivity improving** if $f \geq 0$ implies $Af > 0$ for $f \not\equiv 0$. Clearly A is positivity preserving (improving) if and only if $\langle g, Af \rangle \geq 0$ (> 0) for $f, g \geq 0$ and $f, g \not\equiv 0$.

Example. Multiplication by a positive function is positivity preserving (but not improving). Convolution with a strictly positive function is positivity improving. $\diamond$

We first show that positivity improving operators have positive eigenfunctions.

Theorem 10.12. *Suppose $A \in \mathfrak{L}(L^2(\mathbb{R}^n))$ is a self-adjoint, positivity improving and real (i.e., it maps real functions to real functions) operator. If $\|A\|$ is an eigenvalue, then it is simple and the corresponding eigenfunction is strictly positive.*

Proof. Let ψ be an eigenfunction. It is no restriction to assume that ψ is real (since A is real, both real and imaginary parts of ψ are eigenfunctions as well). We assume $\|\psi\| = 1$ and denote by $\psi_\pm = \frac{|\psi| \pm \psi}{2}$ the positive and negative parts of ψ. Then by $|A\psi| = |A\psi_+ - A\psi_-| \leq A\psi_+ + A\psi_- = A|\psi|$ we have

$$\|A\| = \langle \psi, A\psi \rangle \leq \langle |\psi|, |A\psi| \rangle \leq \langle |\psi|, A|\psi| \rangle \leq \|A\|;$$

10.5. Nondegeneracy of the ground state

that is, $\langle \psi, A\psi \rangle = \langle |\psi|, A|\psi| \rangle$ and thus

$$\langle \psi_+, A\psi_- \rangle = \frac{1}{4}(\langle |\psi|, A|\psi| \rangle - \langle \psi, A\psi \rangle) = 0.$$

Consequently $\psi_- = 0$ or $\psi_+ = 0$ since otherwise $A\psi_- > 0$ and hence also $\langle \psi_+, A\psi_- \rangle > 0$. Without restriction, $\psi = \psi_+ \geq 0$ and since A is positivity increasing, we even have $\psi = \|A\|^{-1} A\psi > 0$.

If there were a second eigenfunction it could also be chosen positive by the above argument and hence could not be orthogonal to the first. □

So we need a positivity improving operator. By (7.49) and (7.50) both e^{-tH_0}, $t > 0$, and $R_\lambda(H_0)$, $\lambda < 0$, are since they are given by convolution with a strictly positive function. Our hope is that this property carries over to $H = H_0 + V$.

Theorem 10.13. *Suppose $H = H_0 + V$ is self-adjoint and bounded from below with $C_c^\infty(\mathbb{R}^n)$ as a core. If $E_0 = \min \sigma(H)$ is an eigenvalue, it is simple and the corresponding eigenfunction is strictly positive.*

Proof. We first show that e^{-tH}, $t > 0$, is positivity preserving. If we set $H_n = H_0 + V_n$ with $V_n = V \chi_{\{x \mid |V(x)| \leq n\}}$, then V_n is bounded and e^{-tH_n} is positivity preserving by the Trotter product formula (Theorem 5.12) since both e^{-tH_0} and e^{-tV_n} are. Moreover, we have $H_n \psi \to H\psi$ for $\psi \in C_c^\infty(\mathbb{R}^n)$ (note that necessarily $V \in L^2_{loc}$) and hence $H_n \xrightarrow{sr} H$ in the strong resolvent sense by Lemma 6.36. Hence $e^{-tH_n} \xrightarrow{s} e^{-tH}$ by Theorem 6.31, which shows that e^{-tH} is at least positivity preserving (since 0 cannot be an eigenvalue of e^{-tH}, it cannot map a positive function to 0).

Next I claim that for ψ positive the closed set

$$N(\psi) = \{\varphi \in L^2(\mathbb{R}^n) \mid \varphi \geq 0, \langle \varphi, e^{-sH} \psi \rangle = 0 \,\forall s \geq 0\}$$

is just $\{0\}$. If $\varphi \in N(\psi)$, we have by $e^{-sH} \psi \geq 0$ that $\varphi e^{-sH} \psi = 0$. Hence $e^{tV_n} \varphi e^{-sH} \psi = 0$; that is, $e^{tV_n} \varphi \in N(\psi)$. In other words, both e^{tV_n} and e^{-tH} leave $N(\psi)$ invariant, and invoking Trotter's formula again, the same is true for

$$e^{-t(H-V_n)} = \underset{k\to\infty}{\text{s-lim}} \left(e^{-\frac{t}{k}H} e^{\frac{t}{k}V_n}\right)^k.$$

Since $e^{-t(H-V_n)} \xrightarrow{s} e^{-tH_0}$, we finally obtain that e^{-tH_0} leaves $N(\psi)$ invariant, but this operator is positivity increasing and thus $N(\psi) = \{0\}$.

Now it remains to use (7.48), which shows

$$\langle \varphi, R_H(\lambda)\psi \rangle = \int_0^\infty e^{\lambda t} \langle \varphi, e^{-tH} \psi \rangle dt > 0, \quad \lambda < E_0,$$

for φ, ψ positive. So $R_H(\lambda)$ is positivity increasing for $\lambda < E_0$.

If ψ is an eigenfunction of H corresponding to E_0, it is an eigenfunction of $R_H(\lambda)$ corresponding to $\frac{1}{E_0-\lambda}$, and the claim follows since $\|R_H(\lambda)\| = \frac{1}{E_0-\lambda}$. $\square$

The assumptions are, for example, satisfied for the potentials V considered in Theorem 10.2.

Problem 10.12. *Suppose A is a bounded integral operator in $L^2(\mathbb{R}^n)$. Show that it is positivity preserving if and only if its kernel $A(x,y)$ is positive. (Hint: Problem 0.41.)*

Chapter 11

Atomic Schrödinger operators

11.1. Self-adjointness

In this section, we want to have a look at the Hamiltonian corresponding to more than one interacting particle. It is given by

$$(11.1) \qquad H = -\sum_{j=1}^{N} \Delta_j + \sum_{j<k}^{N} V_{j,k}(x_j - x_k).$$

We first consider the case of two particles, which will give us a feeling for how the many-particle case differs from the one-particle case and how the difficulties can be overcome.

We denote the coordinates corresponding to the first particle by $x_1 = (x_{1,1}, x_{1,2}, x_{1,3})$ and those corresponding to the second particle by $x_2 = (x_{2,1}, x_{2,2}, x_{2,3})$. If we assume that the interaction is again of the Coulomb type, the Hamiltonian is given by

$$(11.2) \qquad H = -\Delta_1 - \Delta_2 - \frac{\gamma}{|x_1 - x_2|}, \quad \mathfrak{D}(H) = H^2(\mathbb{R}^6).$$

Since Theorem 10.2 does not allow singularities for $n \geq 3$, it does *not* tell us whether H is self-adjoint. Let

$$(11.3) \qquad (y_1, y_2) = \frac{1}{\sqrt{2}} \begin{pmatrix} \mathbb{I} & \mathbb{I} \\ -\mathbb{I} & \mathbb{I} \end{pmatrix} (x_1, x_2).$$

Then H reads in this new coordinate system as

$$
(11.4) \qquad H = (-\Delta_1) + (-\Delta_2 - \frac{\gamma/\sqrt{2}}{|y_2|}).
$$

In particular, it is the sum of a free particle plus a particle in an external Coulomb field. From a physics point of view, the first part corresponds to the center of mass motion and the second part to the relative motion.

Using that $\gamma/(\sqrt{2}|y_2|)$ has $(-\Delta_2)$-bound 0 in $L^2(\mathbb{R}^3)$, it is not hard to see that the same is true for the $(-\Delta_1 - \Delta_2)$-bound in $L^2(\mathbb{R}^6)$ (details will follow in the next section). In particular, H is self-adjoint and semi-bounded for every $\gamma \in \mathbb{R}$. Moreover, you might suspect that $\gamma/(\sqrt{2}|y_2|)$ is relatively compact with respect to $-\Delta_1 - \Delta_2$ in $L^2(\mathbb{R}^6)$ since it is with respect to $-\Delta_2$ in $L^2(\mathbb{R}^6)$. However, this is *not* true! This is due to the fact that $\gamma/(\sqrt{2}|y_2|)$ does not vanish as $|y| \to \infty$.

Let us look at this problem from the physical viewpoint. If $\lambda \in \sigma_{ess}(H)$, this means that the movement of the whole system is somehow unbounded. There are two possibilities for this.

First, both particles are far away from each other (such that we can neglect the interaction) and the energy corresponds to the sum of the kinetic energies of both particles. Since both can be arbitrarily small (but positive), we expect $[0, \infty) \subseteq \sigma_{ess}(H)$.

Secondly, both particles remain close to each other and move together. In the last set of coordinates, this corresponds to a bound state of the second operator. Hence we expect $[\lambda_0, \infty) \subseteq \sigma_{ess}(H)$, where $\lambda_0 = -\gamma^2/8$ is the smallest eigenvalue of the second operator if the forces are attracting ($\gamma \geq 0$) and $\lambda_0 = 0$ if they are repelling ($\gamma \leq 0$).

It is not hard to translate this intuitive idea into a rigorous proof. Let $\psi_1(y_1)$ be a Weyl sequence corresponding to $\lambda \in [0, \infty)$ for $-\Delta_1$ and let $\psi_2(y_2)$ be a Weyl sequence corresponding to λ_0 for $-\Delta_2 - \gamma/(\sqrt{2}|y_2|)$. Then, $\psi_1(y_1)\psi_2(y_2)$ is a Weyl sequence corresponding to $\lambda + \lambda_0$ for H and thus $[\lambda_0, \infty) \subseteq \sigma_{ess}(H)$. Conversely, we have $-\Delta_1 \geq 0$, respectively, $-\Delta_2 - \gamma/(\sqrt{2}|y_2|) \geq \lambda_0$, and hence $H \geq \lambda_0$. Thus we obtain

$$
(11.5) \qquad \sigma(H) = \sigma_{ess}(H) = [\lambda_0, \infty), \quad \lambda_0 = \begin{cases} -\gamma^2/8, & \gamma \geq 0, \\ 0, & \gamma \leq 0. \end{cases}
$$

Clearly, the physically relevant information is the spectrum of the operator $-\Delta_2 - \gamma/(\sqrt{2}|y_2|)$ which is hidden by the spectrum of $-\Delta_1$. Hence, in order to reveal the physics, one first has to *remove* the center of mass motion.

To avoid clumsy notation, we will restrict ourselves to the case of one atom with N electrons whose nucleus is fixed at the origin. In particular, this implies that we do not have to deal with the center of mass motion

11.1. Self-adjointness

encountered in our example above. In this case, the Hamiltonian is given by

$$H^{(N)} = -\sum_{j=1}^{N} \Delta_j - \sum_{j=1}^{N} V_{ne}(x_j) + \sum_{j=1}^{N}\sum_{j<k}^{N} V_{ee}(x_j - x_k),$$

(11.6) $\quad \mathfrak{D}(H^{(N)}) = H^2(\mathbb{R}^{3N}),$

where V_{ne} describes the interaction of one electron with the nucleus and V_{ee} describes the interaction of two electrons. Explicitly, we have

(11.7) $\quad V_j(x) = \dfrac{\gamma_j}{|x|}, \qquad \gamma_j > 0, \; j = ne, ee.$

We first need to establish the self-adjointness of $H^{(N)} = H_0 + V^{(N)}$. This will follow from Kato's theorem.

Theorem 11.1 (Kato). *Let $V_k \in L^\infty(\mathbb{R}^d) + L^2(\mathbb{R}^d)$, $d \leq 3$, be real-valued and let $V_k(y^{(k)})$ be the multiplication operator in $L^2(\mathbb{R}^n)$, $n = Nd$, obtained by letting $y^{(k)}$ be the first d coordinates of a unitary transform of $\mathbb{R}^n$. Then V_k is H_0 bounded with H_0-bound 0. In particular,*

(11.8) $\quad H = H_0 + \sum_{k} V_k(y^{(k)}), \qquad \mathfrak{D}(H) = H^2(\mathbb{R}^n),$

is self-adjoint and $C_0^\infty(\mathbb{R}^n)$ is a core.

Proof. It suffices to consider one k. After a unitary transform of $\mathbb{R}^n$ we can assume $y^{(1)} = (x_1, \ldots, x_d)$ since such transformations leave both the scalar product of $L^2(\mathbb{R}^n)$ and H_0 invariant. Now let $\psi \in \mathcal{S}(\mathbb{R}^n)$. Then

$$\|V_k\psi\|^2 \leq a^2 \int_{\mathbb{R}^n} |\Delta_1\psi(x)|^2 d^n x + b^2 \int_{\mathbb{R}^n} |\psi(x)|^2 d^n x,$$

where $\Delta_1 = \sum_{j=1}^{d} \partial^2/\partial^2 x_j$, by Theorem 10.2. Hence we obtain

$$\|V_k\psi\|^2 \leq a^2 \int_{\mathbb{R}^n} |\sum_{j=1}^{d} p_j^2 \hat\psi(p)|^2 d^n p + b^2 \|\psi\|^2$$

$$\leq a^2 \int_{\mathbb{R}^n} |\sum_{j=1}^{n} p_j^2 \hat\psi(p)|^2 d^n p + b^2 \|\psi\|^2$$

$$= a^2 \|H_0\psi\|^2 + b^2 \|\psi\|^2,$$

which implies that V_k is relatively bounded with bound 0. The rest follows from the Kato–Rellich theorem. $\square$

So $V^{(N)}$ is H_0 bounded with H_0-bound 0 and thus $H^{(N)} = H_0 + V^{(N)}$ is self-adjoint on $\mathfrak{D}(H^{(N)}) = \mathfrak{D}(H_0)$.

11.2. The HVZ theorem

The considerations of the beginning of this section show that it is not so easy to determine the essential spectrum of $H^{(N)}$ since the potential does not decay in all directions as $|x| \to \infty$. However, there is still something we can do. Denote the infimum of the spectrum of $H^{(N)}$ by λ^N. Then, let us split the system into $H^{(N-1)}$ plus a single electron. If the single electron is far away from the remaining system such that there is little interaction, the energy should be the sum of the kinetic energy of the single electron and the energy of the remaining system. Hence, arguing as in the two-electron example of the previous section, we expect

Theorem 11.2 (HVZ). *Let $H^{(N)}$ be the self-adjoint operator given in (11.6). Then $H^{(N)}$ is bounded from below and*

$$\sigma_{ess}(H^{(N)}) = [\lambda^{N-1}, \infty), \tag{11.9}$$

where $\lambda^{N-1} = \min \sigma(H^{(N-1)}) < 0$.

In particular, the ionization energy (i.e., the energy needed to remove one electron from the atom in its ground state) of an atom with N electrons is given by $\lambda^N - \lambda^{N-1}$.

Our goal for the rest of this section is to prove this result which is due to Zhislin, van Winter, and Hunziker and is known as the HVZ theorem. In fact there is a version which holds for general N-body systems. The proof is similar but involves some additional notation.

The idea of proof is the following. To prove $[\lambda^{N-1}, \infty) \subseteq \sigma_{ess}(H^{(N)})$, we choose Weyl sequences for $H^{(N-1)}$ and $-\Delta_N$ and proceed according to our intuitive picture from above. To prove $\sigma_{ess}(H^{(N)}) \subseteq [\lambda^{N-1}, \infty)$, we will *localize* $H^{(N)}$ on sets where one electron is far away from the nucleus whenever some of the others are. On these sets, the interaction term between this electron and the nucleus is decaying and hence does not contribute to the essential spectrum. So it remains to estimate the infimum of the spectrum of a system where one electron does not interact with the nucleus. Since the interaction term with the other electrons is positive, we can finally estimate this infimum by the infimum of the case where one electron is completely decoupled from the rest.

We begin with the first inclusion. Let $\psi^{N-1}(x_1, \ldots, x_{N-1}) \in H^2(\mathbb{R}^{3(N-1)})$ such that $\|\psi^{N-1}\| = 1$, $\|(H^{(N-1)} - \lambda^{N-1})\psi^{N-1}\| \leq \varepsilon$ and $\psi^1 \in H^2(\mathbb{R}^3)$ such that $\|\psi^1\| = 1$, $\|(-\Delta_N - \lambda)\psi^1\| \leq \varepsilon$ for some $\lambda \geq 0$. Now consider

11.2. The HVZ theorem

$\psi_r(x_1, \ldots, x_N) = \psi^{N-1}(x_1, \ldots, x_{N-1})\psi_r^1(x_N)$, $\psi_r^1(x_N) = \psi^1(x_N - r)$. Then

$$\|(H^{(N)} - \lambda - \lambda^{N-1})\psi_r\| \leq \|(H^{(N-1)} - \lambda^{N-1})\psi^{N-1}\|\|\psi_r^1\|$$
$$+ \|\psi^{N-1}\|\|(-\Delta_N - \lambda)\psi_r^1\|$$

(11.10)
$$+ \|(V_N - \sum_{j=1}^{N-1} V_{N,j})\psi_r\|,$$

where $V_N = V_{ne}(x_N)$ and $V_{N,j} = V_{ee}(x_N - x_j)$. Using the fact that $(V_N - \sum_{j=1}^{N-1} V_{N,j})\psi^{N-1} \in L^2(\mathbb{R}^{3N})$ and $|\psi_r^1| \to 0$ pointwise as $|r| \to \infty$ (by Lemma 10.1), the third term can be made smaller than ε by choosing $|r|$ large (dominated convergence). In summary,

(11.11) $$\|(H^{(N)} - \lambda - \lambda^{N-1})\psi_r\| \leq 3\varepsilon,$$

proving $[\lambda^{N-1}, \infty) \subseteq \sigma_{ess}(H^{(N)})$.

The second inclusion is more involved. We begin with a **localization formula**.

Lemma 11.3 (IMS localization formula). *Suppose $\phi_j \in C^\infty(\mathbb{R}^n)$, $1 \leq j \leq m$, is such that*

(11.12) $$\sum_{j=1}^m \phi_j(x)^2 = 1, \quad x \in \mathbb{R}^n.$$

Then

(11.13) $$\Delta \psi = \sum_{j=1}^m \left(\phi_j \Delta(\phi_j \psi) + |\partial \phi_j|^2 \psi \right), \quad \psi \in H^2(\mathbb{R}^n).$$

Proof. The proof follows from a straightforward computation using the identities $\sum_j \phi_j \partial_k \phi_j = 0$ and $\sum_j ((\partial_k \phi_j)^2 + \phi_j \partial_k^2 \phi_j) = 0$, which follow by differentiating (11.12). □

Now we will choose ϕ_j, $1 \leq j \leq N$, in such a way that, for x outside some ball, $x \in \text{supp}(\phi_j)$ implies that the j'th particle is far away from the nucleus.

Lemma 11.4. *Fix some $C \in (0, \frac{1}{\sqrt{N}})$. There exist smooth functions $\phi_j \in C^\infty(\mathbb{R}^n, [0,1])$, $1 \leq j \leq N$, such that (11.12) holds,*

(11.14) $$\text{supp}(\phi_j) \cap \{x | \ |x| \geq 1\} \subseteq \{x | \ |x_j| \geq C|x|\},$$

and $|\partial \phi_j(x)| \to 0$ as $|x| \to \infty$.

Proof. The open sets

$$U_j = \{x \in S^{3N-1} | \ |x_j| > C\}$$

cover the unit sphere in $\mathbb{R}^N$; that is,

$$\bigcup_{j=1}^N U_j = S^{3N-1}.$$

By Lemma 0.17, there is a partition of unity $\tilde{\phi}_j(x)$ subordinate to this cover. Extend $\tilde{\phi}_j(x)$ to a smooth function from $\mathbb{R}^{3N}\backslash\{0\}$ to $[0,1]$ by

$$\tilde{\phi}_j(\lambda x) = \tilde{\phi}_j(x), \qquad x \in S^{3N-1}, \lambda > 0,$$

and pick a function $\tilde{\phi} \in C^\infty(\mathbb{R}^{3N}, [0,1])$ with support inside the unit ball which is 1 in a neighborhood of the origin. Then the

$$\phi_j = \frac{\tilde{\phi} + (1-\tilde{\phi})\tilde{\phi}_j}{\sqrt{\sum_{\ell=1}^N(\tilde{\phi} + (1-\tilde{\phi})\tilde{\phi}_\ell)^2}}$$

are the desired functions. The gradient tends to zero since $\phi_j(\lambda x) = \phi_j(x)$ for $\lambda \geq 1$ and $|x| \geq 1$ which implies $(\partial \phi_j)(\lambda x) = \lambda^{-1}(\partial \phi_j)(x)$. □

By our localization formula, we have

$$H^{(N)} = \sum_{j=1}^N \phi_j H^{(N,j)} \phi_j + P - K,$$

(11.15) $$K = \sum_{j=1}^N \left(\phi_j^2 V_j + |\partial\phi_j|^2\right), \qquad P = \sum_{j=1}^N \phi_j^2 \sum_{\ell \neq j}^N V_{j,\ell},$$

where

(11.16) $$H^{(N,j)} = -\sum_{\ell=1}^N \Delta_\ell - \sum_{\ell \neq j}^N V_\ell + \sum_{k<\ell,\,k,\ell\neq j}^N V_{k,\ell}$$

is the Hamiltonian with the j'th electron decoupled from the rest of the system. Here we have abbreviated $V_j(x) = V_{ne}(x_j)$ and $V_{j,\ell} = V_{ee}(x_j - x_\ell)$.

Since K vanishes as $|x| \to \infty$, we expect it to be relatively compact with respect to the rest. By Lemma 6.23, it suffices to check that it is relatively compact with respect to H_0. The terms $|\partial\phi_j|^2$ are bounded and vanish at ∞; hence they are H_0 compact by Lemma 7.21. However, the terms $\phi_j^2 V_j$ have singularities and will be covered by the following lemma.

Lemma 11.5. *Let V be a multiplication operator which is H_0 bounded with H_0-bound 0 and suppose that $\|\chi_{\{x||x|\geq R\}} V R_{H_0}(z)\| \to 0$ as $R \to \infty$. Then V is relatively compact with respect to H_0.*

Proof. Let ψ_n converge to 0 weakly. Note that $\|\psi_n\| \leq M$ for some $M > 0$. It suffices to show that $\|V R_{H_0}(z)\psi_n\|$ converges to 0. Choose

11.2. The HVZ theorem

$\phi \in C_0^\infty(\mathbb{R}^n, [0,1])$ such that it is one for $|x| \le R$. Note $\phi \mathfrak{D}(H_0) \subset \mathfrak{D}(H_0)$. Then

$$\|VR_{H_0}(z)\psi_n\| \le \|(1-\phi)VR_{H_0}(z)\psi_n\| + \|V\phi R_{H_0}(z)\psi_n\|$$
$$\le \|(1-\phi)VR_{H_0}(z)\| \|\psi_n\|$$
$$+ a\|H_0 \phi R_{H_0}(z)\psi_n\| + b\|\phi R_{H_0}(z)\psi_n\|.$$

By assumption, the first term can be made smaller than ε by choosing R large. Next, the same is true for the second term choosing a small since $H_0 \phi R_{H_0}(z)$ is bounded (by Problem 2.9 and the closed graph theorem). Finally, the last term can also be made smaller than ε by choosing n large since ϕ is H_0 compact. $\square$

So K is relatively compact with respect to $H^{(N)}$. In particular, $H^{(N)} + K$ is self-adjoint on $H^2(\mathbb{R}^{3N})$ and $\sigma_{ess}(H^{(N)}) = \sigma_{ess}(H^{(N)} + K)$. Since the operators $H^{(N,j)}$, $1 \le j \le N$, are all of the form $H^{(N-1)}$ plus one particle which does not interact with the others and the nucleus, we have $H^{(N,j)} - \lambda^{N-1} \ge 0$, $1 \le j \le N$. Moreover, we have $P \ge 0$ and hence

$$\langle \psi, (H^{(N)} + K - \lambda^{N-1})\psi \rangle = \sum_{j=1}^{N} \langle \phi_j \psi, (H^{(N,j)} - \lambda^{N-1})\phi_j \psi \rangle$$
(11.17)
$$+ \langle \psi, P\psi \rangle \ge 0.$$

Thus we obtain the remaining inclusion

(11.18) $\quad \sigma_{ess}(H^{(N)}) = \sigma_{ess}(H^{(N)} + K) \subseteq \sigma(H^{(N)} + K) \subseteq [\lambda^{N-1}, \infty)$,

which finishes the proof of the HVZ theorem.

Note that the same proof works if we add additional nuclei at fixed locations. That is, we can also treat molecules if we assume that the nuclei are fixed in space.

Finally, let us consider the example of helium-like atoms ($N = 2$). By the HVZ theorem and the considerations of the previous section, we have

(11.19) $\quad \sigma_{ess}(H^{(2)}) = [-\frac{\gamma_{ne}^2}{4}, \infty)$.

Moreover, if $\gamma_{ee} = 0$ (no electron interaction), we can take products of one-particle eigenfunctions to show that

(11.20) $\quad -\gamma_{ne}^2 \left(\frac{1}{4n^2} + \frac{1}{4m^2} \right) \in \sigma_p(H^{(2)}(\gamma_{ee} = 0)), \quad n, m \in \mathbb{N}.$

In particular, there are eigenvalues embedded in the essential spectrum in this case. Moreover, since the electron interaction term is positive, we see

(11.21) $\quad H^{(2)} \ge -\frac{\gamma_{ne}^2}{2}.$

Note that there can be no positive eigenvalues by the virial theorem. This even holds for arbitrary N,

(11.22) $$\sigma_p(H^{(N)}) \subset (-\infty, 0).$$

Chapter 12

Scattering theory

12.1. Abstract theory

In physical measurements, one often has the following situation. A particle is shot into a region where it interacts with some forces and then leaves the region again. Outside this region the forces are negligible and hence the time evolution should be asymptotically free. Hence one expects asymptotic states $\psi_\pm(t) = \exp(-\mathrm{i}tH_0)\psi_\pm(0)$ to exist such that

(12.1) $$\|\psi(t) - \psi_\pm(t)\| \to 0 \quad \text{as} \quad t \to \pm\infty.$$

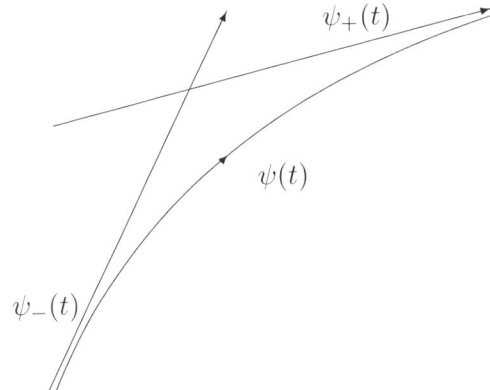

Rewriting this condition, we see
(12.2)
$$0 = \lim_{t\to\pm\infty} \|\mathrm{e}^{-\mathrm{i}tH}\psi(0) - \mathrm{e}^{-\mathrm{i}tH_0}\psi_\pm(0)\| = \lim_{t\to\pm\infty} \|\psi(0) - \mathrm{e}^{\mathrm{i}tH}\mathrm{e}^{-\mathrm{i}tH_0}\psi_\pm(0)\|$$

and motivated by this, we define the **wave operators** by

(12.3) $$\begin{aligned}\mathfrak{D}(\Omega_\pm) &= \{\psi \in \mathfrak{H} | \exists \lim_{t\to\pm\infty} \mathrm{e}^{\mathrm{i}tH}\mathrm{e}^{-\mathrm{i}tH_0}\psi\}, \\ \Omega_\pm \psi &= \lim_{t\to\pm\infty} \mathrm{e}^{\mathrm{i}tH}\mathrm{e}^{-\mathrm{i}tH_0}\psi.\end{aligned}$$

The set $\mathfrak{D}(\Omega_\pm)$ is the set of all incoming/outgoing asymptotic states $\psi_\pm$, and $\mathrm{Ran}(\Omega_\pm)$ is the set of all states which have an incoming/outgoing asymptotic state. If a state ψ has both, that is, $\psi \in \mathrm{Ran}(\Omega_+) \cap \mathrm{Ran}(\Omega_-)$, it is called a **scattering state**.

By construction we have

$$\text{(12.4)} \qquad \|\Omega_\pm \psi\| = \lim_{t\to\pm\infty} \|e^{itH} e^{-itH_0} \psi\| = \lim_{t\to\pm\infty} \|\psi\| = \|\psi\|,$$

and it is not hard to see that $\mathfrak{D}(\Omega_\pm)$ is closed. Moreover, interchanging the roles of H_0 and H amounts to replacing $\Omega_\pm$ by $\Omega_\pm^{-1}$ and hence $\mathrm{Ran}(\Omega_\pm)$ is also closed. In summary,

Lemma 12.1. *The sets $\mathfrak{D}(\Omega_\pm)$ and $\mathrm{Ran}(\Omega_\pm)$ are closed and $\Omega_\pm : \mathfrak{D}(\Omega_\pm) \to \mathrm{Ran}(\Omega_\pm)$ is unitary.*

Next, observe that

$$\text{(12.5)} \qquad \lim_{t\to\pm\infty} e^{itH} e^{-itH_0}(e^{-isH_0}\psi) = \lim_{t\to\pm\infty} e^{-isH}(e^{i(t+s)H} e^{-i(t+s)H_0}\psi)$$

and hence

$$\text{(12.6)} \qquad \Omega_\pm e^{-itH_0}\psi = e^{-itH}\Omega_\pm \psi, \quad \psi \in \mathfrak{D}(\Omega_\pm).$$

In addition, $\mathfrak{D}(\Omega_\pm)$ is invariant under $\exp(-itH_0)$ and $\mathrm{Ran}(\Omega_\pm)$ is invariant under $\exp(-itH)$. Moreover, if $\psi \in \mathfrak{D}(\Omega_\pm)^\perp$, then

$$\text{(12.7)} \qquad \langle \varphi, \exp(-itH_0)\psi \rangle = \langle \exp(itH_0)\varphi, \psi \rangle = 0, \quad \varphi \in \mathfrak{D}(\Omega_\pm).$$

Hence $\mathfrak{D}(\Omega_\pm)^\perp$ is invariant under $\exp(-itH_0)$ and $\mathrm{Ran}(\Omega_\pm)^\perp$ is invariant under $\exp(-itH)$. Consequently, $\mathfrak{D}(\Omega_\pm)$ reduces $\exp(-itH_0)$ and $\mathrm{Ran}(\Omega_\pm)$ reduces $\exp(-itH)$. Moreover, differentiating (12.6) with respect to t, we obtain from Theorem 5.1 the **intertwining property** of the wave operators.

Theorem 12.2. *The subspaces $\mathfrak{D}(\Omega_\pm)$, respectively, $\mathrm{Ran}(\Omega_\pm)$, reduce H_0, respectively, H, and the operators restricted to these subspaces are unitarily equivalent:*

$$\text{(12.8)} \qquad \Omega_\pm H_0 \psi = H \Omega_\pm \psi, \quad \psi \in \mathfrak{D}(\Omega_\pm) \cap \mathfrak{D}(H_0).$$

It is interesting to know the correspondence between incoming and outgoing states. Hence we define the **scattering operator**

$$\text{(12.9)} \qquad S = \Omega_+^{-1} \Omega_-, \quad \mathfrak{D}(S) = \{\psi \in \mathfrak{D}(\Omega_-) | \Omega_- \psi \in \mathrm{Ran}(\Omega_+)\}.$$

Note that we have $\mathfrak{D}(S) = \mathfrak{D}(\Omega_-)$ if and only if $\mathrm{Ran}(\Omega_-) \subseteq \mathrm{Ran}(\Omega_+)$ and $\mathrm{Ran}(S) = \mathfrak{D}(\Omega_+)$ if and only if $\mathrm{Ran}(\Omega_+) \subseteq \mathrm{Ran}(\Omega_-)$. Moreover, S is unitary from $\mathfrak{D}(S)$ onto $\mathrm{Ran}(S)$ and we have

$$\text{(12.10)} \qquad H_0 S\psi = S H_0 \psi, \quad \mathfrak{D}(H_0) \cap \mathfrak{D}(S).$$

However, note that this whole theory is meaningless until we can show that the domains $\mathfrak{D}(\Omega_\pm)$ are nontrivial. We first show a criterion due to Cook.

12.1. Abstract theory

Lemma 12.3 (Cook). *Suppose $\mathfrak{D}(H_0) \subseteq \mathfrak{D}(H)$. If*

(12.11) $$\int_0^\infty \|(H - H_0)\exp(\mp itH_0)\psi\|dt < \infty, \quad \psi \in \mathfrak{D}(H_0),$$

then $\psi \in \mathfrak{D}(\Omega_\pm)$, respectively. Moreover, we even have

(12.12) $$\|(\Omega_\pm - \mathbb{I})\psi\| \leq \int_0^\infty \|(H - H_0)\exp(\mp itH_0)\psi\|dt$$

in this case.

Proof. The result follows from

(12.13) $$e^{itH}e^{-itH_0}\psi = \psi + i\int_0^t \exp(isH)(H - H_0)\exp(-isH_0)\psi ds,$$

which holds for $\psi \in \mathfrak{D}(H_0)$. $\square$

As a simple consequence, we obtain the following result for Schrödinger operators in $\mathbb{R}^3$:

Theorem 12.4. *Suppose H_0 is the free Schrödinger operator and $H = H_0 + V$ with $V \in L^2(\mathbb{R}^3)$. Then the wave operators exist and $\mathfrak{D}(\Omega_\pm) = \mathfrak{H}$.*

Proof. Since we want to use Cook's lemma, we need to estimate

$$\|V\psi(s)\|^2 = \int_{\mathbb{R}^3} |V(x)\psi(s,x)|^2 dx, \quad \psi(s) = \exp(isH_0)\psi,$$

for given $\psi \in \mathfrak{D}(H_0)$. Invoking (7.42), we get

$$\|V\psi(s)\| \leq \|\psi(s)\|_\infty \|V\| \leq \frac{1}{(4\pi s)^{3/2}}\|\psi\|_1\|V\|, \quad s > 0,$$

at least for $\psi \in L^1(\mathbb{R}^3)$. Moreover, this implies

$$\int_1^\infty \|V\psi(s)\|ds \leq \frac{1}{4\pi^{3/2}}\|\psi\|_1\|V\|$$

and thus every such ψ is in $\mathfrak{D}(\Omega_+)$. Since such functions are dense, we obtain $\mathfrak{D}(\Omega_+) = \mathfrak{H}$, and similarly for Ω_-. $\square$

By the intertwining property, $\psi \in \mathfrak{D}(\Omega_\pm)$ is an eigenfunction of H_0 if it is an eigenfunction of H corresponding to the same eigenvalue. Hence for $\psi \in \mathfrak{H}_{pp}(H_0)$ it is easy to check whether it is in $\mathfrak{D}(\Omega_\pm)$, and only the continuous subspace is of interest. We will say that the **wave operators exist** if all elements of $\mathfrak{H}_{ac}(H_0)$ are asymptotic states, that is,

(12.14) $$\mathfrak{H}_{ac}(H_0) \subseteq \mathfrak{D}(\Omega_\pm),$$

and that they are **complete** if, in addition, all elements of $\mathfrak{H}_{ac}(H)$ are scattering states, that is,

(12.15) $$\mathfrak{H}_{ac}(H) \subseteq \mathrm{Ran}(\Omega_\pm).$$

If we even have

(12.16) $$\mathfrak{H}_c(H) \subseteq \mathrm{Ran}(\Omega_\pm),$$

they are called **asymptotically complete**.

We will be mainly interested in the case where H_0 is the free Schrödinger operator and hence $\mathfrak{H}_{ac}(H_0) = \mathfrak{H}$. In this latter case, the wave operators exist if $\mathfrak{D}(\Omega_\pm) = \mathfrak{H}$, they are complete if $\mathfrak{H}_{ac}(H) = \mathrm{Ran}(\Omega_\pm)$, and they are asymptotically complete if $\mathfrak{H}_c(H) = \mathrm{Ran}(\Omega_\pm)$. In particular, asymptotic completeness implies $\mathfrak{H}_{sc}(H) = \{0\}$ since H restricted to $\mathrm{Ran}(\Omega_\pm)$ is unitarily equivalent to H_0. Completeness implies that the scattering operator is unitary. Hence, by the intertwining property, kinetic energy is preserved during scattering:

(12.17) $$\langle \psi_-, H_0 \psi_- \rangle = \langle S\psi_-, SH_0\psi_- \rangle = \langle S\psi_-, H_0 S\psi_- \rangle = \langle \psi_+, H_0\psi_+ \rangle$$

for $\psi_- \in \mathfrak{D}(H_0)$ and $\psi_+ = S\psi_-$.

12.2. Incoming and outgoing states

In the remaining sections, we want to apply this theory to Schrödinger operators. Our first goal is to give a precise meaning to some terms in the intuitive picture of scattering theory introduced in the previous section.

This physical picture suggests that we should be able to decompose $\psi \in \mathfrak{H}$ into an incoming and an outgoing part. But how should incoming, respectively, outgoing, be defined for $\psi \in \mathfrak{H}$? Well, incoming (outgoing) means that the expectation of x^2 should decrease (increase). Set $x(t)^2 = \exp(iH_0 t) x^2 \exp(-iH_0 t)$. Then, abbreviating $\psi(t) = e^{-itH_0}\psi$,

(12.18) $$\frac{d}{dt}\mathbb{E}_\psi(x(t)^2) = \langle \psi(t), i[H_0, x^2]\psi(t)\rangle = 4\langle \psi(t), D\psi(t)\rangle, \quad \psi \in \mathcal{S}(\mathbb{R}^n),$$

where D is the dilation operator introduced in (10.9). Hence it is natural to consider $\psi \in \mathrm{Ran}(P_\pm)$,

(12.19) $$P_\pm = P_D((0, \pm\infty)),$$

as outgoing, respectively, incoming, states. If we project a state in $\mathrm{Ran}(P_\pm)$ to energies in the interval (a^2, b^2), we expect that it cannot be found in a ball of radius proportional to $a|t|$ as $t \to \pm\infty$ (a is the minimal velocity of the particle, since we have assumed the mass to be two). In fact, we will show below that the tail decays faster then any inverse power of $|t|$.

We first collect some properties of D which will be needed later on. Note

(12.20) $$\mathcal{F}D = -D\mathcal{F}$$

and hence $\mathcal{F}f(D) = f(-D)\mathcal{F}$. Additionally, we will look for a transformation which maps D to a multiplication operator.

12.2. Incoming and outgoing states

Since the dilation group acts on $|x|$ only, it seems reasonable to switch to polar coordinates $x = r\omega$, $(t, \omega) \in \mathbb{R}^+ \times S^{n-1}$. Since $U(s)$ essentially transforms r into $r\exp(s)$, we will replace r by $\rho = \log(r)$. In these coordinates we have

$$(12.21) \qquad U(s)\psi(e^\rho \omega) = e^{-ns/2}\psi(e^{(\rho-s)}\omega)$$

and hence $U(s)$ corresponds to a shift of ρ (the constant in front is absorbed by the volume element). Thus D corresponds to differentiation with respect to this coordinate and all we have to do to make it a multiplication operator is to take the Fourier transform with respect to ρ.

This leads us to the **Mellin transform**

$$(12.22) \qquad \mathcal{M}: L^2(\mathbb{R}^n) \to L^2(\mathbb{R} \times S^{n-1}),$$
$$\psi(r\omega) \to (\mathcal{M}\psi)(\lambda, \omega) = \frac{1}{\sqrt{2\pi}} \int_0^\infty r^{-i\lambda}\psi(r\omega) r^{\frac{n}{2}-1} dr.$$

By construction, $\mathcal{M}$ is unitary; that is,
$$(12.23) \qquad \int_\mathbb{R} \int_{S^{n-1}} |(\mathcal{M}\psi)(\lambda,\omega)|^2 d\lambda d^{n-1}\omega = \int_{\mathbb{R}^+} \int_{S^{n-1}} |\psi(r\omega)|^2 r^{n-1} dr d^{n-1}\omega,$$

where $d^{n-1}\omega$ is the normalized surface measure on S^{n-1}. Moreover,

$$(12.24) \qquad \mathcal{M}^{-1} U(s) \mathcal{M} = e^{-is\lambda}$$

and hence

$$(12.25) \qquad \mathcal{M}^{-1} D \mathcal{M} = \lambda.$$

From this it is straightforward to show that

$$(12.26) \qquad \sigma(D) = \sigma_{ac}(D) = \mathbb{R}, \qquad \sigma_{sc}(D) = \sigma_{pp}(D) = \emptyset$$

and that $\mathcal{S}(\mathbb{R}^n)$ is a core for D. In particular, we have $P_+ + P_- = \mathbb{I}$.

Using the Mellin transform, we can now prove Perry's estimate.

Lemma 12.5. *Suppose $f \in C_c^\infty(\mathbb{R})$ with $\mathrm{supp}(f) \subset (a^2, b^2)$ for some $a, b > 0$. For every $R \in \mathbb{R}$, $N \in \mathbb{N}$ there is a constant C such that*

$$(12.27) \qquad \|\chi_{\{x \mid |x| < 2a|t|\}} e^{-itH_0} f(H_0) P_D((\pm R, \pm\infty))\| \leq \frac{C}{(1+|t|)^N}, \qquad \pm t \geq 0,$$

respectively.

Proof. We prove only the $+$ case, the remaining one being similar. Consider $\psi \in \mathcal{S}(\mathbb{R}^n)$. Introducing

$$\psi(t, x) = e^{-itH_0} f(H_0) P_D((R, \infty))\psi(x) = \langle K_{t,x}, \mathcal{F} P_D((R, \infty))\psi\rangle$$
$$= \langle K_{t,x}, P_D((-\infty, -R))\hat\psi\rangle,$$

where
$$K_{t,x}(p) = \frac{1}{(2\pi)^{n/2}} e^{i(tp^2 - px)} f(p^2)^*,$$
we see that it suffices to show
$$\|P_D((-\infty, -R))K_{t,x}\|^2 \leq \frac{const}{(1+|t|)^{2N}}, \quad \text{for } |x| < 2a|t|, \; t > 0.$$
Now we invoke the Mellin transform to estimate this norm:
$$\|P_D((-\infty, -R))K_{t,x}\|^2 = \int_{-\infty}^{-R} \int_{S^{n-1}} |(\mathcal{M}K_{t,x})(\lambda, \omega)|^2 d\lambda d^{n-1}\omega.$$
Since
(12.28) $$(\mathcal{M}K_{t,x})(\lambda, \omega) = \frac{1}{(2\pi)^{(n+1)/2}} \int_0^\infty \tilde{f}(r) e^{i\alpha(r)} dr$$
with $\tilde{f}(r) = f(r^2)^* r^{n/2-1} \in C_c^\infty((a,b))$, $\alpha(r) = tr^2 + r\omega x - \lambda \log(r)$. Estimating the derivative of α, we see
$$\alpha'(r) = 2tr + \omega x - \lambda/r > 0, \quad r \in (a,b),$$
for $\lambda \leq -R$ and $t > -R(2\varepsilon a)^{-1}$, where ε is the distance of a to the support of $\tilde{f}$. Hence we can find a constant such that
$$|\alpha'(r)| \geq const(1 + |\lambda| + |t|), \quad r \in \text{supp}(\tilde{f}),$$
for $\lambda \leq -R$, $t > -R(\varepsilon a)^{-1}$. Now the method of stationary phase (Problem 12.1) implies
$$|(\mathcal{M}K_{t,x})(\lambda, \omega)| \leq \frac{const}{(1+|\lambda|+|t|)^N}$$
for λ, t as before. By increasing the constant, we can even assume that it holds for $t \geq 0$ and $\lambda \leq -R$. This finishes the proof. $\square$

Corollary 12.6. *Suppose that $f \in C_c^\infty((0, \infty))$ and $R \in \mathbb{R}$. Then the operator $P_D((\pm R, \pm \infty)) f(H_0) \exp(-itH_0)$ converges strongly to 0 as $t \to \mp \infty$.*

Proof. Abbreviating $P_D = P_D((\pm R, \pm \infty))$ and $\chi = \chi_{\{x|\,|x|<2a|t|\}}$, we have
$$\|P_D f(H_0) e^{-itH_0} \psi\| \leq \|\chi e^{itH_0} f(H_0)^* P_D\| \|\psi\| + \|f(H_0)\| \|(\mathbb{I} - \chi)\psi\|$$
since $\|A\| = \|A^*\|$. Taking $t \to \mp\infty$, the first term goes to zero by our lemma and the second goes to zero since $\chi\psi \to \psi$. $\square$

Problem 12.1 (Method of stationary phase). *Consider the integral*
$$I(t) = \int_{-\infty}^\infty f(r) e^{it\phi(r)} dr$$
with $f \in C_c^\infty(\mathbb{R})$ and a real-valued phase $\phi \in C^\infty(\mathbb{R})$. Show that $|I(t)| \leq C_N t^{-N}$ for every $N \in \mathbb{N}$ if $|\phi'(r)| \geq 1$ for $r \in \text{supp}(f)$. (Hint: Make a

12.3. Schrödinger operators with short range potentials

change of variables $\rho = \phi(r)$ and conclude that it suffices to show the case $\phi(r) = r$. Now use integration by parts.)

12.3. Schrödinger operators with short range potentials

By the RAGE theorem we know that for $\psi \in \mathfrak{H}_c$, $\psi(t)$ will eventually leave every compact ball (at least on the average). Hence we expect that the time evolution will asymptotically look like the free one for $\psi \in \mathfrak{H}_c$ if the potential decays sufficiently fast. In other words, we expect such potentials to be asymptotically complete.

Suppose V is relatively bounded with bound less than one. Introduce

(12.29) $\qquad h_1(r) = \|V R_{H_0}(z)\chi_r\|, \quad h_2(r) = \|\chi_r V R_{H_0}(z)\|, \quad r \geq 0,$

where

(12.30) $\qquad \chi_r = \chi_{\{x|\, |x|\geq r\}}.$

The potential V will be called **short range** if these quantities are integrable. We first note that it suffices to check this for h_1 or h_2 and for one $z \in \rho(H_0)$.

Lemma 12.7. *The function h_1 is integrable if and only if h_2 is. Moreover, h_j integrable for one $z_0 \in \rho(H_0)$ implies h_j integrable for all $z \in \rho(H_0)$.*

Proof. Pick $\phi \in C^\infty(\mathbb{R}^n, [0,1])$ such that $\phi(x) = 0$ for $0 \leq |x| \leq 1/2$ and $\phi(x) = 1$ for $1 \leq |x|$. Then it is not hard to see that h_j is integrable if and only if $\tilde{h}_j$ is integrable, where

$$\tilde{h}_1(r) = \|V R_{H_0}(z)\phi_r\|, \quad \tilde{h}_2(r) = \|\phi_r V R_{H_0}(z)\|, \quad r \geq 1,$$

and $\phi_r(x) = \phi(x/r)$. Using

$$[R_{H_0}(z), \phi_r] = -R_{H_0}(z)[H_0(z), \phi_r]R_{H_0}(z)$$
$$= -R_{H_0}(z)(\Delta\phi_r + 2(\partial\phi_r)\partial)R_{H_0}(z)$$

and $\Delta\phi_r = \phi_{r/2}\Delta\phi_r$, $\|\Delta\phi_r\|_\infty \leq \|\Delta\phi\|_\infty/r^2$, respectively, $(\partial\phi_r) = \phi_{r/2}(\partial\phi_r)$, $\|\partial\phi_r\|_\infty \leq \|\partial\phi\|_\infty/r$, we see

$$|\tilde{h}_1(r) - \tilde{h}_2(r)| \leq \frac{c}{r}\tilde{h}_1(r/2), \quad r \geq 1.$$

Hence $\tilde{h}_2$ is integrable if $\tilde{h}_1$ is. Conversely,

$$\tilde{h}_1(r) \leq \tilde{h}_2(r) + \frac{c}{r}\tilde{h}_1(r/2) \leq \tilde{h}_2(r) + \frac{c}{r}\tilde{h}_2(r/2) + \frac{2c}{r^2}\tilde{h}_1(r/4)$$

shows that $\tilde{h}_2$ is integrable if $\tilde{h}_1$ is.

Invoking the first resolvent formula

$$\|\phi_r V R_{H_0}(z)\| \leq \|\phi_r V R_{H_0}(z_0)\|\|\mathbb{I} - (z - z_0)R_{H_0}(z)\|$$

finishes the proof. $\square$

As a first consequence, note

Lemma 12.8. *If V is short range, then $R_H(z) - R_{H_0}(z)$ is compact.*

Proof. The operator $R_H(z)V(\mathbb{I}-\chi_r)R_{H_0}(z)$ is compact since $(\mathbb{I}-\chi_r)R_{H_0}(z)$ is by Lemma 7.21 and $R_H(z)V$ is bounded by Lemma 6.23. Moreover, by our short range condition, it converges in norm to
$$R_H(z)VR_{H_0}(z) = R_H(z) - R_{H_0}(z)$$
as $r \to \infty$ (at least for some subsequence). $\square$

In particular, by Weyl's theorem, we have $\sigma_{ess}(H) = [0,\infty)$. Moreover, V short range implies that H and H_0 look alike far outside.

Lemma 12.9. *Suppose $R_H(z) - R_{H_0}(z)$ is compact. Then so is $f(H) - f(H_0)$ for every $f \in C_\infty(\mathbb{R})$ and*

(12.31)
$$\lim_{r \to \infty} \|(f(H) - f(H_0))\chi_r\| = 0.$$

Proof. The first part is Lemma 6.21 and the second part follows from part (ii) of Lemma 6.9 since χ_r converges strongly to 0. $\square$

However, this is clearly not enough to prove asymptotic completeness, and we need a more careful analysis.

We begin by showing that the wave operators exist. By Cook's criterion (Lemma 12.3), we need to show that

(12.32)
$$\|V\exp(\mp itH_0)\psi\| \leq \|VR_{H_0}(-1)\|\|(\mathbb{I}-\chi_{2a|t|})\exp(\mp itH_0)(H_0+\mathbb{I})\psi\|$$
$$+ \|VR_{H_0}(-1)\chi_{2a|t|}\|\|(H_0+\mathbb{I})\psi\|$$

is integrable for a dense set of vectors ψ. The second term is integrable by our short range assumption. The same is true by Perry's estimate (Lemma 12.5) for the first term if we choose $\psi = f(H_0)P_D((\pm R, \pm\infty))\varphi$. Since vectors of this form are dense, we see that the wave operators exist,

(12.33)
$$\mathfrak{D}(\Omega_\pm) = \mathfrak{H}.$$

Since H restricted to $\text{Ran}(\Omega_\pm^*)$ is unitarily equivalent to H_0, we obtain $[0,\infty) = \sigma_{ac}(H_0) \subseteq \sigma_{ac}(H)$. Furthermore, by $\sigma_{ac}(H) \subseteq \sigma_{ess}(H) = [0,\infty)$, we even have $\sigma_{ac}(H) = [0,\infty)$.

To prove asymptotic completeness of the wave operators, we will need to show that the $(\Omega_\pm - \mathbb{I})f(H_0)P_\pm$ are compact.

Lemma 12.10. *Suppose V is short range. Let $f \in C_c^\infty((0,\infty))$ and suppose ψ_n converges weakly to 0. Then*

(12.34)
$$\lim_{n \to \infty} \|(\Omega_\pm - \mathbb{I})f(H_0)P_\pm\psi_n\| = 0;$$

that is, $(\Omega_\pm - \mathbb{I})f(H_0)P_\pm$ is compact.

12.3. Schrödinger operators with short range potentials

Proof. By (12.13) we see

$$\|R_H(z)(\Omega_\pm - \mathbb{I})f(H_0)P_\pm\psi_n\| \leq \int_0^\infty \|R_H(z)V\exp(-isH_0)f(H_0)P_\pm\psi_n\|dt.$$

Since $R_H(z)VR_{H_0}$ is compact, we see that the integrand

$$R_H(z)V\exp(-isH_0)f(H_0)P_\pm\psi_n$$
$$= R_H(z)VR_{H_0}\exp(-isH_0)(H_0+1)f(H_0)P_\pm\psi_n$$

converges pointwise to 0. Moreover, arguing as in (12.32), the integrand is bounded by an L^1 function depending only on $\|\psi_n\|$. Thus $R_H(z)(\Omega_\pm - \mathbb{I})f(H_0)P_\pm$ is compact by the dominated convergence theorem. Furthermore, using the intertwining property, we see that

$$(\Omega_\pm - \mathbb{I})f(H_0)P_\pm = R_H(z)(\Omega_\pm - \mathbb{I})\tilde{f}(H_0)P_\pm$$
$$+ (R_H(z) - R_{H_0}(z))\tilde{f}(H_0)P_\pm$$

is compact by Lemma 12.8, where $\tilde{f}(\lambda) = (\lambda - z)f(\lambda) \in C_c^\infty((0,\infty))$. □

Now we have gathered enough information to tackle the problem of asymptotic completeness.

We first show that the singular continuous spectrum is absent. This is not really necessary, but it avoids the use of Cesàro means in our main argument.

Abbreviate $P = P_H^{sc}P_H((a,b))$, $0 < a < b$. Since H restricted to $\operatorname{Ran}(\Omega_\pm)$ is unitarily equivalent to H_0 (which has purely absolutely continuous spectrum), the singular part must live on $\operatorname{Ran}(\Omega_\pm)^\perp$; that is, $P_H^{sc}\Omega_\pm = 0$. Thus $Pf(H_0) = P(\mathbb{I}-\Omega_+)f(H_0)P_+ + P(\mathbb{I}-\Omega_-)f(H_0)P_-$ is compact. Since $f(H) - f(H_0)$ is compact, it follows that $Pf(H)$ is also compact. Choosing f such that $f(\lambda) = 1$ for $\lambda \in [a,b]$, we see that $P = Pf(H)$ is compact and hence finite dimensional. In particular, $\sigma_{sc}(H) \cap (a,b)$ is a finite set. But a continuous measure cannot be supported on a finite set, showing $\sigma_{sc}(H) \cap (a,b) = \emptyset$. Since $0 < a < b$ are arbitrary, we even have $\sigma_{sc}(H) \cap (0,\infty) = \emptyset$ and by $\sigma_{sc}(H) \subseteq \sigma_{ess}(H) = [0,\infty)$, we obtain $\sigma_{sc}(H) = \emptyset$.

Observe that by replacing P_H^{sc} by P_H^{pp}, the same argument shows that all nonzero eigenvalues are finite dimensional and cannot accumulate in $(0,\infty)$.

In summary, we have shown

Theorem 12.11. *Suppose V is short range. Then*

$$(12.35) \qquad \sigma_{ac}(H) = \sigma_{ess}(H) = [0,\infty), \qquad \sigma_{sc}(H) = \emptyset.$$

All nonzero eigenvalues have finite multiplicity and cannot accumulate in $(0,\infty)$.

Now we come to the anticipated asymptotic completeness result of Enß. Choose

(12.36) $$\psi \in \mathfrak{H}_c(H) = \mathfrak{H}_{ac}(H) \quad \text{such that} \quad \psi = f(H)\psi$$

for some $f \in C_c^\infty((0, \infty))$. By the RAGE theorem, the sequence $\psi(t)$ converges weakly to zero as $t \to \pm\infty$. Abbreviate $\psi(t) = \exp(-itH)\psi$. Introduce

(12.37) $$\varphi_\pm(t) = f(H_0)P_\pm \psi(t),$$

which satisfy

(12.38) $$\lim_{t \to \pm\infty} \|\psi(t) - \varphi_+(t) - \varphi_-(t)\| = 0.$$

Indeed, this follows from

(12.39) $$\psi(t) = \varphi_+(t) + \varphi_-(t) + (f(H) - f(H_0))\psi(t)$$

and Lemma 6.21. Moreover, we even have

(12.40) $$\lim_{t \to \pm\infty} \|(\Omega_\pm - \mathbb{I})\varphi_\pm(t)\| = 0$$

by Lemma 12.10. Now suppose $\psi \in \operatorname{Ran}(\Omega_\pm)^\perp$. Then

(12.41) $$\begin{aligned} \|\psi\|^2 &= \lim_{t \to \pm\infty} \langle \psi(t), \psi(t) \rangle \\ &= \lim_{t \to \pm\infty} \langle \psi(t), \varphi_+(t) + \varphi_-(t) \rangle \\ &= \lim_{t \to \pm\infty} \langle \psi(t), \Omega_+\varphi_+(t) + \Omega_-\varphi_-(t) \rangle. \end{aligned}$$

By Theorem 12.2, $\operatorname{Ran}(\Omega_\pm)^\perp$ is invariant under H and thus $\psi(t) \in \operatorname{Ran}(\Omega_\pm)^\perp$ implying

(12.42) $$\begin{aligned} \|\psi\|^2 &= \lim_{t \to \pm\infty} \langle \psi(t), \Omega_\mp \varphi_\mp(t) \rangle \\ &= \lim_{t \to \pm\infty} \langle P_\mp f(H_0)^* \Omega_\mp^* \psi(t), \psi(t) \rangle. \end{aligned}$$

Invoking the intertwining property, we see

(12.43) $$\|\psi\|^2 = \lim_{t \to \pm\infty} \langle P_\mp f(H_0)^* e^{-itH_0} \Omega_\mp^* \psi, \psi(t) \rangle = 0$$

by Corollary 12.6. Hence $\operatorname{Ran}(\Omega_\pm) = \mathfrak{H}_{ac}(H) = \mathfrak{H}_c(H)$ and we thus have shown

Theorem 12.12 (Enß)**.** *Suppose V is short range. Then the wave operators are asymptotically complete.*

Part 3

Appendix

Appendix A

Almost everything about Lebesgue integration

In this appendix I give a brief introduction to measure theory. Good references are [**8**], [**39**], or [**56**].

A.1. Borel measures in a nutshell

The first step in defining the Lebesgue integral is extending the notion of size from intervals to arbitrary sets. Unfortunately, this turns out to be too much, since a classical paradox by Banach and Tarski shows that one can break the unit ball in $\mathbb{R}^3$ into a finite number of wild (choosing the pieces uses the Axiom of Choice and cannot be done with a jigsaw) pieces, rotate and translate them, and reassemble them to get two copies of the unit ball (compare Problem A.5). Hence any reasonable notion of size (i.e., one which is translation and rotation invariant) cannot be defined for all sets!

A collection of subsets $\mathcal{A}$ of a given set X such that

- $X \in \mathcal{A}$,
- $\mathcal{A}$ is closed under finite unions,
- $\mathcal{A}$ is closed under complements,

is called an **algebra**. Note that $\emptyset \in \mathcal{A}$ and that $\mathcal{A}$ is also closed under finite intersections and relative complements: $\emptyset = X'$, $A \cap B = (A' \cup B')'$ (de Morgan), and $A \backslash B = A \cap B'$, where $A' = X \backslash A$ denotes the complement.

If an algebra is closed under countable unions (and hence also countable intersections), it is called a **σ-algebra**.

Example. Let $X = \{1, 2, 3\}$; then $\mathcal{A} = \{\emptyset, \{1\}, \{2,3\}, X\}$ is an algebra. ◇

Moreover, the intersection of any family of (σ-)algebras $\{\mathcal{A}_\alpha\}$ is again a (σ-)algebra, and for any collection S of subsets there is a unique smallest (σ-)algebra $\Sigma(S)$ containing S (namely the intersection of all (σ-)algebras containing S). It is called the (σ-)algebra generated by S.

Example. For a given set X, the power set $\mathfrak{P}(X)$ is clearly the largest σ-algebra and $\{\emptyset, X\}$ is the smallest. ◇

Example. Let X be some set with a σ-algebra Σ. Then every subset $Y \subseteq X$ has a natural σ-algebra $\Sigma \cap Y = \{A \cap Y | A \in \Sigma\}$ (show that this is indeed a σ-algebra) known as the **relative σ-algebra**.

Note that if S generates Σ, then $S \cap Y$ generates $\Sigma \cap Y$: $\Sigma(S) \cap Y = \Sigma(S \cap Y)$. Indeed, since $\Sigma \cap Y$ is a σ-algebra containing $S \cap Y$, we have $\Sigma(S \cap Y) \subseteq \Sigma(S) \cap Y = \Sigma \cap Y$. Conversely, consider $\{A \in \Sigma | A \cap Y \in \Sigma(S \cap Y)\}$ which is a σ-algebra (check this). Since this last σ-algebra contains S it must be equal to $\Sigma = \Sigma(S)$ and thus $\Sigma \cap Y \subseteq \Sigma(S \cap Y)$. ◇

If X is a topological space, the **Borel σ-algebra** $\mathfrak{B}(X)$ of X is defined to be the σ-algebra generated by all open (respectively, all closed) sets. In fact, if X is second countable, any countable base will suffice to generate the Borel σ-algebra (recall Lemma 0.1).

Sets in the Borel σ-algebra are called **Borel sets**.

Example. In the case $X = \mathbb{R}^n$, the Borel σ-algebra will be denoted by $\mathfrak{B}^n$ and we will abbreviate $\mathfrak{B} = \mathfrak{B}^1$. Note that in order to generate $\mathfrak{B}$, open (or closed) intervals with rational boundary points suffice. ◇

Example. If X is a topological space, then any Borel set $Y \subseteq X$ is also a topological space equipped with the relative topology and its Borel σ-algebra is given by $\mathfrak{B}(Y) = \mathfrak{B}(X) \cap Y = \{A | A \in \mathfrak{B}(X), A \subseteq Y\}$ (show this). ◇

Now let us turn to the definition of a measure: A set X together with a σ-algebra Σ is called a **measurable space**. A **measure** μ is a map $\mu : \Sigma \to [0, \infty]$ on a σ-algebra Σ such that

- $\mu(\emptyset) = 0$,

- $\mu(\bigcup_{j=1}^\infty A_j) = \sum_{j=1}^\infty \mu(A_j)$ if $A_j \cap A_k = \emptyset$ for all $j \neq k$ (σ-additivity).

Here the sum is set equal to ∞ if one of the summands is ∞ or if it diverges.

A.1. Borel measures in a nutshell

The measure μ is called **σ-finite** if there is a countable cover $\{X_j\}_{j=1}^\infty$ of X such that $X_j \in \Sigma$ and $\mu(X_j) < \infty$ for all j. (Note that it is no restriction to assume $X_j \subseteq X_{j+1}$.) It is called **finite** if $\mu(X) < \infty$ and a **probability measure** if $\mu(X) = 1$. The sets in Σ are called **measurable sets** and the triple (X, Σ, μ) is referred to as a **measure space**.

Example. Take a set X and $\Sigma = \mathfrak{P}(X)$ and set $\mu(A)$ to be the number of elements of A (respectively, ∞ if A is infinite). This is the so-called **counting measure**. It will be finite if and only if X is finite and σ-finite if and only if X is countable. ◇

Example. Take a set X and $\Sigma = \mathfrak{P}(X)$. Fix a point $x \in X$ and set $\mu(A) = 1$ if $x \in A$ and $\mu(A) = 0$ otherwise. This is the **Dirac measure** centered at x. ◇

Example. Let μ_1, μ_2 be two measures on (X, Σ) and $\alpha_1, \alpha_2 \geq 0$. Then $\mu = \alpha_1 \mu_1 + \alpha_2 \mu_2$ defined via
$$\mu(A) = \alpha_1 \mu_1(A) + \alpha_2 \mu_2(A)$$
is again a measure. Furthermore, given a countable number of measures μ_n and numbers $\alpha_n \geq 0$, $\mu = \sum_n \alpha_n \mu_n$ is again a measure (show this). ◇

Example. Let μ be a measure on (X, Σ) and $Y \subseteq X$ a measurable subset. Then
$$\nu(A) = \mu(A \cap Y)$$
is again a measure on (X, Σ) (show this). ◇

Example. If $Y \in \Sigma$, we can restrict the σ-algebra $\Sigma|_Y = \{A \in \Sigma | A \subseteq Y\}$ such that $(Y, \Sigma|_Y, \mu|_Y)$ is again a measurable space. It will be σ-finite if (X, Σ, μ) is. ◇

If we replace the σ-algebra by an algebra $\mathcal{A}$, then μ is called a **premeasure**. In this case, σ-additivity clearly only needs to hold for disjoint sets A_n for which $\bigcup_n A_n \in \mathcal{A}$.

We will write $A_n \nearrow A$ if $A_n \subseteq A_{n+1}$ with $A = \bigcup_n A_n$ and $A_n \searrow A$ if $A_{n+1} \subseteq A_n$ with $A = \bigcap_n A_n$.

Theorem A.1. *Any measure μ satisfies the following properties:*

(i) $A \subseteq B$ implies $\mu(A) \leq \mu(B)$ (monotonicity).
(ii) $\mu(A_n) \to \mu(A)$ if $A_n \nearrow A$ (continuity from below).
(iii) $\mu(A_n) \to \mu(A)$ if $A_n \searrow A$ and $\mu(A_1) < \infty$ (continuity from above).

Proof. The first claim is obvious from $\mu(B) = \mu(A) + \mu(B \setminus A)$. To see the second, define $\tilde{A}_1 = A_1$, $\tilde{A}_n = A_n \setminus A_{n-1}$ and note that these sets are disjoint and satisfy $A_n = \bigcup_{j=1}^n \tilde{A}_j$. Hence $\mu(A_n) = \sum_{j=1}^n \mu(\tilde{A}_j) \to \sum_{j=1}^\infty \mu(\tilde{A}_j) = \mu(\bigcup_{j=1}^\infty \tilde{A}_j) = \mu(A)$ by σ-additivity. The third follows from the second using $\tilde{A}_n = A_1 \backslash A_n \nearrow A_1 \backslash A$ implying $\mu(\tilde{A}_n) = \mu(A_1) - \mu(A_n) \to \mu(A_1 \backslash A) = \mu(A_1) - \mu(A)$. □

Example. Consider the counting measure on $X = \mathbb{N}$ and let $A_n = \{j \in \mathbb{N} | j \geq n\}$. Then $\mu(A_n) = \infty$, but $\mu(\bigcap_n A_n) = \mu(\emptyset) = 0$, which shows that the requirement $\mu(A_1) < \infty$ in item (iii) of Theorem A.1 is not superfluous. ◇

A measure on the Borel σ-algebra is called a **Borel measure** if $\mu(K) < \infty$ for every compact set K. Note that some authors do not require this last condition.

Example. Let $X = \mathbb{R}$ and $\Sigma = \mathfrak{B}$. The Dirac measure is a Borel measure. The counting measure is no Borel measure since $\mu([a, b]) = \infty$ for $a < b$. ◇

A measure on the Borel σ-algebra is called **outer regular** if

(A.1) $$\mu(A) = \inf_{O \supseteq A, O \text{ open}} \mu(O)$$

and **inner regular** if

(A.2) $$\mu(A) = \sup_{K \subseteq A, K \text{ compact}} \mu(K).$$

It is called **regular** if it is both outer and inner regular.

Example. Let $X = \mathbb{R}$ and $\Sigma = \mathfrak{B}$. The counting measure is inner regular but not outer regular (every nonempty open set has infinite measure). The Dirac measure is a regular Borel measure. ◇

But how can we obtain some more interesting Borel measures? We will restrict ourselves to the case of $X = \mathbb{R}$ for simplicity, in which case Borel measures are also known as **Lebesgue–Stieltjes measures**. Then the strategy is as follows: Start with the algebra of finite unions of disjoint intervals and define μ for those sets (as the sum over the intervals). This yields a premeasure. Extend this to an *outer measure* for all subsets of $\mathbb{R}$. Show that the restriction to the Borel sets is a measure.

Let us first show how we should define μ for intervals: To every Borel measure on $\mathfrak{B}$ we can assign its **distribution function**

(A.3) $$\mu(x) = \begin{cases} -\mu((x, 0]), & x < 0, \\ 0, & x = 0, \\ \mu((0, x]), & x > 0, \end{cases}$$

which is right continuous and nondecreasing.

A.1. Borel measures in a nutshell

Example. The distribution function of the Dirac measure centered at 0 is

$$\mu(x) = \begin{cases} -1, & x < 0, \\ 0, & x \geq 0. \end{cases}$$

◇

For a finite measure, the alternate normalization $\tilde{\mu}(x) = \mu((-\infty, x])$ can be used. The resulting distribution function differs from our above definition by a constant $\mu(x) = \tilde{\mu}(x) - \mu((-\infty, 0])$. In particular, this is the normalization used in probability theory.

Conversely, to obtain a measure from a nondecreasing function $m : \mathbb{R} \to \mathbb{R}$, we proceed as follows: Recall that an interval is a subset of the real line of the form

(A.4) $\quad I = (a, b], \quad I = [a, b], \quad I = (a, b), \quad \text{or} \quad I = [a, b),$

with $a \leq b$, $a, b \in \mathbb{R} \cup \{-\infty, \infty\}$. Note that (a, a), $[a, a)$, and $(a, a]$ denote the empty set, whereas $[a, a]$ denotes the singleton $\{a\}$. For any proper interval with different endpoints (i.e. $a < b$), we can define its measure to be

(A.5) $\quad \mu(I) = \begin{cases} m(b+) - m(a+), & I = (a, b], \\ m(b+) - m(a-), & I = [a, b], \\ m(b-) - m(a+), & I = (a, b), \\ m(b-) - m(a-), & I = [a, b), \end{cases}$

where $m(a\pm) = \lim_{\varepsilon \downarrow 0} m(a \pm \varepsilon)$ (which exist by monotonicity). If one of the endpoints is infinite we agree to use $m(\pm\infty) = \lim_{x \to \pm\infty} m(x)$. For the empty set we of course set $\mu(\emptyset) = 0$ and for the singletons we set

(A.6) $\quad \mu(\{a\}) = m(a+) - m(a-)$

(which agrees with (A.5) except for the case $I = (a, a)$, which would give a negative value for the empty set if μ jumps at a). Note that $\mu(\{a\}) = 0$ if and only if $m(x)$ is continuous at a and that there can be only countably many points with $\mu(\{a\}) > 0$ since a nondecreasing function can have at most countably many jumps. Moreover, observe that the definition of μ does not involve the actual value of m at a jump. Hence any function $\tilde{m}$ with $m(x-) \leq \tilde{m}(x) \leq m(x+)$ gives rise to the same μ. We will frequently assume that m is right continuous such that it coincides with the distribution function up to a constant, $\mu(x) = m(x+) - m(0+)$. In particular, μ determines m up to a constant and the value at the jumps.

Now we can consider the algebra of finite unions of disjoint intervals (check that this is indeed an algebra) and extend (A.5) to finite unions of disjoint intervals by summing over all intervals. It is straightforward to verify that μ is well defined (one set can be represented by different unions of intervals) and by construction additive. In fact, it is even a premeasure.

Lemma A.2. *The interval function μ defined in (A.5) gives rise to a unique σ-finite premeasure on the algebra $\mathcal{A}$ of finite unions of disjoint intervals.*

Proof. It remains to verify σ-additivity. We need to show that, for any disjoint union,
$$\mu(\bigcup_k I_k) = \sum_k \mu(I_k)$$
whenever $I_k \in \mathcal{A}$ and $I = \bigcup_k I_k \in \mathcal{A}$. Since each I_k is a finite union of intervals, we can as well assume each I_k is just one interval (just split I_k into its subintervals and note that the sum does not change by additivity). Similarly, we can assume that I is just one interval (just treat each subinterval separately).

By additivity, μ is monotone and hence
$$\sum_{k=1}^{n} \mu(I_k) = \mu(\bigcup_{k=1}^{n} I_k) \leq \mu(I)$$
which shows
$$\sum_{k=1}^{\infty} \mu(I_k) \leq \mu(I).$$
To get the converse inequality, we need to work harder.

We can cover each I_k by some slightly larger open interval J_k such that $\mu(J_k) \leq \mu(I_k) + \frac{\varepsilon}{2^k}$ (only closed endpoints need extension). First suppose I is compact. Then finitely many of the J_k, say the first n, cover I and we have
$$\mu(I) \leq \mu(\bigcup_{k=1}^{n} J_k) \leq \sum_{k=1}^{n} \mu(J_k) \leq \sum_{k=1}^{\infty} \mu(I_k) + \varepsilon.$$
Since $\varepsilon > 0$ is arbitrary, this shows σ-additivity for compact intervals. By additivity we can always add/subtract the endpoints of I and hence σ-additivity holds for any bounded interval. If I is unbounded we can again assume that it is closed by adding an endpoint if necessary. Then for any $x > 0$ we can find an n such that $\{J_k\}_{k=1}^{n}$ cover at least $I \cap [-x, x]$ and hence
$$\sum_{k=1}^{\infty} \mu(I_k) \geq \sum_{k=1}^{n} \mu(I_k) \geq \sum_{k=1}^{n} \mu(J_k) - \varepsilon \geq \mu([-x, x] \cap I) - \varepsilon.$$
Since $x > 0$ and $\varepsilon > 0$ are arbitrary, we are done. $\square$

In particular, this is a premeasure on the algebra of finite unions of intervals which can be extended to a measure:

A.1. Borel measures in a nutshell

Theorem A.3. *For every nondecreasing function $m : \mathbb{R} \to \mathbb{R}$ there exists a unique Borel measure μ which extends (A.5). Two different functions generate the same measure if and only if the difference is a constant away from the discontinuities.*

Since the proof of this theorem is rather involved, we defer it to the next section and look at some examples first.

Example. Suppose $\Theta(x) = 0$ for $x < 0$ and $\Theta(x) = 1$ for $x \geq 0$. Then we obtain the so-called **Dirac measure** at 0, which is given by $\Theta(A) = 1$ if $0 \in A$ and $\Theta(A) = 0$ if $0 \notin A$. ◇

Example. Suppose $\lambda(x) = x$. Then the associated measure is the ordinary **Lebesgue measure** on $\mathbb{R}$. We will abbreviate the Lebesgue measure of a Borel set A by $\lambda(A) = |A|$. ◇

A set $A \in \Sigma$ is called a **support** for μ if $\mu(X \backslash A) = 0$. Note that a support is not unique (see the examples below). If X is a topological space and $\Sigma = \mathfrak{B}(X)$, one defines **the support** (also **topological support**) of μ via

$$(A.7) \quad \operatorname{supp}(\mu) = \{x \in X | \mu(O) > 0 \text{ for every open neighborhood } O \text{ of } x\}.$$

Equivalently, one obtains $\operatorname{supp}(\mu)$ by removing all points which have an open neighborhood of measure zero. In particular, this shows that $\operatorname{supp}(\mu)$ is closed. If X is second countable, then $\operatorname{supp}(\mu)$ is indeed a support for μ: For every point $x \notin \operatorname{supp}(\mu)$, let O_x be an open neighborhood of measure zero. These sets cover $X \backslash \operatorname{supp}(\mu)$ and by the Lindelöf theorem there is a countable subcover, which shows that $X \backslash \operatorname{supp}(\mu)$ has measure zero.

Example. Let $X = \mathbb{R}$, $\Sigma = \mathfrak{B}$. The support of the Lebesgue measure λ is all of $\mathbb{R}$. However, every single point has Lebesgue measure zero and so has every countable union of points (by σ-additivity). Hence any set whose complement is countable is a support. There are even uncountable sets of Lebesgue measure zero (see the Cantor set below) and hence a support might even lack an uncountable number of points.

The support of the Dirac measure centered at 0 is the single point 0. Any set containing 0 is a support of the Dirac measure.

In general, the support of a Borel measure on $\mathbb{R}$ is given by

$$\operatorname{supp}(d\mu) = \{x \in \mathbb{R} | \mu(x - \varepsilon) < \mu(x + \varepsilon), \forall \varepsilon > 0\}.$$

Here we have used $d\mu$ to emphasize that we are interested in the support of the measure $d\mu$ which is different from the support of its distribution function $\mu(x)$. ◇

A property is said to hold μ-**almost everywhere** (a.e.) if it holds on a support for μ or, equivalently, if the set where it does not hold is contained in a set of measure zero.

Example. The set of rational numbers is countable and hence has Lebesgue measure zero, $\lambda(\mathbb{Q}) = 0$. So, for example, the characteristic function of the rationals $\mathbb{Q}$ is zero almost everywhere with respect to Lebesgue measure.

Any function which vanishes at 0 is zero almost everywhere with respect to the Dirac measure centered at 0. ◇

Example. The **Cantor set** is an example of a closed uncountable set of Lebesgue measure zero. It is constructed as follows: Start with $C_0 = [0, 1]$ and remove the middle third to obtain $C_1 = [0, \frac{1}{3}] \cup [\frac{2}{3}, 1]$. Next, again remove the middle thirds of the remaining sets to obtain $C_2 = [0, \frac{1}{9}] \cup [\frac{2}{9}, \frac{1}{3}] \cup [\frac{2}{3}, \frac{7}{9}] \cup [\frac{8}{9}, 1]$:

$$
\begin{array}{l}
\underline{\hspace{4cm}} \quad C_0 \\
\underline{\hspace{1.7cm}} \quad \underline{\hspace{1.7cm}} \quad C_1 \\
\text{—} \quad \text{—} \quad \text{—} \quad \text{—} \quad C_2 \\
\text{- -} \quad \text{- -} \quad \text{- -} \quad \text{- -} \quad C_3 \\
\vdots
\end{array}
$$

Proceeding like this, we obtain a sequence of nesting sets C_n and the limit $C = \bigcap_n C_n$ is the Cantor set. Since C_n is compact, so is C. Moreover, C_n consists of 2^n intervals of length 3^{-n}, and thus its Lebesgue measure is $\lambda(C_n) = (2/3)^n$. In particular, $\lambda(C) = \lim_{n \to \infty} \lambda(C_n) = 0$. Using the ternary expansion, it is extremely simple to describe: C is the set of all $x \in [0, 1]$ whose ternary expansion contains no ones, which shows that C is uncountable (why?). It has some further interesting properties: it is totally disconnected (i.e., it contains no subintervals) and perfect (it has no isolated points). ◇

Problem A.1. Find all algebras over $X = \{1, 2, 3\}$.

Problem A.2. Show that $\mathcal{A} = \{A \subseteq X | A \text{ or } X \backslash A \text{ is finite}\}$ is an algebra (with X some fixed set). Show that $\Sigma = \{A \subseteq X | A \text{ or } X \backslash A \text{ is countable}\}$ is a σ-algebra. (Hint: To verify closedness under unions, consider the cases where all sets are finite and where one set has finite complement.)

Problem A.3. Take some set X and $\Sigma = \{A \subseteq X | A \text{ or } X \backslash A \text{ is countable}\}$. Show that
$$\nu(A) = \begin{cases} 0, & \text{if } A \text{ is countable}, \\ 1, & \text{otherwise} \end{cases}$$
is a measure.

Problem A.4. Show that if X is finite, then every algebra is a σ-algebra. Show that this is not true in general if X is countable.

Problem A.5 (Vitali set). *Call two numbers $x, y \in [0, 1)$ equivalent if $x - y$ is rational. Construct the set V by choosing one representative from each equivalence class. Show that V cannot be measurable with respect to any nontrivial finite translation invariant measure on $[0, 1)$. (Hint: How can you build up $[0, 1)$ from translations of V?)*

A.2. Extending a premeasure to a measure

The purpose of this section is to prove Theorem A.3. It is rather technical and can be skipped on first reading.

In order to prove Theorem A.3, we need to show how a premeasure can be extended to a measure. To show that the extension is unique, we need a better criterion to check when a given system of sets is in fact a σ-algebra. In many situations it is easy to show that a given set is closed under complements and under countable unions of disjoint sets. Hence we call a collection of sets $\mathcal{D}$ with these properties a **Dynkin system** (also **λ-system**) if it also contains X.

Note that a Dynkin system is closed under proper relative complements since $A, B \in \mathcal{D}$ implies $B \setminus A = (B' \cup A)' \in \mathcal{D}$ provided $A \subseteq B$. Moreover, if it is also closed under finite intersections (or arbitrary finite unions), then it is an algebra and hence also a σ-algebra. To see the last claim, note that if $A = \bigcup_j A_j$ then also $A = \bigcup_j B_j$ where the sets $B_j = A_j \setminus \bigcup_{k<j} A_k$ are disjoint.

As with σ-algebras, the intersection of Dynkin systems is a Dynkin system and every collection of sets S generates a smallest Dynkin system $\mathcal{D}(S)$. The important observation is that if S is closed under finite intersections (in which case it is sometimes called a **π-system**), then so is $\mathcal{D}(S)$ and hence it will be a σ-algebra.

Lemma A.4 (Dynkin's π-λ theorem). *Let S be a collection of subsets of X which is closed under finite intersections (or unions). Then $\mathcal{D}(S) = \Sigma(S)$.*

Proof. It suffices to show that $\mathcal{D} = \mathcal{D}(S)$ is closed under finite intersections. To this end, consider the set $D(A) = \{B \in \mathcal{D} | A \cap B \in \mathcal{D}\}$ for $A \in \mathcal{D}$. I claim that $D(A)$ is a Dynkin system.

First of all, $X \in D(A)$ since $A \cap X = A \in \mathcal{D}$. Next, if $B \in D(A)$ then $A \cap B' = A \setminus (B \cap A) \in \mathcal{D}$ (since $\mathcal{D}$ is closed under proper relative complements) implying $B' \in D(A)$. Finally, if $B = \bigcup_j B_j$ with $B_j \in D(A)$ disjoint, then $A \cap B = \bigcup_j (A \cap B_j) \in \mathcal{D}$ with $B_j \in \mathcal{D}$ disjoint, implying $B \in D(A)$.

Now if $A \in S$, we have $S \subseteq D(A)$, implying $D(A) = \mathcal{D}$. Consequently $A \cap B \in \mathcal{D}$ if at least one of the sets is in S. But this shows $S \subseteq D(A)$ and hence $D(A) = \mathcal{D}$ for every $A \in \mathcal{D}$. So $\mathcal{D}$ is closed under finite intersections and thus is a σ-algebra. The case of unions is analogous. □

The typical use of this lemma is as follows: First verify some property for sets in a set S which is closed under finite intersections and generates the σ-algebra. In order to show that it holds for every set in $\Sigma(S)$, it suffices to show that the collection of sets for which it holds is a Dynkin system.

As an application, we show that a premeasure determines the corresponding measure μ uniquely (if there is one at all):

Theorem A.5 (Uniqueness of measures). *Let $S \subseteq \Sigma$ be a collection of sets which generates Σ, which is closed under finite intersections, and which contains a sequence of increasing sets $X_n \nearrow X$ of finite measure $\mu(X_n) < \infty$. Then μ is uniquely determined by the values on S.*

Proof. Let $\tilde{\mu}$ be a second measure and note $\mu(X) = \lim_{n\to\infty} \mu(X_n) = \lim_{n\to\infty} \tilde{\mu}(X_n) = \tilde{\mu}(X)$. We first suppose $\mu(X) < \infty$.

Then
$$\mathcal{D} = \{A \in \Sigma | \mu(A) = \tilde{\mu}(A)\}$$
is a Dynkin system. In fact, by $\mu(A') = \mu(X) - \mu(A) = \tilde{\mu}(X) - \tilde{\mu}(A) = \tilde{\mu}(A')$ for $A \in \mathcal{D}$, we see that $\mathcal{D}$ is closed under complements. Furthermore, by continuity of measures from below, it is also closed under countable disjoint unions. Since $\mathcal{D}$ contains S by assumption, we conclude $\mathcal{D} = \Sigma(S) = \Sigma$ from Lemma A.4. This finishes the finite case.

To extend our result to the general case, observe that the finite case implies $\mu(A \cap X_j) = \tilde{\mu}(A \cap X_j)$ (just restrict μ, $\tilde{\mu}$ to X_j). Hence
$$\mu(A) = \lim_{j\to\infty} \mu(A \cap X_j) = \lim_{j\to\infty} \tilde{\mu}(A \cap X_j) = \tilde{\mu}(A)$$
and we are done. □

Corollary A.6. *Let μ be a σ-finite premeasure on an algebra $\mathcal{A}$. Then there is at most one extension to $\Sigma(\mathcal{A})$.*

So it remains to ensure that there is an extension at all. For any premeasure μ we define

(A.8) $$\mu^*(A) = \inf\left\{\sum_{n=1}^{\infty} \mu(A_n) \,\Big|\, A \subseteq \bigcup_{n=1}^{\infty} A_n, \, A_n \in \mathcal{A}\right\}$$

where the infimum extends over all countable covers from $\mathcal{A}$. Then the function $\mu^* : \mathfrak{P}(X) \to [0, \infty]$ is an **outer measure**; that is, it has the properties (Problem A.6)

- $\mu^*(\emptyset) = 0$,
- $A_1 \subseteq A_2 \Rightarrow \mu^*(A_1) \le \mu^*(A_2)$, and

A.2. Extending a premeasure to a measure

- $\mu^*(\bigcup_{n=1}^\infty A_n) \le \sum_{n=1}^\infty \mu^*(A_n)$ (subadditivity).

Note that $\mu^*(A) = \mu(A)$ for $A \in \mathcal{A}$ (Problem A.7).

Theorem A.7 (Extensions via outer measures). *Let μ^* be an outer measure. Then the set Σ of all sets A satisfying the Carathéodory condition*

(A.9) $\qquad \mu^*(E) = \mu^*(A \cap E) + \mu^*(A' \cap E), \quad \forall E \subseteq X$

(where $A' = X \setminus A$ is the complement of A) forms a σ-algebra and μ^* restricted to this σ-algebra is a measure.

Proof. We first show that Σ is an algebra. It clearly contains X and is closed under complements. Concerning unions, let $A, B \in \Sigma$. Applying Carathéodory's condition twice shows

$$\mu^*(E) = \mu^*(A \cap B \cap E) + \mu^*(A' \cap B \cap E) + \mu^*(A \cap B' \cap E)$$
$$+ \mu^*(A' \cap B' \cap E)$$
$$\ge \mu^*((A \cup B) \cap E) + \mu^*((A \cup B)' \cap E),$$

where we have used de Morgan and

$$\mu^*(A \cap B \cap E) + \mu^*(A' \cap B \cap E) + \mu^*(A \cap B' \cap E) \ge \mu^*((A \cup B) \cap E),$$

which follows from subadditivity and $(A \cup B) \cap E = (A \cap B \cap E) \cup (A' \cap B \cap E) \cup (A \cap B' \cap E)$. Since the reverse inequality is just subadditivity, we conclude that Σ is an algebra.

Next, let A_n be a sequence of sets from Σ. Without restriction we can assume that they are disjoint (compare the argument for item (ii) in the proof of Theorem A.1). Abbreviate $\tilde{A}_n = \bigcup_{k \le n} A_k$, $A = \bigcup_n A_n$. Then for every set E we have

$$\mu^*(\tilde{A}_n \cap E) = \mu^*(A_n \cap \tilde{A}_n \cap E) + \mu^*(A_n' \cap \tilde{A}_n \cap E)$$
$$= \mu^*(A_n \cap E) + \mu^*(\tilde{A}_{n-1} \cap E)$$
$$= \ldots = \sum_{k=1}^n \mu^*(A_k \cap E).$$

Using $\tilde{A}_n \in \Sigma$ and monotonicity of μ^*, we infer

$$\mu^*(E) = \mu^*(\tilde{A}_n \cap E) + \mu^*(\tilde{A}_n' \cap E)$$
$$\ge \sum_{k=1}^n \mu^*(A_k \cap E) + \mu^*(A' \cap E).$$

Letting $n \to \infty$ and using subadditivity finally gives

$$\mu^*(E) \geq \sum_{k=1}^{\infty} \mu^*(A_k \cap E) + \mu^*(A' \cap E)$$

(A.10)
$$\geq \mu^*(A \cap E) + \mu^*(A' \cap E) \geq \mu^*(E)$$

and we infer that Σ is a σ-algebra.

Finally, setting $E = A$ in (A.10), we have

$$\mu^*(A) = \sum_{k=1}^{\infty} \mu^*(A_k \cap A) + \mu^*(A' \cap A) = \sum_{k=1}^{\infty} \mu^*(A_k)$$

and we are done. $\square$

Remark: The constructed measure μ is **complete**; that is, for every measurable set A of measure zero, every subset of A is again measurable (Problem A.8).

The only remaining question is whether there are any nontrivial sets satisfying the Carathéodory condition.

Lemma A.8. *Let μ be a premeasure on $\mathcal{A}$ and let μ^* be the associated outer measure. Then every set in $\mathcal{A}$ satisfies the Carathéodory condition.*

Proof. Let $A_n \in \mathcal{A}$ be a countable cover for E. Then for every $A \in \mathcal{A}$ we have

$$\sum_{n=1}^{\infty} \mu(A_n) = \sum_{n=1}^{\infty} \mu(A_n \cap A) + \sum_{n=1}^{\infty} \mu(A_n \cap A') \geq \mu^*(E \cap A) + \mu^*(E \cap A')$$

since $A_n \cap A \in \mathcal{A}$ is a cover for $E \cap A$ and $A_n \cap A' \in \mathcal{A}$ is a cover for $E \cap A'$. Taking the infimum, we have $\mu^*(E) \geq \mu^*(E \cap A) + \mu^*(E \cap A')$, which finishes the proof. $\square$

Concerning regularity, we note:

Lemma A.9. *Suppose outer regularity (A.1) holds for every set in the algebra; then μ is outer regular.*

Proof. By assumption, we can replace each set A_n in (A.8) by a possibly slightly larger open set and hence the infimum in (A.8) can be realized with open sets. $\square$

Thus, as a consequence we obtain Theorem A.3 except for regularity. Outer regularity is easy to see for a finite union of intervals since we can replace each interval by a possibly slightly larger open interval with only slightly larger measure. Inner regularity will be postponed until Lemma A.14.

Problem A.6. *Show that μ^* defined in (A.8) is an outer measure. (Hint for the last property: Take a cover $\{B_{nk}\}_{k=1}^\infty$ for A_n such that $\mu^*(A_n) = \frac{\varepsilon}{2^n} + \sum_{k=1}^\infty \mu(B_{nk})$ and note that $\{B_{nk}\}_{n,k=1}^\infty$ is a cover for $\bigcup_n A_n$.)*

Problem A.7. *Show that μ^* defined in (A.8) extends μ. (Hint: For the cover A_n it is no restriction to assume $A_n \cap A_m = \emptyset$ and $A_n \subseteq A$.)*

Problem A.8. *Show that the measure constructed in Theorem A.7 is complete.*

Problem A.9. *Let μ be a finite measure. Show that*

$$(A.11) \qquad d(A, B) = \mu(A \Delta B), \qquad A \Delta B = (A \cup B) \setminus (A \cap B)$$

is a metric on Σ if we identify sets of measure zero. Show that if $\mathcal{A}$ is an algebra, then it is dense in $\Sigma(\mathcal{A})$. (Hint: Show that the sets which can be approximated by sets in $\mathcal{A}$ form a Dynkin system.)

A.3. Measurable functions

The Riemann integral works by splitting the x coordinate into small intervals and approximating $f(x)$ on each interval by its minimum and maximum. The problem with this approach is that the difference between maximum and minimum will only tend to zero (as the intervals get smaller) if $f(x)$ is sufficiently nice. To avoid this problem, we can force the difference to go to zero by considering, instead of an interval, the set of x for which $f(x)$ lies between two given numbers $a < b$. Now we need the size of the set of these x, that is, the size of the preimage $f^{-1}((a, b))$. For this to work, preimages of intervals must be measurable.

Let (X, Σ_X) and (Y, Σ_Y) be measurable spaces. A function $f : X \to Y$ is called **measurable** if $f^{-1}(A) \in \Sigma_X$ for every $A \in \Sigma_Y$. Clearly it suffices to check this condition for every set A in a collection of sets which generate Σ_Y, since the collection of sets for which it holds forms a σ-algebra by $f^{-1}(Y \setminus A) = X \setminus f^{-1}(A)$ and $f^{-1}(\bigcup_j A_j) = \bigcup_j f^{-1}(A_j)$.

We will be mainly interested in the case where $(Y, \Sigma_Y) = (\mathbb{R}^n, \mathfrak{B}^n)$.

Lemma A.10. *A function $f : X \to \mathbb{R}^n$ is measurable if and only if*

$$(A.12) \qquad f^{-1}(I) \in \Sigma \qquad \forall I = \prod_{j=1}^n (a_j, \infty).$$

In particular, a function $f : X \to \mathbb{R}^n$ is measurable if and only if every component is measurable, and a complex-valued function $f : X \to \mathbb{C}^n$ is measurable if and only if both its real and imaginary parts are.

Proof. We need to show that $\mathfrak{B}^n$ is generated by rectangles of the above form. The σ-algebra generated by these rectangles also contains all open rectangles of the form $I = \prod_{j=1}^n (a_j, b_j)$, which form a base for the topology. $\square$

Clearly the intervals (a_j, ∞) can also be replaced by $[a_j, \infty)$, $(-\infty, a_j)$, or $(-\infty, a_j]$.

If X is a topological space and Σ the corresponding Borel σ-algebra, we will also call a measurable function a **Borel function**. Note that, in particular,

Lemma A.11. *Let (X, Σ_X), (Y, Σ_Y), (Z, Σ_Z) be topological spaces with their corresponding Borel σ-algebras. Any continuous function $f : X \to Y$ is measurable. Moreover, if $f : X \to Y$ and $g : Y \to Z$ are measurable functions, then the composition $g \circ f$ is again measurable.*

The set of all measurable functions forms an algebra.

Lemma A.12. *Let X be a topological space and Σ its Borel σ-algebra. Suppose $f, g : X \to \mathbb{R}$ are measurable functions. Then the sum $f + g$ and the product fg are measurable.*

Proof. Note that addition and multiplication are continuous functions from $\mathbb{R}^2 \to \mathbb{R}$ and hence the claim follows from the previous lemma. $\square$

Sometimes it is also convenient to allow $\pm\infty$ as possible values for f, that is, functions $f : X \to \overline{\mathbb{R}}$, $\overline{\mathbb{R}} = \mathbb{R} \cup \{-\infty, \infty\}$. In this case, $A \subseteq \overline{\mathbb{R}}$ is called Borel if $A \cap \mathbb{R}$ is. This implies that $f : X \to \overline{\mathbb{R}}$ will be Borel if and only if $f^{-1}(\pm\infty)$ are Borel and $f : X \setminus f^{-1}(\{-\infty, \infty\}) \to \mathbb{R}$ is Borel. Since

$$(A.13) \qquad \{+\infty\} = \bigcap_n (n, +\infty], \qquad \{-\infty\} = \overline{\mathbb{R}} \setminus \bigcup_n (-n, +\infty],$$

we see that $f : X \to \overline{\mathbb{R}}$ is measurable if and only if

$$(A.14) \qquad f^{-1}((a, \infty]) \in \Sigma \qquad \forall\, a \in \mathbb{R}.$$

Again the intervals $(a, \infty]$ can also be replaced by $[a, \infty]$, $[-\infty, a)$, or $[-\infty, a]$.

Hence it is not hard to check that the previous lemma still holds if one either avoids undefined expressions of the type $\infty - \infty$ and $\pm\infty \cdot 0$ or makes a definite choice, e.g., $\infty - \infty = 0$ and $\pm\infty \cdot 0 = 0$.

Moreover, the set of all measurable functions is closed under all important limiting operations.

Lemma A.13. *Suppose $f_n : X \to \overline{\mathbb{R}}$ is a sequence of measurable functions. Then*

$$(A.15) \qquad \inf_{n \in \mathbb{N}} f_n, \quad \sup_{n \in \mathbb{N}} f_n, \quad \liminf_{n \to \infty} f_n, \quad \limsup_{n \to \infty} f_n$$

are measurable as well.

Proof. It suffices to prove that $\sup f_n$ is measurable since the rest follows from $\inf f_n = -\sup(-f_n)$, $\liminf f_n = \sup_n \inf_{k\geq n} f_k$, and $\limsup f_n = \inf_n \sup_{k\geq n} f_k$. But $(\sup f_n)^{-1}((a,\infty]) = \bigcup_n f_n^{-1}((a,\infty])$ are Borel and we are done. $\square$

A few immediate consequences are worthwhile noting: It follows that if f and g are measurable functions, so are $\min(f,g)$, $\max(f,g)$, $|f| = \max(f,-f)$, and $f^{\pm} = \max(\pm f, 0)$. Furthermore, the pointwise limit of measurable functions is again measurable.

Sometimes the case of arbitrary suprema and infima is also of interest. In this respect, the following observation is useful: Recall that a function $f : X \to \overline{\mathbb{R}}$ is **lower semicontinuous** if the set $f^{-1}((a,\infty])$ is open for every $a \in \mathbb{R}$. Then it follows from the definition that the sup over an arbitrary collection of lower semicontinuous functions

$$\overline{f}(x) = \sup_\alpha f_\alpha(x) \tag{A.16}$$

is again lower semicontinuous. Similarly, f is **upper semicontinuous** if the set $f^{-1}([-\infty,a))$ is open for every $a \in \mathbb{R}$. In this case the infimum

$$\underline{f}(x) = \inf_\alpha f_\alpha(x) \tag{A.17}$$

is again upper semicontinuous. Note that f is lower semicontinuous if and only if $-f$ is upper semicontinuous.

Problem A.10. *Show that the supremum over lower semicontinuous functions is again lower semicontinuous.*

Problem A.11. *Let X be a metric space. Show that f is lower semicontinuous if and only if*

$$\liminf_{x \to x_0} f(x) \geq f(x_0), \qquad x_0 \in X.$$

Similarly, f is upper semicontinuous if and only if

$$\limsup_{x \to x_0} f(x) \leq f(x_0), \qquad x_0 \in X.$$

A.4. How wild are measurable objects?

In this section, we want to investigate how far measurable objects are away from well-understood ones. As our first task, we want to show that measurable sets can be well approximated by using closed sets from the inside and open sets from the outside in nice spaces like $\mathbb{R}^n$.

Lemma A.14. *Let X be a metric space and μ a Borel measure which is finite on finite balls. Then μ is σ-finite and for every $A \in \mathfrak{B}(X)$ and any given $\varepsilon > 0$ there exists an open set O and a closed set F such that*

(A.18) $$F \subseteq A \subseteq O \quad \text{and} \quad \mu(O \setminus F) \leq \varepsilon.$$

Proof. That μ is σ-finite is immediate from the definitions since for any fixed $x_0 \in X$, the open balls $X_n = B_n(x_0)$ have finite measure and satisfy $X_n \nearrow X$.

To see that (A.18) holds, we begin with the case when μ is finite. Denote by $\mathcal{A}$ the set of all Borel sets satisfying (A.18). Then $\mathcal{A}$ contains every closed set F: Given F, define $O_n = \{x \in X \mid d(x, F) < 1/n\}$ and note that O_n are open sets which satisfy $O_n \searrow F$. Thus, by Theorem A.1 (iii), $\mu(O_n \setminus F) \to 0$ and hence $F \in \mathcal{A}$.

Moreover, $\mathcal{A}$ is even a σ-algebra. That it is closed under complements is easy to see (note that $\tilde{O} = X \setminus F$ and $\tilde{F} = X \setminus O$ are the required sets for $\tilde{A} = X \setminus A$). To see that it is closed under countable unions, consider $A = \bigcup_{n=1}^\infty A_n$ with $A_n \in \mathcal{A}$. Then there are F_n, O_n such that $\mu(O_n \setminus F_n) \leq \varepsilon 2^{-n-1}$. Now $O = \bigcup_{n=1}^\infty O_n$ is open and $F = \bigcup_{n=1}^N F_n$ is closed for any finite N. Since $\mu(A)$ is finite, we can choose N sufficiently large such that $\mu(\bigcup_{N+1}^\infty F_n \setminus F) \leq \varepsilon/2$. Then we have found two sets of the required type: $\mu(O \setminus F) \leq \sum_{n=1}^\infty \mu(O_n \setminus F_n) + \mu(\bigcup_{n=N+1}^\infty F_n \setminus F) \leq \varepsilon$. Thus $\mathcal{A}$ is a σ-algebra containing the open sets, and hence it is the entire Borel algebra.

Now suppose μ is not finite. Pick some $x_0 \in X$ and set $X_0 = B_{2/3}(x_0)$ and $X_n = B_{n+2/3}(x_0) \setminus \overline{B_{n-2/3}(x_0)}$, $n \in \mathbb{N}$. Let $A_n = A \cap X_n$ and note that $A = \bigcup_{n=0}^\infty A_n$. By the finite case we can choose $F_n \subseteq A_n \subseteq O_n \subseteq X_n$ such that $\mu(O_n \setminus F_n) \leq \varepsilon 2^{-n-1}$. Now set $F = \bigcup_n F_n$ and $O = \bigcup_n O_n$ and observe that F is closed. Indeed, let $x \in \overline{F}$ and let x_j be some sequence from F converging to x. Since $d(x_0, x_j) \to d(x_0, x)$ this sequence must eventually lie in $F_n \cup F_{n+1}$ for some fixed n, implying $x \in \overline{F_n \cup F_{n+1}} = F_n \cup F_{n+1} \subseteq F$. Finally, $\mu(O \setminus F) \leq \sum_{n=0}^\infty \mu(O_n \setminus F_n) \leq \varepsilon$ as required. $\square$

This result immediately gives us outer regularity and, if we strengthen our assumption, also inner regularity.

Corollary A.15. *Under the assumptions of the previous lemma,*

(A.19) $$\mu(A) = \inf_{O \supseteq A,\, O \text{ open}} \mu(O) = \sup_{F \subseteq A,\, F \text{ closed}} \mu(F)$$

and μ is outer regular. If X is proper (i.e., every closed ball is compact), then μ is also inner regular.

(A.20) $$\mu(A) = \sup_{K \subseteq A,\, K \text{ compact}} \mu(K).$$

Proof. Finally, (A.19) follows from $\mu(A) = \mu(O) - \mu(O\setminus A) = \mu(F) + \mu(A\setminus F)$ and if every finite ball is compact, for every sequence of closed sets F_n with $\mu(F_n) \to \mu(A)$ we also have compact sets $K_n = F_n \cap \overline{B_n(x_0)}$ with $\mu(K_n) \to \mu(A)$. □

By the Heine–Borel theorem, every bounded closed ball in $\mathbb{R}^n$ (or $\mathbb{C}^n$) is compact and thus has finite measure by the very definition of a Borel measure. Hence every Borel measure on $\mathbb{R}^n$ (or $\mathbb{C}^n$) satisfies the assumptions of Lemma A.14.

An inner regular measure on a Hausdorff space which is locally finite (every point has a neighborhood of finite measure) is called a **Radon measure**. Accordingly, every Borel measure on $\mathbb{R}^n$ (or $\mathbb{C}^n$) is automatically a Radon measure.

Example. Since Lebesgue measure on $\mathbb{R}$ is regular, we can cover the rational numbers by an open set of arbitrary small measure (it is also not hard to find such a set directly) but we cannot cover it by an open set of measure zero (since any open set contains an interval and hence has positive measure). However, if we slightly extend the family of admissible sets, this will be possible. ⋄

Looking at the Borel σ-algebra, the next general sets after open sets are countable intersections of open sets, known as G_δ sets (here G and δ stand for the German words *Gebiet* and *Durchschnitt*, respectively). The next general sets after closed sets are countable unions of closed sets, known as F_σ sets (here F and σ stand for the French words *fermé* and *somme*, respectively).

Example. The irrational numbers are a G_δ set in $\mathbb{R}$. To see this, let x_n be an enumeration of the rational numbers and consider the intersection of the open sets $O_n = \mathbb{R}\setminus\{x_n\}$. The rational numbers are hence an F_σ set. ⋄

Corollary A.16. *A set in $\mathbb{R}^n$ is Borel if and only if it differs from a G_δ set by a Borel set of measure zero. Similarly, a set in $\mathbb{R}^n$ is Borel if and only if it differs from an F_σ set by a Borel set of measure zero.*

Proof. Since G_δ sets are Borel, only the converse direction is nontrivial. By Lemma A.14 we can find open sets O_n such that $\mu(O_n\setminus A) \le 1/n$. Now let $G = \bigcap_n O_n$. Then $\mu(G\setminus A) \le \mu(O_n\setminus A) \le 1/n$ for any n and thus $\mu(G\setminus A) = 0$. The second claim is analogous. □

Problem A.12. *Show directly (without using regularity) that for every $\varepsilon > 0$ there is an open set O of Lebesgue measure $|O| < \varepsilon$ which covers the rational numbers.*

Problem A.13. *Show that a Borel set $A \subseteq \mathbb{R}$ has Lebesgue measure zero if and only if for every ε there exists a countable set of intervals I_j which cover A and satisfy $\sum_j |I_j| < \varepsilon$.*

A.5. Integration — Sum me up, Henri

Throughout this section, (X, Σ, μ) will be a measure space. A measurable function $s : X \to \mathbb{R}$ is called **simple** if its image is finite; that is, if

$$\text{(A.21)} \qquad s = \sum_{j=1}^{p} \alpha_j \chi_{A_j}, \qquad s(X) = \{\alpha_j\}_{j=1}^{p}, \qquad A_j = s^{-1}(\alpha_j) \in \Sigma.$$

Here χ_A is the **characteristic function** of A; that is, $\chi_A(x) = 1$ if $x \in A$ and $\chi_A(x) = 0$ otherwise. Note that the set of simple functions is a vector space, and while there are different ways of writing a simple function as a linear combination of characteristic functions, the representation (A.21) is unique.

For a nonnegative simple function s as in (A.21) we define its **integral** as

$$\text{(A.22)} \qquad \int_A s \, d\mu = \sum_{j=1}^{p} \alpha_j \mu(A_j \cap A).$$

Here we use the convention $0 \cdot \infty = 0$.

Lemma A.17. *The integral has the following properties:*

(i) $\int_A s \, d\mu = \int_X \chi_A s \, d\mu$.
(ii) $\int_{\bigcup_{j=1}^{\infty} A_j} s \, d\mu = \sum_{j=1}^{\infty} \int_{A_j} s \, d\mu$, $\qquad A_j \cap A_k = \emptyset$ for $j \neq k$.
(iii) $\int_A \alpha s \, d\mu = \alpha \int_A s \, d\mu$, $\alpha \geq 0$.
(iv) $\int_A (s+t) d\mu = \int_A s \, d\mu + \int_A t \, d\mu$.
(v) $A \subseteq B \Rightarrow \int_A s \, d\mu \leq \int_B s \, d\mu$.
(vi) $s \leq t \Rightarrow \int_A s \, d\mu \leq \int_A t \, d\mu$.

Proof. (i) is clear from the definition. (ii) follows from σ-additivity of μ. (iii) is obvious. (iv) Let $s = \sum_j \alpha_j \chi_{A_j}$, $t = \sum_j \beta_j \chi_{B_j}$ as in (A.21) and abbreviate $C_{jk} = (A_j \cap B_k) \cap A$. Then, by (ii),

$$\int_A (s+t) d\mu = \sum_{j,k} \int_{C_{jk}} (s+t) d\mu = \sum_{j,k} (\alpha_j + \beta_k) \mu(C_{jk})$$

$$= \sum_{j,k} \left(\int_{C_{jk}} s \, d\mu + \int_{C_{jk}} t \, d\mu \right) = \int_A s \, d\mu + \int_A t \, d\mu.$$

A.5. Integration — Sum me up, Henri

(v) follows from monotonicity of μ. (vi) follows since by (iv) we can write $s = \sum_j \alpha_j \chi_{C_j}$, $t = \sum_j \beta_j \chi_{C_j}$ where, by assumption, $\alpha_j \leq \beta_j$. □

Our next task is to extend this definition to nonnegative measurable functions by

$$\text{(A.23)} \qquad \int_A f\, d\mu = \sup_{s \leq f} \int_A s\, d\mu,$$

where the supremum is taken over all simple functions $s \leq f$. Note that, except for possibly (ii) and (iv), Lemma A.17 still holds for arbitrary nonnegative functions s, t.

Theorem A.18 (Monotone convergence, Beppo Levi's theorem). *Let f_n be a monotone nondecreasing sequence of nonnegative measurable functions, $f_n \nearrow f$. Then*

$$\text{(A.24)} \qquad \int_A f_n\, d\mu \to \int_A f\, d\mu.$$

Proof. By property (vi), $\int_A f_n\, d\mu$ is monotone and converges to some number α. By $f_n \leq f$ and again (vi) we have

$$\alpha \leq \int_A f\, d\mu.$$

To show the converse, let s be simple such that $s \leq f$ and let $\theta \in (0,1)$. Put $A_n = \{x \in A \,|\, f_n(x) \geq \theta s(x)\}$ and note $A_n \nearrow A$ (show this). Then

$$\int_A f_n\, d\mu \geq \int_{A_n} f_n\, d\mu \geq \theta \int_{A_n} s\, d\mu.$$

Letting $n \to \infty$, we see that

$$\alpha \geq \theta \int_A s\, d\mu.$$

Since this is valid for every $\theta < 1$, it still holds for $\theta = 1$. Finally, since $s \leq f$ is arbitrary, the claim follows. □

In particular,

$$\text{(A.25)} \qquad \int_A f\, d\mu = \lim_{n \to \infty} \int_A s_n\, d\mu$$

for every monotone sequence $s_n \nearrow f$ of simple functions. Note that there is always such a sequence, for example,

$$\text{(A.26)} \qquad s_n(x) = \sum_{k=0}^{n 2^n} \frac{k}{2^n} \chi_{f^{-1}(A_k)}(x), \quad A_k = [\frac{k}{2^n}, \frac{k+1}{2^n}), \quad A_{n 2^n} = [n, \infty).$$

By construction, s_n converges uniformly if f is bounded, since $0 \leq f(x) - s_n(x) < \frac{1}{2^n}$ if $f(x) \leq n$.

Now what about the missing items (ii) and (iv) from Lemma A.17? Since limits can be spread over sums, item (iv) holds, and (ii) also follows directly from the monotone convergence theorem. We even have the following result:

Lemma A.19. *If $f \geq 0$ is measurable, then $d\nu = f\, d\mu$ defined via*

(A.27) $$\nu(A) = \int_A f\, d\mu$$

is a measure such that

(A.28) $$\int g\, d\nu = \int gf\, d\mu$$

for every measurable function g.

Proof. As already mentioned, additivity of ν is equivalent to linearity of the integral and σ-additivity follows from Theorem A.18:

$$\nu\left(\bigcup_{n=1}^{\infty} A_n\right) = \int \left(\sum_{n=1}^{\infty} \chi_{A_n}\right) f\, d\mu = \sum_{n=1}^{\infty} \int \chi_{A_n} f\, d\mu = \sum_{n=1}^{\infty} \nu(A_n).$$

The second claim holds for simple functions and hence for all functions by construction of the integral. □

If f_n is not necessarily monotone, we have at least

Theorem A.20 (Fatou's lemma). *If f_n is a sequence of nonnegative measurable function, then*

(A.29) $$\int_A \liminf_{n \to \infty} f_n\, d\mu \leq \liminf_{n \to \infty} \int_A f_n\, d\mu.$$

Proof. Set $g_n = \inf_{k \geq n} f_k$. Then $g_n \leq f_n$, implying

$$\int_A g_n\, d\mu \leq \int_A f_n\, d\mu.$$

Now take the lim inf on both sides and note that by the monotone convergence theorem,

$$\liminf_{n \to \infty} \int_A g_n\, d\mu = \lim_{n \to \infty} \int_A g_n\, d\mu = \int_A \lim_{n \to \infty} g_n\, d\mu = \int_A \liminf_{n \to \infty} f_n\, d\mu,$$

proving the claim. □

Example. Consider $f_n = \chi_{[n,n+1]}$. Then $\lim_{n \to \infty} f_n(x) = 0$ for every $x \in \mathbb{R}$. However, $\int_\mathbb{R} f_n(x)\, dx = 1$. This shows that the inequality in Fatou's lemma cannot be replaced by equality in general. ◇

A.5. Integration — Sum me up, Henri

If the integral is finite for both the positive and negative part $f^{\pm} = \max(\pm f, 0)$ of an arbitrary measurable function f, we call f **integrable** and set

(A.30) $$\int_A f\, d\mu = \int_A f^{+}\, d\mu - \int_A f^{-}\, d\mu.$$

Similarly, we handle the case where f is complex-valued by calling f integrable if both the real and imaginary parts are and setting

(A.31) $$\int_A f\, d\mu = \int_A \operatorname{Re}(f)\, d\mu + i \int_A \operatorname{Im}(f)\, d\mu.$$

Clearly, f is integrable if and only if $|f|$ is. The set of all integrable functions is denoted by $\mathcal{L}^1(X, d\mu)$.

Lemma A.21. *The integral is linear and Lemma A.17 holds for integrable functions s, t.*

Furthermore, for all integrable functions f, g, we have

(A.32) $$\left| \int_A f\, d\mu \right| \leq \int_A |f|\, d\mu$$

and (triangle inequality)

(A.33) $$\int_A |f + g|\, d\mu \leq \int_A |f|\, d\mu + \int_A |g|\, d\mu.$$

Proof. Linearity and Lemma A.17 are straightforward to check. To see (A.32) put $\alpha = \frac{z^*}{|z|}$, where $z = \int_A f\, d\mu$ (without restriction $z \neq 0$). Then

$$\left| \int_A f\, d\mu \right| = \alpha \int_A f\, d\mu = \int_A \alpha f\, d\mu = \int_A \operatorname{Re}(\alpha f)\, d\mu \leq \int_A |f|\, d\mu,$$

proving (A.32). The last claim follows from $|f + g| \leq |f| + |g|$. $\square$

Lemma A.22. *Let f be measurable. Then*

(A.34) $$\int_X |f|\, d\mu = 0 \quad \Leftrightarrow \quad f(x) = 0 \quad \mu - a.e.$$

Moreover, suppose f is nonnegative or integrable. Then

(A.35) $$\mu(A) = 0 \quad \Rightarrow \quad \int_A f\, d\mu = 0.$$

Proof. Observe that we have $A = \{x | f(x) \neq 0\} = \bigcup_n A_n$, where $A_n = \{x \mid |f(x)| > \frac{1}{n}\}$. If $\int_X |f|\, d\mu = 0$, we must have $\mu(A_n) = 0$ for every n and hence $\mu(A) = \lim_{n \to \infty} \mu(A_n) = 0$.

The converse will follow from (A.35) since $\mu(A) = 0$ (with A as before) implies $\int_X |f|\, d\mu = \int_A |f|\, d\mu = 0$.

Finally, to see (A.35) note that by our convention $0 \cdot \infty = 0$ it holds for any simple function and hence for any nonnegative f by definition of the integral (A.23). Since any function can be written as a linear combination of four nonnegative functions, this also implies the case when f is integrable. $\square$

Note that the proof also shows that if f is not 0 almost everywhere, there is an $\varepsilon > 0$ such that $\mu(\{x \mid |f(x)| \geq \varepsilon\}) > 0$.

In particular, the integral does not change if we restrict the domain of integration to a support of μ or if we change f on a set of measure zero. In particular, functions which are equal a.e. have the same integral.

Finally, our integral is well behaved with respect to limiting operations. We first state a simple generalization of Fatou's lemma.

Lemma A.23 (generalized Fatou lemma). *If f_n is a sequence of real-valued measurable functions and g some integrable function, then*

$$(A.36) \qquad \int_A \liminf_{n \to \infty} f_n \, d\mu \leq \liminf_{n \to \infty} \int_A f_n \, d\mu$$

if $g \leq f_n$ and

$$(A.37) \qquad \limsup_{n \to \infty} \int_A f_n \, d\mu \leq \int_A \limsup_{n \to \infty} f_n \, d\mu$$

if $f_n \leq g$.

Proof. To see the first, apply Fatou's lemma to $f_n - g$ and subtract $\int_A g \, d\mu$ on both sides of the result. The second follows from the first using $\liminf(-f_n) = -\limsup f_n$. $\square$

If in the last lemma we even have $|f_n| \leq g$, we can combine both estimates to obtain
(A.38)
$$\int_A \liminf_{n \to \infty} f_n \, d\mu \leq \liminf_{n \to \infty} \int_A f_n \, d\mu \leq \limsup_{n \to \infty} \int_A f_n \, d\mu \leq \int_A \limsup_{n \to \infty} f_n \, d\mu,$$

which is known as the **Fatou–Lebesgue theorem**. In particular, in the special case where f_n converges, we obtain

Theorem A.24 (Dominated convergence). *Let f_n be a convergent sequence of measurable functions and set $f = \lim_{n \to \infty} f_n$. Suppose there is an integrable function g such that $|f_n| \leq g$. Then f is integrable and*

$$(A.39) \qquad \lim_{n \to \infty} \int f_n \, d\mu = \int f \, d\mu.$$

Proof. The real and imaginary parts satisfy the same assumptions, and hence it suffices to prove the case where f_n and f are real-valued. Moreover, since $\liminf f_n = \limsup f_n = f$, equation (A.38) establishes the claim. $\square$

Remark: Since sets of measure zero do not contribute to the value of the integral, it clearly suffices if the requirements of the dominated convergence theorem are satisfied almost everywhere (with respect to μ).

Example. Note that the existence of g is crucial: The functions $f_n(x) = \frac{1}{2n}\chi_{[-n,n]}(x)$ on $\mathbb{R}$ converge uniformly to 0 but $\int_\mathbb{R} f_n(x)dx = 1$. ◇

Example. If $\mu(x) = \Theta(x)$ is the Dirac measure at 0, then

$$\int_\mathbb{R} f(x)d\mu(x) = f(0).$$

In fact, the integral can be restricted to any support and hence to $\{0\}$.

If $\mu(x) = \sum_n \alpha_n \Theta(x - x_n)$ is a sum of Dirac measures, then (Problem A.14)

$$\int_\mathbb{R} f(x)d\mu(x) = \sum_n \alpha_n f(x_n).$$

Hence our integral contains sums as special cases. ◇

Finally, let me conclude this section with a remark on how to compute Lebesgue integrals in the classical case of Lebesgue measure on some interval $(a,b) \subseteq \mathbb{R}$. Given a continuous function $f \in C(a,b)$ which is integrable over (a,b), we can introduce

(A.40) $$F(x) = \int_{(a,x]} f(y)dy, \qquad x \in (a,b).$$

Then one has

$$\frac{F(x+\varepsilon) - F(x)}{\varepsilon} = f(x) + \frac{1}{\varepsilon}\int_{(x,x+\varepsilon]} (f(y) - f(x))dy$$

(where $(x, x+\varepsilon]$ has to be understood as $(x+\varepsilon, x]$ if $\varepsilon < 0$) and

$$\limsup_{\varepsilon \to 0} \frac{1}{\varepsilon} \int_{(x,x+\varepsilon]} |f(y) - f(x)|dy \leq \limsup_{\varepsilon \to 0} \sup_{y \in (x,x+\varepsilon]} |f(y) - f(x)| = 0$$

by the continuity of f at x. Thus $F \in C^1(a,b)$ and

$$F'(x) = f(x),$$

which is a variant of the **fundamental theorem of calculus**. This tells us that the integral of a continuous function f can be computed in terms of its antiderivative and, in particular, all tools from calculus like integration by parts or integration by substitution are readily available for the Lebesgue integral on $\mathbb{R}$. Moreover, the Lebesgue integral must coincide with the Riemann integral for (piecewise) continuous functions.

Problem A.14. *Consider a countable set of measures μ_n and numbers $\alpha_n \geq 0$. Let $\mu = \sum_n \alpha_n \mu_n$ and show*

(A.41) $$\int_A f\, d\mu = \sum_n \alpha_n \int_A f\, d\mu_n$$

for any measurable function which is either nonnegative or integrable.

Problem A.15. *Show that the set $B(X)$ of bounded measurable functions with the sup norm is a Banach space. Show that the set $S(X)$ of simple functions is dense in $B(X)$. Show that the integral is a bounded linear functional on $B(X)$ if $\mu(X) < \infty$. (Hence Theorem 0.29 could be used to extend the integral from simple to bounded measurable functions.)*

Problem A.16. *Show that the monotone convergence holds for nondecreasing sequences of real-valued measurable functions $f_n \nearrow f$ provided f_1 is integrable.*

Problem A.17. *Show that the dominated convergence theorem implies (under the same assumptions)*

$$\lim_{n \to \infty} \int |f_n - f|\, d\mu = 0.$$

Problem A.18. *Let f be an integrable function satisfying $f(x) \leq M$. Show that*

$$\int_A f\, d\mu \leq M\mu(A)$$

with equality if and only if $f(x) = M$ for a.e. $x \in A$.

Problem A.19. *Let $X \subseteq \mathbb{R}$, Y be some measure space, and $f : X \times Y \to \mathbb{C}$. Suppose $y \mapsto f(x, y)$ is measurable for every x and $x \mapsto f(x, y)$ is continuous for every y. Show that*

(A.42) $$F(x) = \int_A f(x, y)\, d\mu(y)$$

is continuous if there is an integrable function $g(y)$ such that $|f(x, y)| \leq g(y)$.

Problem A.20. *Let $X \subseteq \mathbb{R}$, Y be some measure space, and $f : X \times Y \to \mathbb{C}$. Suppose $y \mapsto f(x, y)$ is integrable for all x and $x \mapsto f(x, y)$ is differentiable for a.e. y. Show that*

(A.43) $$F(x) = \int_A f(x, y)\, d\mu(y)$$

is differentiable if there is an integrable function $g(y)$ such that $|\frac{\partial}{\partial x} f(x, y)| \leq g(y)$. Moreover, $y \mapsto \frac{\partial}{\partial x} f(x, y)$ is measurable and

(A.44) $$F'(x) = \int_A \frac{\partial}{\partial x} f(x, y)\, d\mu(y)$$

in this case.

A.6. Product measures

Let μ_1 and μ_2 be two measures on Σ_1 and Σ_2, respectively. Let $\Sigma_1 \otimes \Sigma_2$ be the σ-algebra generated by **rectangles** of the form $A_1 \times A_2$.

Example. Let $\mathfrak{B}$ be the Borel sets in $\mathbb{R}$. Then $\mathfrak{B}^2 = \mathfrak{B} \otimes \mathfrak{B}$ are the Borel sets in $\mathbb{R}^2$ (since the rectangles are a basis for the product topology). ◇

Any set in $\Sigma_1 \otimes \Sigma_2$ has the **section property**; that is,

Lemma A.25. *Suppose $A \in \Sigma_1 \otimes \Sigma_2$. Then its sections*

$$(A.45) \quad A_1(x_2) = \{x_1 | (x_1, x_2) \in A\} \quad \text{and} \quad A_2(x_1) = \{x_2 | (x_1, x_2) \in A\}$$

are measurable.

Proof. Denote all sets $A \in \Sigma_1 \otimes \Sigma_2$ with the property that $A_1(x_2) \in \Sigma_1$ by S. Clearly all rectangles are in S and it suffices to show that S is a σ-algebra. Now, if $A \in S$, then $(A')_1(x_2) = (A_1(x_2))' \in \Sigma_1$ and thus S is closed under complements. Similarly, if $A_n \in S$, then $(\bigcup_n A_n)_1(x_2) = \bigcup_n (A_n)_1(x_2)$ shows that S is closed under countable unions. □

This implies that if f is a measurable function on $X_1 \times X_2$, then $f(., x_2)$ is measurable on X_1 for every x_2 and $f(x_1, .)$ is measurable on X_2 for every x_1 (observe $A_1(x_2) = \{x_1 | f(x_1, x_2) \in B\}$, where $A = \{(x_1, x_2) | f(x_1, x_2) \in B\}$).

Given two measures μ_1 on Σ_1 and μ_2 on Σ_2, we now want to construct the **product measure** $\mu_1 \otimes \mu_2$ on $\Sigma_1 \otimes \Sigma_2$ such that

$$(A.46) \quad \mu_1 \otimes \mu_2(A_1 \times A_2) = \mu_1(A_1)\mu_2(A_2), \quad A_j \in \Sigma_j, \ j = 1, 2.$$

Since the rectangles are closed under intersection, Theorem A.5 implies that there is at most one measure on $\Sigma_1 \otimes \Sigma_2$ provided μ_1 and μ_2 are σ-finite.

Theorem A.26. *Let μ_1 and μ_2 be two σ-finite measures on Σ_1 and Σ_2, respectively. Let $A \in \Sigma_1 \otimes \Sigma_2$. Then $\mu_2(A_2(x_1))$ and $\mu_1(A_1(x_2))$ are measurable and*

$$(A.47) \quad \int_{X_1} \mu_2(A_2(x_1)) d\mu_1(x_1) = \int_{X_2} \mu_1(A_1(x_2)) d\mu_2(x_2).$$

Proof. As usual, we begin with the case where μ_1 and μ_2 are finite. Let $\mathcal{D}$ be the set of all subsets for which our claim holds. Note that $\mathcal{D}$ contains at least all rectangles. Thus it suffices to show that $\mathcal{D}$ is a Dynkin system by Lemma A.4. To see this, note that measurability and equality of both integrals follow from $A_1(x_2)' = A'_1(x_2)$ (implying $\mu_1(A'_1(x_2)) = \mu_1(X_1) - \mu_1(A_1(x_2))$) for complements and from the monotone convergence theorem for disjoint unions of sets.

If μ_1 and μ_2 are σ-finite, let $X_{i,j} \nearrow X_i$ with $\mu_i(X_{i,j}) < \infty$ for $i = 1, 2$. Now $\mu_2((A \cap X_{1,j} \times X_{2,j})_2(x_1)) = \mu_2(A_2(x_1) \cap X_{2,j})\chi_{X_{1,j}}(x_1)$ and similarly with 1 and 2 exchanged. Hence by the finite case,

$$(A.48) \quad \int_{X_1} \mu_2(A_2 \cap X_{2,j})\chi_{X_{1,j}} d\mu_1 = \int_{X_2} \mu_1(A_1 \cap X_{1,j})\chi_{X_{2,j}} d\mu_2,$$

and the σ-finite case follows from the monotone convergence theorem. $\square$

Hence for given $A \in \Sigma_1 \otimes \Sigma_2$, we can define

$$(A.49) \quad \mu_1 \otimes \mu_2(A) = \int_{X_1} \mu_2(A_2(x_1)) d\mu_1(x_1) = \int_{X_2} \mu_1(A_1(x_2)) d\mu_2(x_2)$$

or equivalently, since $\chi_{A_1(x_2)}(x_1) = \chi_{A_2(x_1)}(x_2) = \chi_A(x_1, x_2)$,

$$\mu_1 \otimes \mu_2(A) = \int_{X_1} \left(\int_{X_2} \chi_A(x_1, x_2) d\mu_2(x_2) \right) d\mu_1(x_1)$$

$$(A.50) \quad = \int_{X_2} \left(\int_{X_1} \chi_A(x_1, x_2) d\mu_1(x_1) \right) d\mu_2(x_2).$$

Then $\mu_1 \otimes \mu_2$ gives rise to a unique measure on $A \in \Sigma_1 \otimes \Sigma_2$ since σ-additivity follows from the monotone convergence theorem.

Finally we have

Theorem A.27 (Fubini). *Let f be a measurable function on $X_1 \times X_2$ and let μ_1, μ_2 be σ-finite measures on X_1, X_2, respectively.*

(i) *If $f \geq 0$, then $\int f(., x_2) d\mu_2(x_2)$ and $\int f(x_1, .) d\mu_1(x_1)$ are both measurable and*

$$\iint f(x_1, x_2) d\mu_1 \otimes \mu_2(x_1, x_2) = \int \left(\int f(x_1, x_2) d\mu_1(x_1) \right) d\mu_2(x_2)$$

$$(A.51) \quad = \int \left(\int f(x_1, x_2) d\mu_2(x_2) \right) d\mu_1(x_1).$$

(ii) *If f is complex-valued, then*

$$(A.52) \quad \int |f(x_1, x_2)| d\mu_1(x_1) \in \mathcal{L}^1(X_2, d\mu_2),$$

respectively,

$$(A.53) \quad \int |f(x_1, x_2)| d\mu_2(x_2) \in \mathcal{L}^1(X_1, d\mu_1),$$

if and only if $f \in \mathcal{L}^1(X_1 \times X_2, d\mu_1 \otimes d\mu_2)$. In this case, (A.51) holds.

Proof. By Theorem A.26 and linearity, the claim holds for simple functions. To see (i), let $s_n \nearrow f$ be a sequence of nonnegative simple functions. Then it

A.6. Product measures

follows by applying the monotone convergence theorem (twice for the double integrals).

For (ii) we can assume that f is real-valued by considering its real and imaginary parts separately. Moreover, splitting $f = f^+ - f^-$ into its positive and negative parts, the claim reduces to (i). □

In particular, if $f(x_1, x_2)$ is either nonnegative or integrable, then the order of integration can be interchanged. The case of nonnegative functions is also called **Tonelli's theorem**. In the general case the integrability condition is crucial, as the following example shows.

Example. Let $X = [0, 1] \times [0, 1]$ with Lebesgue measure and consider
$$f(x, y) = \frac{x - y}{(x + y)^3}.$$
Then
$$\int_0^1 \int_0^1 f(x, y) dx \, dy = -\int_0^1 \frac{1}{(1 + y)^2} dy = -\frac{1}{2}$$
but (by symmetry)
$$\int_0^1 \int_0^1 f(x, y) dy \, dx = \int_0^1 \frac{1}{(1 + x)^2} dx = \frac{1}{2}.$$
Consequently, f cannot be integrable over X (verify this directly). ◇

Lemma A.28. *If μ_1 and μ_2 are outer regular measures, then so is $\mu_1 \otimes \mu_2$.*

Proof. Outer regularity holds for every rectangle and hence also for the algebra of finite disjoint unions of rectangles (Problem A.21). Thus the claim follows from Lemma A.9. □

In connection with Theorem A.5, the following observation is of interest:

Lemma A.29. *If S_1 generates Σ_1 and S_2 generates Σ_2, then $S_1 \times S_2 = \{A_1 \times A_2 | A_j \in S_j, j = 1, 2\}$ generates $\Sigma_1 \otimes \Sigma_2$.*

Proof. Denote the σ-algebra generated by $S_1 \times S_2$ by Σ. Consider the set $\{A_1 \in \Sigma_1 | A_1 \times X_2 \in \Sigma\}$ which is clearly a σ-algebra containing S_1 and thus equal to Σ_1. In particular, $\Sigma_1 \times X_2 \subset \Sigma$ and similarly $X_1 \times \Sigma_2 \subset \Sigma$. Hence also $(\Sigma_1 \times X_2) \cap (X_1 \times \Sigma_2) = \Sigma_1 \times \Sigma_2 \subset \Sigma$. □

Finally, note that we can iterate this procedure.

Lemma A.30. *Suppose (X_j, Σ_j, μ_j), $j = 1, 2, 3$, are σ-finite measure spaces. Then $(\Sigma_1 \otimes \Sigma_2) \otimes \Sigma_3 = \Sigma_1 \otimes (\Sigma_2 \otimes \Sigma_3)$ and*

(A.54) $$(\mu_1 \otimes \mu_2) \otimes \mu_3 = \mu_1 \otimes (\mu_2 \otimes \mu_3).$$

Proof. First of all, note that $(\Sigma_1 \otimes \Sigma_2) \otimes \Sigma_3 = \Sigma_1 \otimes (\Sigma_2 \otimes \Sigma_3)$ is the sigma algebra generated by the rectangles $A_1 \times A_2 \times A_3$ in $X_1 \times X_2 \times X_3$. Moreover, since

$$((\mu_1 \otimes \mu_2) \otimes \mu_3)(A_1 \times A_2 \times A_3) = \mu_1(A_1)\mu_2(A_2)\mu_3(A_3)$$
$$= (\mu_1 \otimes (\mu_2 \otimes \mu_3))(A_1 \times A_2 \times A_3),$$

the two measures coincide on rectangles and hence everywhere by Theorem A.5. □

Example. If λ is Lebesgue measure on $\mathbb{R}$, then $\lambda^n = \lambda \otimes \cdots \otimes \lambda$ is Lebesgue measure on $\mathbb{R}^n$. Since λ is outer regular, so is λ^n. Of course regularity also follows from Corollary A.15.

Moreover, Lebesgue measure is translation invariant and up to normalization the only measure with this property. To see this, let μ be a second translation invariant measure. Denote by Q_r a cube with side length $r > 0$. Without loss, we can assume $\mu(Q_1) = 1$. Since we can split Q_1 into m^n cubes of side length $1/m$, we see that $\mu(Q_{1/m}) = m^{-n}$ by translation invariance and additivity. Hence we obtain $\mu(Q_r) = r^n$ for every rational r and thus for every r by continuity from below. Proceeding like this, we see that λ^n and μ coincide on all rectangles which are products of bounded open intervals. Since this set is closed under intersections and generates the Borel algebra $\mathfrak{B}^n$ by Lemma A.29, the claim follows again from Theorem A.5. ◇

Problem A.21. *Show that the set of all finite unions of rectangles $A_1 \times A_2$ forms an algebra. Moreover, every set in this algebra can be written as a finite union of disjoint rectangles.*

Problem A.22. *Let $U \subseteq \mathbb{C}$ be a domain, Y be some measure space, and $f : U \times Y \to \mathbb{C}$. Suppose $y \mapsto f(z,y)$ is measurable for every z and $z \mapsto f(z,y)$ is holomorphic for every y. Show that*

(A.55) $$F(z) = \int_A f(z,y)\,d\mu(y)$$

is holomorphic if for every compact subset $V \subset U$ there is an integrable function $g(y)$ such that $|f(z,y)| \leq g(y)$, $z \in V$. (Hint: Use Fubini and Morera.)

A.7. Transformation of measures and integrals

Finally, we want to transform measures. Let $f : X \to Y$ be a measurable function. Given a measure μ on X we can introduce a measure $f_\star\mu$ on Y via

(A.56) $$(f_\star\mu)(A) = \mu(f^{-1}(A)).$$

A.7. Transformation of measures and integrals

It is straightforward to check that $f_\star\mu$ is indeed a measure. Moreover, note that $f_\star\mu$ is supported on the range of f.

Theorem A.31. *Let $f : X \to Y$ be measurable and let $g : Y \to \mathbb{C}$ be a Borel function. Then the Borel function $g \circ f : X \to \mathbb{C}$ is a.e. nonnegative or integrable if and only if g is, and in both cases,*

$$(A.57) \qquad \int_Y g \, d(f_\star\mu) = \int_X g \circ f \, d\mu.$$

Proof. In fact, it suffices to check this formula for simple functions g, which follows since $\chi_A \circ f = \chi_{f^{-1}(A)}$. $\square$

Example. Let $f(x) = Mx + a$ be an affine transformation, where $M : \mathbb{R}^n \to \mathbb{R}^n$ is some invertible matrix. Then Lebesgue measure transforms according to

$$f_\star \lambda^n = \frac{1}{|\det(M)|} \lambda^n.$$

In fact, it suffices to check $f_\star \lambda^n(R) = |\det(M)|^{-1} \lambda^n(R)$ for finite rectangles R by Theorem A.5. To see this, note that $f_\star \lambda^n$ is translation invariant and hence must be a multiple of λ^n. Moreover, for an orthogonal matrix, this multiple is one (since an orthogonal matrix leaves the unit ball invariant) and for a diagonal matrix it must be the absolute value of the product of the diagonal elements. Finally, since every matrix can be written as $M = O_1 D O_2$, where O_j are orthogonal and D is diagonal (Problem A.24), the claim follows.

As a consequence we obtain

$$\int_A g(Mx + a) d^n x = \frac{1}{|\det(M)|} \int_{MA+a} g(y) d^n y,$$

which applies, for example, to shifts $f(x) = x + a$ or scaling transforms $f(x) = \alpha x$. $\diamond$

This result can be generalized to diffeomorphisms (one-to-one C^1 maps with inverse again C^1):

Lemma A.32. *Let $U, V \subseteq \mathbb{R}^n$ and suppose $f \in C^1(U, V)$ is a diffeomorphism. Then*

$$(A.58) \qquad (f^{-1})_\star d^n x = |J_f(x)| d^n x,$$

where $J_f = \det(\frac{\partial f}{\partial x})$ is the Jacobi determinant of f. In particular,

$$(A.59) \qquad \int_U g(f(x)) |J_f(x)| d^n x = \int_V g(y) d^n y.$$

Proof. It suffices to show

$$\int_{f(R)} d^n y = \int_R |J_f(x)| d^n x$$

for every bounded open rectangle $R \subseteq U$. By Theorem A.5 it will then follow for characteristic functions and thus for arbitrary functions by the very definition of the integral.

To this end we consider the integral

$$I_\varepsilon = \int_{f(R)} \int_R |J_f(f^{-1}(y))| \varphi_\varepsilon(f(z) - y) d^n z \, d^n y$$

Here $\varphi = V_n^{-1} \chi_{B_1(0)}$ and $\varphi_\varepsilon(y) = \varepsilon^{-n} \varphi(\varepsilon^{-1} y)$, where V_n is the volume of the unit ball (cf. below), such that $\int \varphi_\varepsilon(x) d^n x = 1$.

We will evaluate this integral in two ways. To begin with we consider the inner integral

$$h_\varepsilon(y) = \int_R \varphi_\varepsilon(f(z) - y) d^n z.$$

For $\varepsilon < \varepsilon_0$ the integrand is nonzero only for $z \in K = f^{-1}(\overline{B_{\varepsilon_0}(y)})$, where K is some compact set containing $x = f^{-1}(y)$. Using the affine change of coordinates $z = x + \varepsilon w$ we obtain

$$h_\varepsilon(y) = \int_{W_\varepsilon(x)} \varphi\left(\frac{f(x + \varepsilon w) - f(x)}{\varepsilon}\right) d^n w, \qquad W_\varepsilon(x) = \frac{1}{\varepsilon}(K - x).$$

By

$$\left|\frac{f(x + \varepsilon w) - f(x)}{\varepsilon}\right| \geq \frac{1}{C}|w|, \qquad C = \sup_K \|df^{-1}\|$$

the integrand is nonzero only for $w \in B_C(0)$. Hence, as $\varepsilon \to 0$ the domain $W_\varepsilon(x)$ will eventually cover all of $B_C(0)$ and dominated convergence implies

$$\lim_{\varepsilon \downarrow 0} h_\varepsilon(y) = \int_{B_C(0)} \varphi(df(x)w) dw = |J_f(x)|^{-1}.$$

Consequently, $\lim_{\varepsilon \downarrow 0} I_\varepsilon = |f(R)|$ again by dominated convergence. Now we use Fubini to interchange the order of integration

$$I_\varepsilon = \int_R \int_{f(R)} |J_f(f^{-1}(y))| \varphi_\varepsilon(f(z) - y) d^n y \, d^n z.$$

Since $f(z)$ is an interior point of $f(R)$ continuity of $|J_f(f^{-1}(y))|$ implies

$$\lim_{\varepsilon \downarrow 0} \int_{f(R)} |J_f(f^{-1}(y))| \varphi_\varepsilon(f(z) - y) d^n y = |J_f(f^{-1}(f(z)))| = |J_f(z)|$$

and hence dominated convergence shows $\lim_{\varepsilon \downarrow 0} I_\varepsilon = \int_R |J_f(z)| d^n z$. $\square$

A.7. Transformation of measures and integrals

Example. For example we can consider **polar coordinates** $T_2 : [0, \infty) \times [0, 2\pi) \to \mathbb{R}^2$ defined by
$$T_2(\rho, \varphi) = (\rho \cos(\varphi), \rho \sin(\varphi)).$$
Then
$$\det \frac{\partial T_2}{\partial(\rho, \varphi)} = \rho$$
and one has
$$\int_U f(\rho \cos(\varphi), \rho \sin(\varphi)) \rho \, d(\rho, \varphi) = \int_{T_2(U)} f(x) d^2x.$$
Note that T_2 is only bijective when restricted to $(0, \infty) \times [0, 2\pi)$. However, since the set $\{0\} \times [0, 2\pi)$ is of measure zero, it does not contribute to the integral on the left. Similarly, its image $T_2(\{0\} \times [0, 2\pi)) = \{0\}$ does not contribute to the integral on the right. ◇

Example. We can use the previous example to obtain the transformation formula for **spherical coordinates** in $\mathbb{R}^n$ by induction. We illustrate the process for $n = 3$. To this end, let $x = (x_1, x_2, x_3)$ and start with spherical coordinates in $\mathbb{R}^2$ (which are just polar coordinates) for the first two components:
$$x = (\rho \cos(\varphi), \rho \sin(\varphi), x_3), \qquad \rho \in [0, \infty), \varphi \in [0, 2\pi).$$
Next, use polar coordinates for (ρ, x_3):
$$(\rho, x_3) = (r \sin(\theta), r \cos(\theta)), \qquad r \in [0, \infty), \theta \in [0, \pi].$$
Note that the range for θ follows since $\rho \geq 0$. Moreover, observe that $r^2 = \rho^2 + x_3^2 = x_1^2 + x_2^2 + x_3^2 = |x|^2$ as already anticipated by our notation. In summary,
$$x = T_3(r, \varphi, \theta) = (r \sin(\theta) \cos(\varphi), r \sin(\theta) \sin(\varphi), r \cos(\theta)).$$
Furthermore, since T_3 is the composition with T_2 acting on the first two coordinates with the last unchanged and polar coordinates P acting on the first and last coordinate, the chain rule implies
$$\det \frac{\partial T_3}{\partial(r, \varphi, \theta)} = \det \frac{\partial T_2}{\partial(\rho, \varphi, x_3)}\bigg|_{\substack{\rho = r \sin(\theta) \\ x_3 = r \cos(\theta)}} \det \frac{\partial P}{\partial(r, \varphi, \theta)} = r^2 \sin(\theta).$$
Hence one has
$$\int_U f(T_3(r, \varphi, \theta)) r^2 \sin(\theta) d(r, \varphi, \theta) = \int_{T_3(U)} f(x) d^3x.$$
Again T_3 is only bijective on $(0, \infty) \times [0, 2\pi) \times (0, \pi)$.

It is left as an exercise to check that the extension $T_n : [0, \infty) \times [0, 2\pi) \times [0, \pi]^{n-2} \to \mathbb{R}^n$ is given by
$$x = T_n(r, \varphi, \theta_1, \ldots, \theta_{n-2})$$

with
$$\begin{aligned}
x_1 &= r\cos(\varphi)\sin(\theta_1)\sin(\theta_2)\sin(\theta_3)\cdots\sin(\theta_{n-2}),\\
x_2 &= r\sin(\varphi)\sin(\theta_1)\sin(\theta_2)\sin(\theta_3)\cdots\sin(\theta_{n-2}),\\
x_3 &= r\cos(\theta_1)\sin(\theta_2)\sin(\theta_3)\cdots\sin(\theta_{n-2}),\\
x_4 &= r\cos(\theta_2)\sin(\theta_3)\cdots\sin(\theta_{n-2}),\\
&\vdots\\
x_{n-1} &= r\cos(\theta_{n-3})\sin(\theta_{n-2}),\\
x_n &= r\cos(\theta_{n-2}).
\end{aligned}$$

The Jacobi determinant is given by
$$\det\frac{\partial T_n}{\partial(r,\varphi,\theta_1,\ldots,\theta_{n-2})} = r^{n-1}\sin(\theta_1)\sin(\theta_2)^2\cdots\sin(\theta_{n-2})^{n-2}.$$

$\diamond$

Another useful consequence of Theorem A.31 is the following rule for integrating radial functions.

Lemma A.33. *There is a measure σ^{n-1} on the* **unit sphere** $S^{n-1} = \partial B_1(0) = \{x\in\mathbb{R}^n|\,|x|=1\}$, *which is rotation invariant and satisfies*

$$\text{(A.60)}\qquad \int_{\mathbb{R}^n} g(x)d^n x = \int_0^\infty \int_{S^{n-1}} g(r\omega)r^{n-1}d\sigma^{n-1}(\omega)dr,$$

for every integrable (or positive) function g.

Moreover, the surface area of S^{n-1} is given by

$$\text{(A.61)}\qquad S_n = \sigma^{n-1}(S^{n-1}) = nV_n,$$

where $V_n = \lambda^n(B_1(0))$ is the volume of the unit ball in $\mathbb{R}^n$, and if $g(x) = \tilde{g}(|x|)$ is radial, we have

$$\text{(A.62)}\qquad \int_{\mathbb{R}^n} \tilde{g}(|x|)d^n x = S_n\int_0^\infty \tilde{g}(r)r^{n-1}dr.$$

Proof. Consider the transformation $f:\mathbb{R}^n\to[0,\infty)\times S^{n-1}$, $x\mapsto(|x|,\frac{x}{|x|})$ (with $\frac{0}{|0|}=1$). Let $d\mu(r) = r^{n-1}dr$ and

$$\text{(A.63)}\qquad \sigma^{n-1}(A) = n\lambda^n(f^{-1}([0,1)\times A))$$

for every $A\in\mathfrak{B}(S^{n-1}) = \mathfrak{B}^n\cap S^{n-1}$. Note that σ^{n-1} inherits the rotation invariance from λ^n. By Theorem A.31 it suffices to show $f_\star\lambda^n = \mu\otimes\sigma^{n-1}$. This follows from

$$(f_\star\lambda^n)([0,r)\times A) = \lambda^n(f^{-1}([0,r)\times A)) = r^n\lambda^n(f^{-1}([0,1)\times A))$$
$$= \mu([0,r))\sigma^{n-1}(A),$$

since these sets determine the measure uniquely. $\square$

A.7. Transformation of measures and integrals

Example. Let us compute the volume of a ball in $\mathbb{R}^n$:

$$V_n(r) = \int_{\mathbb{R}^n} \chi_{B_r(0)} d^n x.$$

By the simple scaling transform $f(x) = rx$ we obtain $V_n(r) = V_n(1) r^n$ and hence it suffices to compute $V_n = V_n(1)$.

To this end we use (Problem A.26)

$$\pi^n = \int_{\mathbb{R}^n} e^{-|x|^2} d^n x = n V_n \int_0^\infty e^{-r^2} r^{n-1} dr = \frac{nV_n}{2} \int_0^\infty e^{-s} s^{n/2-1} ds$$
$$= \frac{nV_n}{2} \Gamma(\frac{n}{2}) = \frac{V_n}{2} \Gamma(\frac{n}{2}+1)$$

where Γ is the gamma function (Problems A.27). Hence

(A.64) $$V_n = \frac{\pi^{n/2}}{\Gamma(\frac{n}{2}+1)}.$$

By $\Gamma(\frac{1}{2}) = \sqrt{\pi}$ (see Problem A.28), this coincides with the well-known values for $n = 1, 2, 3$. ◇

Example. The above lemma can be used to determine when a radial function is integrable. For example, we obtain

$$|x|^\alpha \in L^1(B_1(0)) \Leftrightarrow \alpha > -n, \qquad |x|^\alpha \in L^1(\mathbb{R}^n \setminus B_1(0)) \Leftrightarrow \alpha < -n.$$

◇

Problem A.23. Let λ be Lebesgue measure on $\mathbb{R}$. Show that if $f \in C^1(\mathbb{R})$ with $f' > 0$, then

$$d(f_\star \lambda)(x) = \frac{1}{f'(f^{-1}(x))} dx.$$

Problem A.24. Show that every invertible matrix M can be written as $M = O_1 D O_2$, where D is diagonal and O_j are orthogonal. (Hint: The matrix $M^* M$ is nonnegative and hence there is an orthogonal matrix U which diagonalizes $M^* M = U^* D^2 U$. Then one can choose $O_1 = MUD^{-1}$ and $O_2 = U^*$.)

Problem A.25. Compute V_n using spherical coordinates.
(Hint: $\int \sin(x)^n dx = -\frac{1}{n} \sin(x)^{n-1} \cos(x) + \frac{n-1}{n} \int \sin(x)^{n-2} dx$.)

Problem A.26. Show

$$I_n = \int_{\mathbb{R}^n} e^{-|x|^2} d^n x = \pi^{n/2}.$$

(Hint: Use Fubini to show $I_n = I_1^n$ and compute I_2 using polar coordinates.)

Problem A.27. *The* **gamma function** *is defined via*

(A.65) $$\Gamma(z) = \int_0^\infty x^{z-1} e^{-x} dx, \qquad \mathrm{Re}(z) > 0.$$

Verify that the integral converges and defines an analytic function in the indicated half-plane (cf. Problem A.22). Use integration by parts to show

(A.66) $$\Gamma(z+1) = z\Gamma(z), \qquad \Gamma(1) = 1.$$

Conclude $\Gamma(n) = (n-1)!$ for $n \in \mathbb{N}$.

Problem A.28. *Show that $\Gamma(\frac{1}{2}) = \sqrt{\pi}$. (Hint: Use the change of coordinates $x = t^2$ and then use Problem A.26.)*

Problem A.29. *Let $U \subseteq \mathbb{R}^m$ be open and let $f : U \to \mathbb{R}^n$ be locally Lipschitz (i.e., for every compact set $K \subset U$ there is some constant L such that $|f(x) - f(y)| \leq L|x - y|$ for all $x, y \in K$). Show that if $A \subset U$ has Lebesgue measure zero, then $f(A)$ is contained in a set of Lebesgue measure zero. (Hint: By Lindelöf it is no restriction to assume that A is contained in a compact ball contained in U.)*

A.8. Vague convergence of measures

Let μ_n be a sequence of Borel measures. We will say that μ_n converges to μ vaguely if

(A.67) $$\int_X f d\mu_n \to \int_X f d\mu$$

for every $f \in C_c(X)$.

We are only interested in the case of Borel measures on $\mathbb{R}$. In this case, we have the following equivalent characterization of vague convergence.

Lemma A.34. *Let μ_n be a sequence of Borel measures on $\mathbb{R}$. Then $\mu_n \to \mu$ vaguely if and only if the distribution functions (normalized at a point of continuity of μ) converge at every point of continuity of μ.*

Proof. Suppose $\mu_n \to \mu$ vaguely. Let I be any bounded interval (closed, half closed, or open) with boundary points x_0, x_1. Moreover, choose continuous functions f, g with compact support such that $f \leq \chi_I \leq g$. Then we have $\int f d\mu \leq \mu(I) \leq \int g d\mu$ and similarly for μ_n. Hence

$$\mu(I) - \mu_n(I) \leq \int g d\mu - \int f d\mu_n \leq \int (g-f) d\mu + \left| \int f d\mu - \int f d\mu_n \right|$$

and

$$\mu(I) - \mu_n(I) \geq \int f d\mu - \int g d\mu_n \geq \int (f-g) d\mu - \left| \int g d\mu - \int g d\mu_n \right|.$$

A.8. Vague convergence of measures

Combining both estimates, we see

$$|\mu(I) - \mu_n(I)| \le \int (g-f)d\mu + \left|\int f d\mu - \int f d\mu_n\right| + \left|\int g d\mu - \int g d\mu_n\right|$$

and so

$$\limsup_{n\to\infty} |\mu(I) - \mu_n(I)| \le \int (g-f)d\mu.$$

Choosing f, g such that $g - f \to \chi_{\{x_0\}} + \chi_{\{x_1\}}$ pointwise, we even get from dominated convergence that

$$\limsup_{n\to\infty} |\mu(I) - \mu_n(I)| \le \mu(\{x_0\}) + \mu(\{x_1\}),$$

which proves that the distribution functions converge at every point of continuity of μ.

Conversely, suppose that the distribution functions converge at every point of continuity of μ. To see that in fact $\mu_n \to \mu$ vaguely, let $f \in C_c(\mathbb{R})$. Fix some $\varepsilon > 0$ and note that, since f is uniformly continuous, there is a $\delta > 0$ such that $|f(x) - f(y)| \le \varepsilon$ whenever $|x - y| \le \delta$. Next, choose some points $x_0 < x_1 < \cdots < x_k$ such that $\mathrm{supp}(f) \subset (x_0, x_k)$, μ is continuous at x_j, and $x_j - x_{j-1} \le \delta$ (recall that a monotone function has at most countable discontinuities). Furthermore, there is some N such that $|\mu_n(x_j) - \mu(x_j)| \le \frac{\varepsilon}{2k}$ for all j and $n \ge N$. Then

$$\left|\int f d\mu_n - \int f d\mu\right| \le \sum_{j=1}^k \int_{(x_{j-1}, x_j]} |f(x) - f(x_j)| d\mu_n(x)$$

$$+ \sum_{j=1}^k |f(x_j)| |\mu((x_{j-1}, x_j]) - \mu_n((x_{j-1}, x_j])|$$

$$+ \sum_{j=1}^k \int_{(x_{j-1}, x_j]} |f(x) - f(x_j)| d\mu(x).$$

Now, for $n \ge N$, the first and the last terms on the right-hand side are both bounded by $(\mu((x_0, x_k]) + \frac{\varepsilon}{k})\varepsilon$ and the middle term is bounded by $\max|f|\varepsilon$. Thus the claim follows. □

Moreover, every bounded sequence of measures has a vaguely convergent subsequence.

Lemma A.35. *Suppose μ_n is a sequence of finite Borel measures on $\mathbb{R}$ such that $\mu_n(\mathbb{R}) \le M$. Then there exists a subsequence which converges vaguely to some measure μ with $\mu(\mathbb{R}) \le M$.*

Proof. Let $\mu_n(x) = \mu_n((-\infty, x])$ be the corresponding distribution functions. By $0 \le \mu_n(x) \le M$ there is a convergent subsequence for fixed x.

Moreover, by the standard diagonal series trick, we can assume that $\mu_n(x)$ converges to some number $\mu(x)$ for each rational x. For irrational x we set $\mu(x) = \inf_{x_0 > x}\{\mu(x_0)|x_0 \text{ rational}\}$. Then $\mu(x)$ is monotone, $0 \le \mu(x_1) \le \mu(x_2) \le M$ for $x_1 \le x_2$. Furthermore,

$$\mu(x-) \le \liminf \mu_n(x) \le \limsup \mu_n(x) \le \mu(x)$$

shows that $\mu_n(x) \to \mu(x)$ at every point of continuity of μ. So we can redefine μ to be right continuous without changing this last fact. $\square$

In the case where the sequence is bounded, (A.67) even holds for a larger class of functions.

Lemma A.36. *Suppose $\mu_n \to \mu$ vaguely and $\mu_n(\mathbb{R}) \le M$. Then (A.67) holds for every $f \in C_\infty(\mathbb{R})$. If in addition $\mu_n(\mathbb{R}) \to \mu(\mathbb{R})$, then (A.67) holds for every $f \in C_b(\mathbb{R})$.*

Proof. Split $f = f_1 + f_2$, where f_1 has compact support and $|f_2| \le \varepsilon$. Then $|\int f d\mu - \int f d\mu_n| \le |\int f_1 d\mu - \int f_1 d\mu_n| + 2\varepsilon M$ and the first claim follows.

Similarly, for the second claim, let $|f| \le C$ and choose R such that $\mu(\mathbb{R}\setminus[-R,R]) < \varepsilon$. Then we have $\mu_n(\mathbb{R}\setminus[-R,R]) < \varepsilon$ for $n \ge N$. Now set $\varphi(x) = 1$ for $|x| \le R$, $\varphi(x) = 1 - |x| - R$ for $R \le |x| \le R+1$, and $\varphi(x) = 0$ for $|x| \ge R$. Then $|\int f d\mu - \int f d\mu_n| \le |\int \varphi f d\mu - \int \varphi f d\mu_n| + 2\varepsilon C$ and the second claim follows. $\square$

Example. The example $d\mu_n(\lambda) = d\Theta(\lambda - n)$ shows that in the above claim f cannot be replaced by a bounded continuous function. Moreover, the example $d\mu_n(\lambda) = n \, d\Theta(\lambda - n)$ also shows that the uniform bound cannot be dropped. $\diamond$

Problem A.30. *Suppose $\mu_n \to \mu$ vaguely and let I be a bounded interval with boundary points x_0 and x_1. Then*

$$\limsup_n \left| \int_I f d\mu_n - \int_I f d\mu \right| \le |f(x_1)|\mu(\{x_1\}) + |f(x_0)|\mu(\{x_0\})$$

for any $f \in C([x_0, x_1])$.

Problem A.31. *Let $\mu_n(X) \le M$ and suppose (A.67) holds for all $f \in U \subseteq C(X)$. Then (A.67) holds for all f in the closed span of U.*

Problem A.32. *Suppose (A.67) holds for all $f \in C_c(\mathbb{R})$ as well as $\mu_n(\mathbb{R}) \to \mu(\mathbb{R}) < \infty$. Then (A.67) holds for all $f \in C_b(\mathbb{R})$. (Hint: Choose some $r > 0$ such that $\mu([-r-1, r+1]) \le \varepsilon$ and let $\phi \in C(\mathbb{R})$ be one for $|x| \ge r+1$ and zero for $|x| \le r$. Now show that $\limsup \mu_n([-r,r]) \le \varepsilon$.)*

A.9. Decomposition of measures

Let μ, ν be two measures on a measurable space (X, Σ). They are called **mutually singular** (in symbols $\mu \perp \nu$) if they are supported on disjoint sets. That is, there is a measurable set N such that $\mu(N) = 0$ and $\nu(X \backslash N) = 0$.

Example. Let λ be the Lebesgue measure and Θ the Dirac measure (centered at 0). Then $\lambda \perp \Theta$: Just take $N = \{0\}$; then $\lambda(\{0\}) = 0$ and $\Theta(\mathbb{R} \backslash \{0\}) = 0$. ◇

On the other hand, ν is called **absolutely continuous** with respect to μ (in symbols $\nu \ll \mu$) if $\mu(A) = 0$ implies $\nu(A) = 0$.

Example. The prototypical example is the measure $d\nu = f\, d\mu$ (compare Lemma A.19). Indeed, by Lemma A.22, $\mu(A) = 0$ implies

$$\nu(A) = \int_A f\, d\mu = 0 \tag{A.68}$$

and shows that ν is absolutely continuous with respect to μ. In fact, we will show below that every absolutely continuous measure is of this form. ◇

The two main results will follow as simple consequence of the following result:

Theorem A.37. *Let μ, ν be σ-finite measures. Then there exists a nonnegative function f and a set N of μ measure zero, such that*

$$\nu(A) = \nu(A \cap N) + \int_A f\, d\mu. \tag{A.69}$$

Proof. We first assume μ, ν to be finite measures. Let $\alpha = \mu + \nu$ and consider the Hilbert space $L^2(X, d\alpha)$. Then

$$\ell(h) = \int_X h\, d\nu$$

is a bounded linear functional on $L^2(X, d\alpha)$ by Cauchy–Schwarz:

$$|\ell(h)|^2 = \left| \int_X 1 \cdot h\, d\nu \right|^2 \leq \left(\int |1|^2 d\nu \right) \left(\int |h|^2 d\nu \right)$$

$$\leq \nu(X) \left(\int |h|^2 d\alpha \right) = \nu(X) \|h\|^2.$$

Hence by the Riesz lemma (Theorem 1.8), there exists a $g \in L^2(X, d\alpha)$ such that

$$\ell(h) = \int_X hg\, d\alpha.$$

By construction,

$$\text{(A.70)} \qquad \nu(A) = \int \chi_A \, d\nu = \int \chi_A g \, d\alpha = \int_A g \, d\alpha.$$

In particular, g must be positive a.e. (take A the set where g is negative). Moreover,

$$\mu(A) = \alpha(A) - \nu(A) = \int_A (1-g) d\alpha$$

which shows that $g \leq 1$ a.e. Now choose $N = \{x | g(x) = 1\}$ such that $\mu(N) = 0$ and set

$$f = \frac{g}{1-g} \chi_{N'}, \qquad N' = X \setminus N.$$

Then, since (A.70) implies $d\nu = g \, d\alpha$, respectively, $d\mu = (1-g) d\alpha$, we have

$$\int_A f d\mu = \int \chi_A \frac{g}{1-g} \chi_{N'} d\mu = \int \chi_{A \cap N'} g \, d\alpha = \nu(A \cap N')$$

as desired.

To see the σ-finite case, observe that $Y_n \nearrow X$, $\mu(Y_n) < \infty$ and $Z_n \nearrow X$, $\nu(Z_n) < \infty$ implies $X_n = Y_n \cap Z_n \nearrow X$ and $\alpha(X_n) < \infty$. Now we set $\tilde{X}_n = X_n \setminus X_{n-1}$ (where $X_0 = \emptyset$) and consider $\mu_n(A) = \mu(A \cap \tilde{X}_n)$ and $\nu_n(A) = \nu(A \cap \tilde{X}_n)$. Then there exist corresponding sets N_n and functions f_n such that

$$\nu_n(A) = \nu_n(A \cap N_n) + \int_A f_n d\mu_n = \nu(A \cap N_n) + \int_A f_n d\mu,$$

where for the last equality we have assumed $N_n \subseteq \tilde{X}_n$ and $f_n(x) = 0$ for $x \in \tilde{X}'_n$ without loss of generality. Now set $N = \bigcup_n N_n$ as well as $f = \sum_n f_n$. Then $\mu(N) = 0$ and

$$\nu(A) = \sum_n \nu_n(A) = \sum_n \nu(A \cap N_n) + \sum_n \int_A f_n d\mu = \nu(A \cap N) + \int_A f d\mu,$$

which finishes the proof. □

Now the anticipated results follow with no effort:

Theorem A.38 (Radon–Nikodym)**.** *Let μ, ν be two σ-finite measures on a measurable space (X, Σ). Then ν is absolutely continuous with respect to μ if and only if there is a nonnegative measurable function f such that*

$$\text{(A.71)} \qquad \nu(A) = \int_A f \, d\mu$$

*for every $A \in \Sigma$. The function f is determined uniquely a.e. with respect to μ and is called the **Radon–Nikodym derivative** $\frac{d\nu}{d\mu}$ of ν with respect to μ.*

A.9. Decomposition of measures

Proof. Just observe that in this case $\nu(A \cap N) = 0$ for every A. Uniqueness will be shown in the next theorem. $\square$

Example. Take $X = \mathbb{R}$. Let μ be the counting measure and ν Lebesgue measure. Then $\nu \ll \mu$ but there is no f with $d\nu = f\, d\mu$. If there were such an f, there must be a point $x_0 \in \mathbb{R}$ with $f(x_0) > 0$ and we have $0 = \nu(\{x_0\}) = \int_{\{x_0\}} f\, d\mu = f(x_0) > 0$, a contradiction. Hence the Radon–Nikodym theorem can fail if μ is not σ-finite. $\diamond$

Theorem A.39 (Lebesgue decomposition). *Let μ, ν be two σ-finite measures on a measurable space (X, Σ). Then ν can be uniquely decomposed as $\nu = \nu_{ac} + \nu_{sing}$, where μ and ν_{sing} are mutually singular and ν_{ac} is absolutely continuous with respect to μ.*

Proof. Taking $\nu_{sing}(A) = \nu(A \cap N)$ and $d\nu_{ac} = f\, d\mu$ from the previous theorem, there is at least one such decomposition. To show uniqueness assume there is another one, $\nu = \tilde{\nu}_{ac} + \tilde{\nu}_{sing}$, and let $\tilde{N}$ be such that $\mu(\tilde{N}) = 0$ and $\tilde{\nu}_{sing}(\tilde{N}') = 0$. Then $\nu_{sing}(A) - \tilde{\nu}_{sing}(A) = \int_A (\tilde{f} - f) d\mu$. In particular, $\int_{A \cap N' \cap \tilde{N}'} (\tilde{f} - f) d\mu = 0$ and hence $\tilde{f} = f$ a.e. away from $N \cup \tilde{N}$. Since $\mu(N \cup \tilde{N}) = 0$, we have $\tilde{f} = f$ a.e. and hence $\tilde{\nu}_{ac} = \nu_{ac}$ as well as $\tilde{\nu}_{sing} = \nu - \tilde{\nu}_{ac} = \nu - \nu_{ac} = \nu_{sing}$. $\square$

Problem A.33. *Let μ be a Borel measure on $\mathfrak{B}$ and suppose its distribution function $\mu(x)$ is continuously differentiable. Show that the Radon–Nikodym derivative equals the ordinary derivative $\mu'(x)$.*

Problem A.34. *Suppose μ is a Borel measure on $\mathbb{R}$ and $f: \mathbb{R} \to \mathbb{R}$ is continuous. Show that $f_* \mu$ is absolutely continuous if μ is. (Hint: Problem A.13.)*

Problem A.35. *Suppose μ and ν are inner regular measures. Show that $\nu \ll \mu$ if and only if $\mu(C) = 0$ implies $\nu(C) = 0$ for every compact set.*

Problem A.36. *Suppose $\nu(A) \leq C\mu(A)$ for all $A \in \Sigma$. Then $d\nu = f\, d\mu$ with $0 \leq f \leq C$ a.e.*

Problem A.37. *Let $d\nu = f\, d\mu$. Suppose $f > 0$ a.e. with respect to μ. Then $\mu \ll \nu$ and $d\mu = f^{-1} d\nu$.*

Problem A.38 (Chain rule). *Show that $\nu \ll \mu$ is a transitive relation. In particular, if $\omega \ll \nu \ll \mu$, show that*

$$\frac{d\omega}{d\mu} = \frac{d\omega}{d\nu} \frac{d\nu}{d\mu}.$$

Problem A.39. *Suppose $\nu \ll \mu$. Show that for every measure ω we have*

$$\frac{d\omega}{d\mu} d\mu = \frac{d\omega}{d\nu} d\nu + d\zeta,$$

where ζ is a positive measure (depending on ω) which is singular with respect to ν. Show that $\zeta = 0$ if and only if $\mu \ll \nu$.

A.10. Derivatives of measures

If μ is a Borel measure on $\mathfrak{B}$ and its distribution function $\mu(x)$ is continuously differentiable, then the Radon–Nikodym derivative is just the ordinary derivative $\mu'(x)$ (Problem A.33). Our aim in this section is to generalize this result to arbitrary Borel measures on $\mathfrak{B}^n$.

Let μ be a Borel measure on $\mathbb{R}^n$. We call

$$(A.72) \qquad (D\mu)(x) = \lim_{\varepsilon \downarrow 0} \frac{\mu(B_\varepsilon(x))}{|B_\varepsilon(x)|}$$

the derivative of μ at $x \in \mathbb{R}^n$ provided the above limit exists. (Here $B_r(x) \subset \mathbb{R}^n$ is a ball of radius r centered at $x \in \mathbb{R}^n$ and $|A|$ denotes the Lebesgue measure of $A \in \mathfrak{B}^n$.)

Example. Consider a Borel measure on $\mathfrak{B}$ and suppose its distribution $\mu(x)$ (as defined in (A.3)) is differentiable at x. Then

$$(D\mu)(x) = \lim_{\varepsilon \downarrow 0} \frac{\mu((x+\varepsilon, x-\varepsilon))}{2\varepsilon} = \lim_{\varepsilon \downarrow 0} \frac{\mu(x+\varepsilon) - \mu(x-\varepsilon)}{2\varepsilon} = \mu'(x).$$

$\diamond$

To compute the derivative of μ, we introduce the **upper** and **lower derivative**,

$$(A.73) \quad (\overline{D}\mu)(x) = \limsup_{\varepsilon \downarrow 0} \frac{\mu(B_\varepsilon(x))}{|B_\varepsilon(x)|} \quad \text{and} \quad (\underline{D}\mu)(x) = \liminf_{\varepsilon \downarrow 0} \frac{\mu(B_\varepsilon(x))}{|B_\varepsilon(x)|}.$$

Clearly, μ is differentiable at x if $(\underline{D}\mu)(x) = (\overline{D}\mu)(x) < \infty$. Next, note that they are measurable: In fact, this follows from

$$(A.74) \qquad (\overline{D}\mu)(x) = \lim_{n \to \infty} \sup_{0 < \varepsilon < 1/n} \frac{\mu(B_\varepsilon(x))}{|B_\varepsilon(x)|}$$

since the supremum on the right-hand side is lower semicontinuous with respect to x (cf. Problem A.10) as $x \mapsto \mu(B_\varepsilon(x))$ is lower semicontinuous (Problem A.40). Similarly for $(\underline{D}\mu)(x)$.

Next, the following geometric fact of $\mathbb{R}^n$ will be needed.

Lemma A.40 (Wiener covering lemma). *Given open balls $B_1 = B_{r_1}(x_1)$, ..., $B_m = B_{r_m}(x_m)$ in $\mathbb{R}^n$, there is a subset of disjoint balls $B_{j_1}, \ldots, B_{j_k}$ such that*

$$(A.75) \qquad \bigcup_{j=1}^m B_j \subseteq \bigcup_{\ell=1}^k B_{3r_{j_\ell}}(x_{j_\ell}).$$

A.10. Derivatives of measures

Proof. Assume that the balls B_j are ordered by decreasing radius. Start with $B_{j_1} = B_1$ and remove all balls from our list which intersect B_{j_1}. Observe that the removed balls are all contained in $B_{3r_1}(x_1)$. Proceeding like this, we obtain the required subset. $\square$

The upshot of this lemma is that we can select a disjoint subset of balls which still controls the Lebesgue volume of the original set up to a universal constant 3^n (recall $|B_{3r}(x)| = 3^n |B_r(x)|$).

Now we can show

Lemma A.41. *Let $\alpha > 0$. For every Borel set A we have*

$$|\{x \in A \,|\, (\overline{D}\mu)(x) > \alpha\}| \leq 3^n \frac{\mu(A)}{\alpha} \tag{A.76}$$

and

$$|\{x \in A \,|\, (\overline{D}\mu)(x) > 0\}| = 0, \text{ whenever } \mu(A) = 0. \tag{A.77}$$

Proof. Let $A_\alpha = \{x \in A | (\overline{D}\mu)(x) > \alpha\}$. We will show

$$|K| \leq 3^n \frac{\mu(O)}{\alpha}$$

for every open set O with $A \subseteq O$ and every compact set $K \subseteq A_\alpha$. The first claim then follows from outer regularity of μ and inner regularity of the Lebesgue measure.

Given fixed K, O, for every $x \in K$ there is some r_x such that $B_{r_x}(x) \subseteq O$ and $|B_{r_x}(x)| < \alpha^{-1}\mu(B_{r_x}(x))$. Since K is compact, we can choose a finite subcover of K from these balls. Moreover, by Lemma A.40 we can refine our set of balls such that

$$|K| \leq 3^n \sum_{i=1}^{k} |B_{r_i}(x_i)| < \frac{3^n}{\alpha} \sum_{i=1}^{k} \mu(B_{r_i}(x_i)) \leq 3^n \frac{\mu(O)}{\alpha}.$$

To see the second claim, observe that $A_0 = \cup_{j=1}^{\infty} A_{1/j}$ and by the first part $|A_{1/j}| = 0$ for every j if $\mu(A) = 0$. $\square$

Theorem A.42 (Lebesgue)**.** *Let f be (locally) integrable. Then for a.e. $x \in \mathbb{R}^n$ we have*

$$\lim_{r \downarrow 0} \frac{1}{|B_r(x)|} \int_{B_r(x)} |f(y) - f(x)| dy = 0. \tag{A.78}$$

The points where (A.78) holds are called **Lebesgue points** of f.

Proof. Decompose f as $f = g + h$, where g is continuous and $\|h\|_1 < \varepsilon$ (Theorem 0.38) and abbreviate

$$D_r(f)(x) = \frac{1}{|B_r(x)|} \int_{B_r(x)} |f(y) - f(x)| dy.$$

Then, since $\lim D_r(g)(x) = 0$ (for every x) and $D_r(f) \leq D_r(g) + D_r(h)$, we have
$$\limsup_{r\downarrow 0} D_r(f)(x) \leq \limsup_{r\downarrow 0} D_r(h)(x) \leq (\overline{D}\mu)(x) + |h(x)|,$$
where $d\mu = |h|d^n x$. This implies
$$\{x \mid \limsup_{r\downarrow 0} D_r(f)(x) \geq 2\alpha\} \subseteq \{x \mid (\overline{D}\mu)(x) \geq \alpha\} \cup \{x \mid |h(x)| \geq \alpha\}$$

and using the first part of Lemma A.41 plus $|\{x \mid |h(x)| \geq \alpha\}| \leq \alpha^{-1}\|h\|_1$ (Problem A.43), we see

$$|\{x \mid \limsup_{r\downarrow 0} D_r(f)(x) \geq 2\alpha\}| \leq (3^n + 1)\frac{\varepsilon}{\alpha}.$$

Since ε is arbitrary, the Lebesgue measure of this set must be zero for every α. That is, the set where the lim sup is positive has Lebesgue measure zero. $\square$

Note that the balls can be replaced by more general sets: A sequence of sets $A_j(x)$ is said to shrink to x nicely if there are balls $B_{r_j}(x)$ with $r_j \to 0$ and a constant $\varepsilon > 0$ such that $A_j(x) \subseteq B_{r_j}(x)$ and $|A_j| \geq \varepsilon |B_{r_j}(x)|$. For example, $A_j(x)$ could be some balls or cubes (not necessarily containing x). However, the portion of $B_{r_j}(x)$ which they occupy must not go to zero! For example, the rectangles $(0, \frac{1}{j}) \times (0, \frac{2}{j}) \subset \mathbb{R}^2$ do shrink nicely to 0, but the rectangles $(0, \frac{1}{j}) \times (0, \frac{2}{j^2})$ do not.

Lemma A.43. *Let f be (locally) integrable. Then at every Lebesgue point we have*

(A.79)
$$f(x) = \lim_{j\to\infty} \frac{1}{|A_j(x)|} \int_{A_j(x)} f(y)dy$$

whenever $A_j(x)$ shrinks to x nicely.

Proof. Let x be a Lebesgue point and choose some nicely shrinking sets $A_j(x)$ with corresponding $B_{r_j}(x)$ and ε. Then

$$\frac{1}{|A_j(x)|} \int_{A_j(x)} |f(y) - f(x)|dy \leq \frac{1}{\varepsilon |B_{r_j}(x)|} \int_{B_{r_j}(x)} |f(y) - f(x)|dy$$

and the claim follows. $\square$

Corollary A.44. *Let μ be a Borel measure on $\mathbb{R}$ which is absolutely continuous with respect to Lebesgue measure. Then its distribution function is differentiable a.e. and $d\mu(x) = \mu'(x)dx$.*

A.10. Derivatives of measures

Proof. By assumption, $d\mu(x) = f(x)dx$ for some locally integrable function f. In particular, the distribution function $\mu(x) = \int_0^x f(y)dy$ is continuous. Moreover, since the sets $(x, x+r)$ shrink nicely to x as $r \to 0$, Lemma A.43 implies

$$\lim_{r \to 0} \frac{\mu((x, x+r))}{r} = \lim_{r \to 0} \frac{\mu(x+r) - \mu(x)}{r} = f(x)$$

at every Lebesgue point of f. Since the same is true for the sets $(x-r, x)$, $\mu(x)$ is differentiable at every Lebesgue point and $\mu'(x) = f(x)$. □

As another consequence we obtain

Theorem A.45. *Let μ be a Borel measure on $\mathbb{R}^n$. The derivative $D\mu$ exists a.e. with respect to Lebesgue measure and equals the Radon–Nikodym derivative of the absolutely continuous part of μ with respect to Lebesgue measure; that is,*

$$(A.80) \qquad \mu_{ac}(A) = \int_A (D\mu)(x) d^n x.$$

Proof. If $d\mu = f d^n x$ is absolutely continuous with respect to Lebesgue measure, then $(D\mu)(x) = f(x)$ at every Lebesgue point of f by Lemma A.43 and the claim follows from Theorem A.42. To see the general case, use the Lebesgue decomposition of μ and let N be a support for the singular part with $|N| = 0$. Then $(\overline{D}\mu_{sing})(x) = 0$ for a.e. $x \in \mathbb{R}^n \backslash N$ by the second part of Lemma A.41. □

In particular, μ is singular with respect to Lebesgue measure if and only if $D\mu = 0$ a.e. with respect to Lebesgue measure.

Using the upper and lower derivatives, we can also give supports for the absolutely and singularly continuous parts.

Theorem A.46. *The set $\{x | 0 < (D\mu)(x) < \infty\}$ is a support for the absolutely continuous and $\{x | (\underline{D}\mu)(x) = \infty\}$ is a support for the singular part.*

Proof. The first part is immediate from the previous theorem. For the second part, first note that by $(\underline{D}\mu)(x) \geq (\underline{D}\mu_{sing})(x)$ we can assume that μ is purely singular. It suffices to show that the set $A_k = \{x \mid (\underline{D}\mu)(x) < k\}$ satisfies $\mu(A_k) = 0$ for every $k \in \mathbb{N}$.

Let $K \subset A_k$ be compact, and let $V_j \supset K$ be some open set such that $|V_j \backslash K| \leq \frac{1}{j}$. For every $x \in K$ there is some $\varepsilon = \varepsilon(x)$ such that $B_\varepsilon(x) \subseteq V_j$ and $\mu(B_{3\varepsilon}(x)) \leq k|B_{3\varepsilon}(x)|$. By compactness, finitely many of these balls cover K and hence

$$\mu(K) \leq \sum_i \mu(B_{\varepsilon_i}(x_i)).$$

Selecting disjoint balls as in Lemma A.40 further shows

$$\mu(K) \leq \sum_\ell \mu(B_{3\varepsilon_{i_\ell}}(x_{i_\ell})) \leq k3^n \sum_\ell |B_{\varepsilon_{i_\ell}}(x_{i_\ell})| \leq k3^n |V_j|.$$

Letting $j \to \infty$, we see $\mu(K) \leq k3^n|K|$ and by regularity we even have $\mu(A) \leq k3^n|A|$ for every $A \subseteq A_k$. Hence μ is absolutely continuous on A_k and since we assumed μ to be singular, we must have $\mu(A_k) = 0$. □

Finally, we note that these supports are minimal. Here a support M of some measure μ is called a **minimal support** (it is sometimes also called an **essential support**) if every subset $M_0 \subseteq M$ which does not support μ (i.e., $\mu(M_0) = 0$) has Lebesgue measure zero.

Example. Let $X = \mathbb{R}$, $\Sigma = \mathfrak{B}$. If $d\mu(x) = \sum_n \alpha_n d\theta(x - x_n)$ is a sum of Dirac measures, then the set $\{x_n\}$ is clearly a minimal support for μ. Moreover, it is clearly the smallest support as none of the x_n can be removed. If we choose $\{x_n\}$ to be the rational numbers, then $\mathrm{supp}(\mu) = \mathbb{R}$, but $\mathbb{R}$ is not a minimal support, as we can remove the irrational numbers.

On the other hand, if we consider the Lebesgue measure λ, then $\mathbb{R}$ is a minimal support. However, the same is true if we remove any set of measure zero, for example, the Cantor set. In particular, since we can remove any single point, we see that, just like supports, minimal supports are not unique. ◇

Lemma A.47. *The set $M_{ac} = \{x | 0 < (D\mu)(x) < \infty\}$ is a minimal support for μ_{ac}.*

Proof. Suppose $M_0 \subseteq M_{ac}$ and $\mu_{ac}(M_0) = 0$. Set $M_\varepsilon = \{x \in M_0 | \varepsilon < (D\mu)(x)\}$ for $\varepsilon > 0$. Then $M_\varepsilon \nearrow M_0$ and

$$|M_\varepsilon| = \int_{M_\varepsilon} d^n x \leq \frac{1}{\varepsilon} \int_{M_\varepsilon} (D\mu)(x) dx = \frac{1}{\varepsilon} \mu_{ac}(M_\varepsilon) \leq \frac{1}{\varepsilon} \mu_{ac}(M_0) = 0$$

shows $|M_0| = \lim_{\varepsilon \downarrow 0} |M_\varepsilon| = 0$. □

Note that the set $M = \{x | 0 < (D\mu)(x)\}$ is a minimal support of μ.

Example. The **Cantor function** is constructed as follows. Take the sets C_n used in the construction of the Cantor set C: C_n is the union of 2^n closed intervals with $2^n - 1$ open gaps in between. Set f_n equal to $j/2^n$ on the j'th gap of C_n and extend it to $[0,1]$ by linear interpolation. Note that, since we are creating precisely one new gap between every old gap when going from C_n to C_{n+1}, the value of f_{n+1} is the same as the value of f_n on the gaps of

A.10. Derivatives of measures

C_n. Explicitly, we have $f_0(x) = x$ and $f_{n+1} = K(f_n)$, where

$$K(f)(x) = \begin{cases} \frac{1}{2}f(3x), & 0 \leq x \leq \frac{1}{3}, \\ \frac{1}{2}f(3x), & \frac{1}{3} \leq x \leq \frac{2}{3}, \\ \frac{1}{2}(1 + f(3x - 2)), & \frac{2}{3} \leq x \leq 1. \end{cases}$$

Since $\|f_{n+1} - f_n\|_\infty \leq \frac{1}{2}\|f_{n+1} - f_n\|_\infty$ we can define the Cantor function as $f = \lim_{n\to\infty} f_n$. By construction, f is a continuous function which is constant on every subinterval of $[0,1]\setminus C$. Since C is of Lebesgue measure zero, this set is of full Lebesgue measure and hence $f' = 0$ a.e. in $[0,1]$. In particular, the corresponding measure, the **Cantor measure**, is supported on C and is purely singular with respect to Lebesgue measure. ◇

Problem A.40. *Show that*

$$\mu(B_\varepsilon(x)) \leq \liminf_{y \to x} \mu(B_\varepsilon(y)) \leq \limsup_{y \to x} \mu(B_\varepsilon(y)) \leq \mu(\overline{B_\varepsilon(x)}).$$

In particular, conclude that $x \mapsto \mu(B_\varepsilon(x))$ is lower semicontinuous for $\varepsilon > 0$.

Problem A.41. *Show that $M = \{x | 0 < (D\mu)(x)\}$ is a minimal support of μ.*

Problem A.42. *Suppose $\overline{D}\mu \leq \alpha$. Show that $d\mu = f\, d^n x$ with $\|f\|_\infty \leq \alpha$.*

Problem A.43 (Chebyshev inequality). *For $f \in L^1(\mathbb{R}^n)$, show*

$$|\{x \in A | |f(x)| > \alpha\}| \leq \frac{1}{\alpha} \int_A |f(x)| d^n x.$$

Problem A.44. *Show that the Cantor function is Hölder continuous, $|f(x) - f(y)| \leq |x - y|^\alpha$, with exponent $\alpha = \log_3(2)$. (Hint: Show that if g satisfies a Hölder estimate $|g(x) - g(y)| \leq M|x - y|^\alpha$, then so does $K(g)$: $|K(g)(x) - K(g)(y)| \leq \frac{3^\alpha}{2} M|x - y|^\alpha$.)*

Bibliographical notes

The aim of this section is not to give a comprehensive guide to the literature, but to document the sources from which I have learned the materials and which I have used during the preparation of this text. In addition, I will point out some standard references for further reading. In some sense, all books on this topic are inspired by von Neumann's celebrated monograph [74] and the present text is no exception.

General references for the first part are Akhiezer and Glazman [1], Berthier (Boutet de Monvel) [10], Blank, Exner, and Havlíček [11], Edmunds and Evans [18], Lax [32], Reed and Simon [49], Weidmann [70], [72], or Yosida [76].

Chapter 0: A first look at Banach and Hilbert spaces

As a reference for general background I can warmly recommend Kelly's classical book [33]. The rest is standard material and can be found in any book on functional analysis.

Chapter 1: Hilbert spaces

The material in this chapter is again classical and can be found in any book on functional analysis. I mainly follow Reed and Simon [49], respectively, Weidmann [70], with the main difference being that I use orthonormal sets and their projections as the central theme from which everything else is derived. For an alternate problem-based approach, see Halmos' book [27].

Chapter 2: Self-adjointness and spectrum

This chapter is still similar in spirit to [49], [70] with some ideas taken from Schechter [57].

Chapter 3: The spectral theorem

The approach via the Herglotz representation theorem follows Weidmann [70]. However, I use projection-valued measures as in Reed and Simon [49] rather than the resolution of the identity. Moreover, I have augmented the discussion by adding material on spectral types and the connections with the boundary values of the resolvent. For a survey containing several recent results, see [35].

Chapter 4: Applications of the spectral theorem

This chapter collects several applications from various sources which I have found useful or which are needed later on. Again, Reed and Simon [49] and Weidmann [70], [73] are the main references here.

Chapter 5: Quantum dynamics

The material is a synthesis of the lecture notes by Enß [20], Reed and Simon [49], [51], and Weidmann [73]. See also the book by Amrein [3]. There are also close connections with operator semigroups and we refer to the classical monograph by Goldstein [25] for further information.

Chapter 6: Perturbation theory for self-adjoint operators

This chapter is similar to [70] (which contains more results) with the main difference being that I have added some material on quadratic forms. In particular, the section on quadratic forms contains, in addition to the classical results, some material which I consider useful but was unable to find (at least not in the present form) in the literature. The prime reference here is Kato's monumental treatise [29] and Simon's book [58]. For further information on trace class operators, see Simon's classic [61]. The idea to extend the usual notion of strong resolvent convergence by allowing the approximating operators to live on subspaces is taken from Weidmann [72].

Chapter 7: The free Schrödinger operator

Most of the material is classical. Much more on the Fourier transform can be found in Reed and Simon [50] or Grafakos [23].

Chapter 8: Algebraic methods

This chapter collects some material which can be found in almost any physics textbook on quantum mechanics. My only contribution is to provide some mathematical details. I recommend the classical book by Thirring [68] and the visual guides by Thaller [66], [67].

Chapter 9: One-dimensional Schrödinger operators

One-dimensional models have always played a central role in understanding quantum mechanical phenomena. In particular, *general wisdom used to say that Schrödinger operators should have absolutely continuous spectrum plus some discrete point spectrum, while singular continuous spectrum is a pathology that should not occur in examples with bounded V* [**16**, Sect. 10.4]. In fact, a large part of [**52**] is devoted to establishing the absence of singular continuous spectrum. This was proven wrong by Pearson, who constructed an explicit one-dimensional example with singular continuous spectrum. Moreover, after the appearance of random models, it became clear that such types of exotic spectra (singular continuous or dense pure point) are frequently generic. The starting point is often the boundary behaviour of the Weyl m-function and its connection with the growth properties of solutions of the underlying differential equation, the latter being known as Gilbert and Pearson or subordinacy theory. One of my main goals is to give a modern introduction to this theory. The section on inverse spectral theory presents a simple proof for the Borg–Marchenko theorem (in the local version of Simon) from Bennewitz [**9**]. Again, this result is the starting point of almost all other inverse spectral results for Sturm–Liouville equations and should enable the reader to start reading research papers in this area.

Other references with further information are the lecture notes by Weidmann [**71**] or the classical books by Coddington and Levinson [**15**], Levitan [**36**], Levitan and Sargsjan [**37**], [**38**], Marchenko [**40**], Naimark [**42**], Pearson [**46**]. See also the recent monographs by Rofe-Betekov and Kholkin [**55**], Zettl [**77**] or the recent collection of historic and survey articles [**4**]. A compilation of exactly solvable potentials can be found in Bagrov and Gitman [**6**, App. I]. For a nice introduction to random models I can recommend the recent notes by Kirsch [**34**] or the classical monographs by Carmona and Lacroix [**13**] or Pastur and Figotin [**45**]. For the discrete analog of Sturm–Liouville and Jacobi operators, see my monograph [**64**].

Chapter 10: One-particle Schrödinger operators

The presentation in the first two sections is influenced by Enß [**20**] and Thirring [**68**]. The solution of the Schrödinger equation in spherical coordinates can be found in any textbook on quantum mechanics. Again I tried to provide some missing mathematical details. Several other explicitly solvable examples can be found in the books by Albeverio et al. [**2**] or Flügge [**22**]. For the formulation of quantum mechanics via path integrals I suggest Roepstorff [**54**] or Simon [**59**].

Chapter 11: Atomic Schrödinger operators

This chapter essentially follows Cycon, Froese, Kirsch, and Simon [**16**]. For a recent review, see Simon [**60**]. For multi-particle operators from the viewpoint of stability of matter, see Lieb and Seiringer [**41**].

Chapter 12: Scattering theory

This chapter follows the lecture notes by Enß [**20**] (see also [**19**]) using some material from Perry [**47**]. Further information on mathematical scattering theory can be found in Amrein, Jauch, and Sinha [**5**], Baumgaertel and Wollenberg [**7**], Chadan and Sabatier [**14**], Cycon, Froese, Kirsch, and Simon [**16**], Komech and Kopylova [**31**], Newton [**43**], Pearson [**46**], Reed and Simon [**51**], or Yafaev [**75**].

Appendix A: Almost everything about Lebesgue integration

Most parts follow Rudin's book [**56**], respectively, Bauer [**8**], with some ideas also taken from Weidmann [**70**]. I have tried to strip everything down to the results needed here while staying self-contained. Another useful reference is the book by Lieb and Loss [**39**]. A comprehensive source are the two volumes by Bogachev [**12**].

Bibliography

[1] N. I. Akhiezer and I. M. Glazman, *Theory of Linear Operators in Hilbert Space*, Vols. I and II, Pitman, Boston, 1981.

[2] S. Albeverio, F. Gesztesy, R. Høegh-Krohn, and H. Holden, *Solvable Models in Quantum Mechanics*, 2nd ed., American Mathematical Society, Providence, 2005.

[3] W. O. Amrein, *Non-Relativistic Quantum Dynamics*, D. Reidel, Dordrecht, 1981.

[4] W. O. Amrein, A. M. Hinz, and D. B. Pearson, *Sturm–Liouville Theory: Past and Present*, Birkhäuser, Basel, 2005.

[5] W. O. Amrein, J. M. Jauch, and K. B. Sinha, *Scattering Theory in Quantum Mechanics*, W. A. Benajmin Inc., New York, 1977.

[6] V. G. Bagrov and D. M. Gitman, *Exact Solutions of Relativistic Wave Equations*, Kluwer Academic Publishers, Dordrecht, 1990.

[7] H. Baumgaertel and M. Wollenberg, *Mathematical Scattering Theory*, Birkhäuser, Basel, 1983.

[8] H. Bauer, *Measure and Integration Theory*, de Gruyter, Berlin, 2001.

[9] C. Bennewitz, *A proof of the local Borg–Marchenko theorem*, Commun. Math. Phys. **218**, 131–132 (2001).

[10] A. M. Berthier, *Spectral Theory and Wave Operators for the Schrödinger Equation*, Pitman, Boston, 1982.

[11] J. Blank, P. Exner, and M. Havlíček, *Hilbert-Space Operators in Quantum Physics*, 2nd ed., Springer, Dordrecht, 2008.

[12] V. I. Bogachev, *Measure Theory*, 2 vols., Springer, Berlin, 2007.

[13] R. Carmona and J. Lacroix, *Spectral Theory of Random Schrödinger Operators*, Birkhäuser, Boston, 1990.

[14] K. Chadan and P. C. Sabatier, *Inverse Problems in Quantum Scattering Theory*, Springer, New York, 1989.

[15] E. A. Coddington and N. Levinson, *Theory of Ordinary Differential Equations*, Krieger, Malabar, 1985.

[16] H. L. Cycon, R. G. Froese, W. Kirsch, and B. Simon, *Schrödinger Operators*, 2nd printing, Springer, Berlin, 2008.

[17] M. Demuth and M. Krishna, *Determining Spectra in Quantum Theory*, Birkhäuser, Boston, 2005.

[18] D. E. Edmunds and W. D. Evans, *Spectral Theory and Differential Operators*, Oxford University Press, Oxford, 1987.

[19] V. Enss, *Asymptotic completeness for quantum mechanical potential scattering*, Comm. Math. Phys. **61**, 285–291 (1978).

[20] V. Enß, *Schrödinger Operators*, lecture notes (unpublished).

[21] L. D. Fadeev and O. A. Yakubovskiĭ, *Lectures on Quantum Mechanics for Mathematics Students*, Amer. Math. Soc., Providence, 2009.

[22] S. Flügge, *Practical Quantum Mechanics*, Springer, Berlin, 1994.

[23] L. Grafakos, *Classical Fourier Analysis*, 2nd ed., Springer, New York, 2008.

[24] I. Gohberg, S. Goldberg, and N. Krupnik, *Traces and Determinants of Linear Operators*, Birkhäuser, Basel, 2000.

[25] J. A. Goldstein, *Semigroups of Linear Operators and Applications*, Oxford University Press, Oxford, 1985.

[26] S. Gustafson and I. M. Sigal, *Mathematical Concepts of Quantum Mechanics*, 2nd ed., Springer, Berlin, 2011.

[27] P. R. Halmos, *A Hilbert Space Problem Book*, 2nd ed., Springer, New York, 1984.

[28] P. D. Hislop and I. M. Sigal, *Introduction to Spectral Theory*, Springer, New York, 1996.

[29] T. Kato, *Perturbation Theory for Linear Operators*, Springer, New York, 1966.

[30] A. Komech, *Quantum Mechanics: Genesis and Achievements*, Springer, Dordrecht, 2013.

[31] A. Komech and E. Kopylova, *Dispersion Decay and Scattering Theory*, John Wiley, Hoboken, 2012.

[32] P. D. Lax, *Functional Analysis*, Wiley-Interscience, New York, 2002.

[33] J. L. Kelly, *General Topology*, Springer, New York, 1955.

[34] W. Kirsch, *An invitation to random Schrödinger operators*, in *Random Schrödinger Operators*, M. Dissertori et al. (eds.), 1–119, Panoramas et Synthèses **25**, Société Mathématique de France, Paris, 2008.

[35] Y. Last, *Quantum dynamics and decompositions of singular continuous spectra*, J. Funct. Anal. **142**, 406–445 (1996).

[36] B. M. Levitan, *Inverse Sturm–Liouville Problems*, VNU Science Press, Utrecht, 1987.

[37] B. M. Levitan and I. S. Sargsjan, *Introduction to Spectral Theory*, American Mathematical Society, Providence, 1975.

[38] B. M. Levitan and I. S. Sargsjan, *Sturm–Liouville and Dirac Operators*, Kluwer Academic Publishers, Dordrecht, 1991.

[39] E. Lieb and M. Loss, *Analysis*, American Mathematical Society, Providence, 1997.

[40] V. A. Marchenko, *Sturm–Liouville Operators and Applications*, Birkhäuser, Basel, 1986.

[41] E. H. Lieb and R. Seiringer, *Stability of Matter*, Cambridge University Press, Cambridge, 2010.

Bibliography

[42] M. A. Naimark, *Linear Differential Operators, Parts I and II*, Ungar, New York, 1967 and 1968.

[43] R. G. Newton, *Scattering Theory of Waves and Particles*, 2nd ed., Dover, New York, 2002.

[44] F. W. J. Olver et al., *NIST Handbook of Mathematical Functions*, Cambridge University Press, Cambridge, 2010.

[45] L. Pastur and A. Figotin, *Spectra of Random and Almost-Periodic Operators*, Springer, Berlin, 1992.

[46] D. Pearson, *Quantum Scattering and Spectral Theory*, Academic Press, London, 1988.

[47] P. Perry, *Mellin transforms and scattering theory*, Duke Math. J. **47**, 187–193 (1987).

[48] E. Prugovečki, *Quantum Mechanics in Hilbert Space*, 2nd ed., Academic Press, New York, 1981.

[49] M. Reed and B. Simon, *Methods of Modern Mathematical Physics I. Functional Analysis*, rev. and enl. ed., Academic Press, San Diego, 1980.

[50] M. Reed and B. Simon, *Methods of Modern Mathematical Physics II. Fourier Analysis, Self-Adjointness*, Academic Press, San Diego, 1975.

[51] M. Reed and B. Simon, *Methods of Modern Mathematical Physics III. Scattering Theory*, Academic Press, San Diego, 1979.

[52] M. Reed and B. Simon, *Methods of Modern Mathematical Physics IV. Analysis of Operators*, Academic Press, San Diego, 1978.

[53] J. R. Retherford, *Hilbert Space: Compact Operators and the Trace Theorem*, Cambridge University Press, Cambridge, 1993.

[54] G. Roepstorff, *Path Integral Approach to Quantum Physics*, Springer, Berlin, 1994.

[55] F. S. Rofe-Beketov and A. M. Kholkin, *Spectral Analysis of Differential Operators. Interplay Between Spectral and Oscillatory Properties*, World Scientific, Hackensack, 2005.

[56] W. Rudin, *Real and Complex Analysis*, 3rd ed., McGraw-Hill, New York, 1987.

[57] M. Schechter, *Operator Methods in Quantum Mechanics*, North Holland, New York, 1981.

[58] B. Simon, *Quantum Mechanics for Hamiltonians Defined as Quadratic Forms*, Princeton University Press, Princeton, 1971.

[59] B. Simon, *Functional Integration and Quantum Physics*, Academic Press, New York, 1979.

[60] B. Simon, *Schrödinger operators in the twentieth century*, J. Math. Phys. **41:6**, 3523–3555 (2000).

[61] B. Simon, *Trace Ideals and Their Applications*, 2nd ed., Amererican Mathematical Society, Providence, 2005.

[62] E. Stein and R. Shakarchi, *Complex Analysis*, Princeton University Press, Princeton, 2003.

[63] L. A. Takhtajan, *Quantum Mechanics for Mathematicians*, Amer. Math. Soc., Providence, 2008.

[64] G. Teschl, *Jacobi Operators and Completely Integrable Nonlinear Lattices*, Math. Surv. and Mon. **72**, Amer. Math. Soc., Rhode Island, 2000.

[65] B. Thaller, *The Dirac Equation*, Springer, Berlin 1992.

[66] B. Thaller, *Visual Quantum Mechanics*, Springer, New York, 2000.

[67] B. Thaller, *Advanced Visual Quantum Mechanics*, Springer, New York, 2005.

[68] W. Thirring, *Quantum Mechanics of Atoms and Molecules*, Springer, New York, 1981.

[69] G. N. Watson, *A Treatise on the Theory of Bessel Functions*, 2nd ed., Cambridge University Press, Cambridge, 1962.

[70] J. Weidmann, *Linear Operators in Hilbert Spaces*, Springer, New York, 1980.

[71] J. Weidmann, *Spectral Theory of Ordinary Differential Operators*, Lecture Notes in Mathematics, **1258**, Springer, Berlin, 1987.

[72] J. Weidmann, *Lineare Operatoren in Hilberträumen, Teil 1: Grundlagen*, B. G. Teubner, Stuttgart, 2000.

[73] J. Weidmann, *Lineare Operatoren in Hilberträumen, Teil 2: Anwendungen*, B. G. Teubner, Stuttgart, 2003.

[74] J. von Neumann, *Mathematical Foundations of Quantum Mechanics*, Princeton University Press, Princeton, 1996.

[75] D. R. Yafaev, *Mathematical Scattering Theory: General Theory*, American Mathematical Society, Providence, 1992.

[76] K. Yosida, *Functional Analysis*, 6th ed., Springer, Berlin, 1980.

[77] A. Zettl, *Sturm–Liouville Theory*, American Mathematical Society, Providence, 2005.

Glossary of notation

$AC(I)$	...absolutely continuous functions, 95
$B_r(x)$	...open ball of radius r around x, 4
$\mathfrak{B}$	$= \mathfrak{B}^1$
$\mathfrak{B}^n$	...Borel σ-field of $\mathbb{R}^n$, 296
$\mathfrak{C}(\mathfrak{H})$	...set of compact operators, 151
$\mathbb{C}$	...the set of complex numbers
$C(U)$	...set of continuous functions from U to $\mathbb{C}$
$C_\infty(U)$	...set of functions in $C(U)$ which vanish at ∞
$C(U,V)$	...set of continuous functions from U to V
$C_c(U,V)$	...set of compactly supported continuous functions
$C^\infty(U,V)$	...set of smooth functions
$C_b(U,V)$	...set of bounded continuous functions
$\chi_\Omega(.)$	...characteristic function of the set Ω
dim	...dimension of a vector space
$\text{dist}(x,Y)$	$= \inf_{y \in Y} \|x-y\|$, distance between x and Y
$\mathfrak{D}(.)$	...domain of an operator
e	...exponential function, $e^z = \exp(z)$
$\mathbb{E}(A)$	...expectation of an operator A, 63
$\mathcal{F}$	...Fourier transform, 187
H	...Schrödinger operator, 257
H_0	...free Schrödinger operator, 197
$H^m(a,b)$	...Sobolev space, 95
$H_0^m(a,b)$	...Sobolev space, 96
$H^m(\mathbb{R}^n)$	...Sobolev space, 194
hull(.)	...convex hull
$\mathfrak{H}$	...a separable Hilbert space

i	...complex unity, $i^2 = -1$
$\mathbb{I}$	...identity operator
Im(.)	...imaginary part of a complex number
inf	...infimum
Ker(A)	...kernel of an operator A, 27
$\mathfrak{L}(X,Y)$	...set of all bounded linear operators from X to Y, 29
$\mathfrak{L}(X)$	$= \mathfrak{L}(X,X)$
$L^p(X, d\mu)$	...Lebesgue space of p integrable functions, 31
$L^p_{loc}(X, d\mu)$	...locally p integrable functions, 36
$L^p_c(X, d\mu)$	...compactly supported p integrable functions
$L^\infty(X, d\mu)$	...Lebesgue space of bounded functions, 32
$L^\infty_\infty(\mathbb{R}^n)$	...Lebesgue space of bounded functions vanishing at ∞
$\ell^p(\mathbb{N})$	...Banach space of p summable sequences, 15
$\ell^2(\mathbb{N})$	...Hilbert space of square summable sequences, 21
$\ell^\infty(\mathbb{N})$	...Banach space of bounded summable sequences, 16
λ	...a real number
$m_a(z)$	...Weyl m-function, 235
$M(z)$	...Weyl M-matrix, 246
max	...maximum
$\mathcal{M}$	...Mellin transform, 287
μ_ψ	...spectral measure, 108
$\mathbb{N}$	...the set of positive integers
$\mathbb{N}_0$	$= \mathbb{N} \cup \{0\}$
$o(x)$	...Landau symbol little-o
$O(x)$	...Landau symbol big-O
Ω	...a Borel set
$\Omega_\pm$	...wave operators, 283
$P_A(.)$	...family of spectral projections of an operator A, 108
$P_\pm$	...projector onto outgoing/incoming states, 286
$\mathbb{Q}$	...the set of rational numbers
$\mathfrak{Q}(.)$	...form domain of an operator, 109
$R(I, X)$	...set of regulated functions, 132
$R_A(z)$	...resolvent of A, 83
Ran(A)	...range of an operator A, 27
rank(A)	$= \dim \text{Ran}(A)$, rank of an operator A, 151
Re(.)	...real part of a complex number
$\rho(A)$	...resolvent set of A, 83
$\mathbb{R}$	...the set of real numbers
$S(I, X)$	...set of simple functions, 132
$\mathcal{S}(\mathbb{R}^n)$	...set of smooth functions with rapid decay, 187
sign(x)	$= x/|x|$ for $x \neq 0$ and 0 for $x = 0$; sign function

Glossary of notation

$\sigma(A)$	...spectrum of an operator A, 83
$\sigma_{ac}(A)$	...absolutely continuous spectrum of A, 119
$\sigma_{sc}(A)$	...singular continuous spectrum of A, 119
$\sigma_{pp}(A)$	...pure point spectrum of A, 119
$\sigma_p(A)$	...point spectrum (set of eigenvalues) of A, 115
$\sigma_d(A)$	...discrete spectrum of A, 170
$\sigma_{ess}(A)$	...essential spectrum of A, 170
$\mathrm{span}(M)$	...set of finite linear combinations from M, 17
$\sup$	...supremum
$\mathrm{supp}(f)$	...support of a function f, 8
$\mathrm{supp}(\mu)$	...support of a measure μ, 301
$\mathbb{Z}$	...the set of integers
z	...a complex number
$\sqrt{z}$	...square root of z with branch cut along $(-\infty, 0]$
z^*	...complex conjugation
A^*	...adjoint of A, 67
$\overline{A}$	...closure of A, 72
$\hat{f}$	$= \mathcal{F}f$, Fourier transform of f, 187
$\check{f}$	$= \mathcal{F}^{-1}f$, inverse Fourier transform of f, 189
$\lvert x \rvert$	$= \sqrt{\sum_{j=1}^n \lvert x_j \rvert^2}$ Euclidean norm in $\mathbb{R}^n$ or $\mathbb{C}^n$
$\lvert \Omega \rvert$	...Lebesgue measure of a Borel set Ω
$\lVert \cdot \rVert$	...norm in the Hilbert space $\mathfrak{H}$, 21
$\lVert \cdot \rVert_p$	...norm in the Banach space L^p, 30
$\langle ., .. \rangle$	...scalar product in $\mathfrak{H}$, 21
$\mathbb{E}_\psi(A)$	$= \langle \psi, A\psi \rangle$, expectation value, 64
$\Delta_\psi(A)$	$= \mathbb{E}_\psi(A^2) - \mathbb{E}_\psi(A)^2$, variance, 64
Δ	...Laplace operator, 197
∂	...gradient, 188
∂_α	...derivative, 187
$\oplus$	...orthogonal sum of vector spaces or operators, 52, 89
$\otimes$	...tensor product, 53, 143
$M^\perp$	...orthogonal complement, 49
A'	...complement of a set
(λ_1, λ_2)	$= \{\lambda \in \mathbb{R} \mid \lambda_1 < \lambda < \lambda_2\}$, open interval
$[\lambda_1, \lambda_2]$	$= \{\lambda \in \mathbb{R} \mid \lambda_1 \leq \lambda \leq \lambda_2\}$, closed interval
$\psi_n \to \psi$	...norm convergence, 14
$\psi_n \rightharpoonup \psi$	...weak convergence, 55

$A_n \to A$ ...norm convergence
$A_n \xrightarrow{s} A$ ...strong convergence, 57
$A_n \rightharpoonup A$ ...weak convergence, 56
$A_n \xrightarrow{nr} A$ ...norm resolvent convergence, 179
$A_n \xrightarrow{sr} A$ ...strong resolvent convergence, 179

Index

a.e., *see also* almost everywhere
absolue value of an operator, 138
absolute convergence, 20
absolutely continuous
　function, 95
　measure, 331
　spectrum, 119
accumulation point, 4
adjoint operator, 54, 67
algebra, 295
almost everywhere, 302
angular momentum operator, 210

B.L.T. theorem, 28
Baire category theorem, 38
ball
　closed, 6
　open, 4
Banach algebra, 29
Banach space, 14
Banach–Steinhaus theorem, 39
base, 5
basis, 17
　orthonormal, 47
　spectral, 106
Bessel function, 204
　modified, 202
　spherical, 267
Bessel inequality, 45
bijective, 8
Bolzano–Weierstraß theorem, 12
Borel
　function, 308

　measure, 298
　　regular, 298
　set, 296
　σ-algebra, 296
　transform, 107, 112
boundary condition
　Dirichlet, 224
　Neumann, 224
　periodic, 224
boundary point, 4
bounded
　operator, 27
　sesquilinear form, 26
　set, 11

C-real, 93
canonical form of compact operators, 161
Cantor
　function, 338
　measure, 339
　set, 302
Cauchy sequence, 7
Cauchy–Schwarz–Bunjakowski
　inequality, 22
Cayley transform, 91
Cesàro average, 150
characteristic function, 312
Chebyshev inequality, 339
closable
　form, 80
　operator, 72
closed

ball, 6
form, 80
operator, 72
set, 6
closed graph theorem, 75
closure, 6
  essential, 117
cluster point, 4
commute, 136
compact, 9
  locally, 12
  sequentially, 11
complete, 7, 14
completion, 26
configuration space, 64
conjugation, 93
conserved quantity, 138
continuous, 8
convergence, 6
convolution, 191
core, 71
cover, 9
C^* algebra, 55
cyclic vector, 106

dense, 7
dilation group, 259
Dirac measure, 301, 317
Dirac operator, 149, 215
Dirichlet boundary condition, 224
discrete set, 4
discrete topology, 4
distance, 3, 12
distribution function, 298
Dollard theorem, 200
domain, 27, 64, 66
dominated convergence theorem, 316
Dynkin system, 303
Dynkin's π-λ theorem, 303

eigenspace, 132
eigenvalue, 83
  multiplicity, 132
eigenvector, 83
element
  adjoint, 55
  normal, 55
  positive, 55
  self-adjoint, 55
  unitary, 55
equivalent norms, 24
essential

closure, 117
range, 84
spectrum, 170
supremum, 32
expectation, 63
Exponential Herglotz representation, 129
extension, 67
Extreme value theorem, 12

finite intersection property, 9
first resolvent formula, 85
form, 80
  bound, 175
  bounded, 26, 82
  closable, 80
  closed, 80
  core, 81
  domain, 77, 109
  hermitian, 80
  nonnegative, 80
  semi-bounded, 80
Fourier
  series, 47
  transform, 150, 187
Friedrichs extension, 80
Fubini theorem, 320
function
  absolutely continuous, 95
  open, 8
fundamental theorem of calculus, 135, 317

gamma function, 328
Gaussian wave packet, 209
gradient, 188
Gram–Schmidt orthogonalization, 48
graph, 72
graph norm, 72
Green's function, 202
ground state, 272

Hamiltonian, 65
Hankel operator, 169
Hankel transform, 203
harmonic oscillator, 212
Hausdorff space, 5
Heine–Borel theorem, 11
Heisenberg picture, 154
Heisenberg uncertainty principle, 193
Hellinger–Toeplitz theorem, 76
Herglotz

Index

function, 107
  representation theorem, 120
Hermite polynomials, 213
hermitian
  form, 80
  operator, 67
Hilbert space, 21, 43
  separable, 47
Hölder's inequality, 16, 32
homeomorphism, 8
HVZ theorem, 278
hydrogen atom, 258

ideal, 55
identity, 29
induced topology, 5
injective, 7
inner product, 21
inner product space, 21
integrable, 315
integral, 312
interior, 6
interior point, 4
intertwining property, 284
involution, 55
ionization, 278
isolated point, 4

Jacobi operator, 76

Kato–Rellich theorem, 159
kernel, 27
KLMN theorem, 175
Kuratowski closure axioms, 6

λ-system, 303
l.c., *see also* limit circle
l.p., *see also* limit point
Lagrange identity, 218
Laguerre polynomial, 267
  generalized, 268
Lebesgue
  decomposition, 333
  measure, 301
  point, 335
Lebesgue–Stieltjes measure, 298
Legendre equation, 262
lemma
  Riemann-Lebesgue, 191
Lidskij trace theorem, 168
limit circle, 223
limit point, 4, 223
Lindelöf theorem, 9

linear
  functional, 29, 50
  operator, 27
linearly independent, 17
Liouville normal form, 222
localization formula, 279
lower semicontinuous, 309

maximum norm, 14
Mean ergodic theorem, 155
mean-square deviation, 64
measurable
  function, 307
  set, 297
  space, 296
measure, 296
  absolutely continuous, 331
  complete, 306
  finite, 297
  growth point, 112
  Lebesgue, 301
  minimal support, 338
  mutually singular, 331
  product, 319
  projection-valued, 100
  space, 297
  spectral, 108
  support, 301
  topological support, 301
Mellin transform, 287
metric space, 3
Minkowski's inequality, 32
mollifier, 35
momentum operator, 208
monotone convergence theorem, 313
Morrey inequality, 196
multi-index, 187
  order, 187
multiplicity
  spectral, 107
mutually singular measures, 331

neighborhood, 4
Neumann
  boundary condition, 224
  function
    spherical, 267
  series, 85
Nevanlinna function, 107
Noether theorem, 208
norm, 14
  operator, 27

norm resolvent convergence, 179
normal, 12, 55, 69, 76, 104
normalized, 22, 44
normed space, 14
nowhere dense, 38
null space, 27

observable, 63
ONB, see also orthonormal basis
one-parameter unitary group, 65
ONS, see also orthonormal set
onto, 8
open
  ball, 4
  function, 8
  set, 4
operator
  adjoint, 54, 67
  bounded, 27
  bounded from below, 79
  closable, 72
  closed, 72
  closure, 72
  compact, 151
  domain, 27, 66
  finite rank, 151
  hermitian, 67
  Hilbert–Schmidt, 163
  linear, 27, 66
  nonnegative, 77
  normal, 69, 76, 104
  positive, 77
  relatively bounded, 157
  relatively compact, 152
  self-adjoint, 68
  semi-bounded, 79
  strong convergence, 56
  symmetric, 67
  unitary, 45, 65
  weak convergence, 57
orthogonal, 22, 44
  complement, 49
  polynomials, 264
  projection, 50
  sum, 52
orthonormal
  basis, 47
  set, 44
orthonormal basis, 47
oscillating, 255
outer measure, 304

parallel, 22, 44
parallelogram law, 23
parity operator, 111
Parseval relation, 47
partial isometry, 139
partition of unity, 13
perpendicular, 22, 44
phase space, 64
π-system, 303
Plücker identity, 222
Plancherel identity, 190
polar coordinates, 325
polar decomposition, 139
polarization identity, 23, 45, 67
position operator, 207
positivity
  improving, 272
  preserving, 272
premeasure, 297
probability density, 63
probability measure, 297
product measure, 319
product topology, 9
projection, 55
proper metric space, 12
pseudometric, 3
pure point spectrum, 119
Pythagorean theorem, 22, 44

quadrangle inequality, 13
quadratic form, 67, see also form
quasinorm, 20

Radon measure, 311
Radon–Nikodym
  derivative, 332
  theorem, 332
RAGE theorem, 153
Rajchman measure, 155
range, 27
  essential, 84
rank, 151
Rayleigh–Ritz method, 140
reducing subspace, 90
regulated function, 132
relative σ-algebra, 296
relative topology, 5
relatively compact, 9, 152
resolution of the identity, 101
resolvent, 83
  convergence, 179
  formula

Index

first, 85
second, 159
Neumann series, 85
set, 83
Riesz lemma, 50
Ritz method, 140

scalar product, 21
scattering operator, 284
scattering state, 284
Schatten p-class, 165
Schauder basis, 17
Schrödinger equation, 65
Schur criterion, 34
Schwartz space, 187
second countable, 5
second resolvent formula, 159
self-adjoint, 55
   essentially, 71
seminorm, 14
separable, 7, 18
series
   absolutely convergent, 20
sesquilinear form, 21
   bounded, 26
   parallelogram law, 25
   polarization identity, 26
short range, 289
σ-algebra, 296
σ-finite, 297
simple function, 132, 312
simple spectrum, 107
singular values, 161
singularly continuous
   spectrum, 119
Sobolev space, 95, 194
span, 17
spectral
   basis, 106
      ordered, 118
   mapping theorem, 118
   measure
      maximal, 118
   theorem, 109
      compact operators, 160
   vector, 106
      maximal, 118
spectrum, 83
   absolutely continuous, 119
   discrete, 170
   essential, 170
   pure point, 119

singularly continuous, 119
spherical coordinates, 260, 325
spherical harmonics, 263
spherically symmetric, 194
∗-ideal, 55
∗-subalgebra, 55
stationary phase, 288
Stieltjes inversion formula, 107, 134
Stone theorem, 147
Stone's formula, 134
Stone–Weierstraß theorem, 60
strong convergence, 56
strong resolvent convergence, 179
Sturm comparison theorem, 254
Sturm–Liouville equation, 217
   regular, 218
subcover, 9
subordinacy, 243
subordinate solution, 243
subspace
   reducing, 90
subspace topology, 5
superposition, 64
supersymmetric quantum mechanics, 215
support, 8
   measure, 301
surjective, 8

Temple's inequality, 142
tensor product, 53
theorem
   B.L.T., 28
   Bair, 38
   Banach–Steinhaus, 39
   Bolzano–Weierstraß, 12
   closed graph, 75
   Dollard, 200
   dominated convergence, 316
   Dynkin's π-λ, 303
   Fatou, 314, 316
   Fatou–Lebesgue, 316
   Fubini, 320
   fundamental thm. of calculus, 317
   Heine–Borel, 11
   Hellinger–Toeplitz, 76
   Herglotz, 120
   HVZ, 278
   Jordan–von Neumann, 23
   Kato–Rellich, 159
   KLMN, 175
   Kneser, 255

Lebesgue, 316
Lebesgue decomposition, 333
Levi, 313
Lindelöf, 9
monotone convergence, 313
Noether, 208
Plancherel, 190
Pythagorean, 22, 44
Radon–Nikodym, 332
RAGE, 153
Riesz, 50
Schur, 34
Sobolev embedding, 196
spectral, 109
spectral mapping, 118
Stone, 147
Stone–Weierstraß, 60
Sturm, 254
Tonelli, 321
Urysohn, 12
virial, 259
Weidmann, 253
Weierstraß, 12, 19
Weyl, 171
Wiener, 150, 194
Tonelli theorem, 321
topological space, 4
topology
  base, 5
  product, 9
total, 18
trace, 167
  class, 167
trace operator, 96
trace topology, 5
triangle inequality, 3, 14
  inverse, 3, 14
trivial topology, 4
Trotter product formula, 155

uncertainty principle, 192, 208
uniform boundedness principle, 39
uniformly convex space, 25
unit sphere, 326
unit vector, 22, 44
unitary, 55, 65
unitary group, 65
  generator, 65
  strongly continuous, 65
  weakly continuous, 147
upper semicontinuous, 309
Urysohn lemma, 12

Vandermonde determinant, 20
variance, 64
virial theorem, 259
Vitali set, 303

wave
  function, 63
  operators, 283
wave equation, 148
weak
  Cauchy sequence, 56
  convergence, 55
  derivative, 96, 195
Weierstraß approximation, 19
Weierstraß theorem, 12
Weyl
  M-matrix, 246
  circle, 230
  relations, 208
  sequence, 86
    singular, 171
  theorem, 171
Weyl–Titchmarsh m-function, 235
Wiener covering lemma, 334
Wiener theorem, 150
Wronskian, 218

Young inequality, 191

Selected Published Titles in This Series

157 Gerald Teschl, Mathematical Methods in Quantum Mechanics: With Applications to Schrödinger Operators, Second Edition, 2014
156 Markus Haase, Functional Analysis, 2014
155 Emmanuel Kowalski, An Introduction to the Representation Theory of Groups, 2014
154 Wilhelm Schlag, A Course in Complex Analysis and Riemann Surfaces, 2014
152 Gábor Székelyhidi, An Introduction to Extremal Kähler Metrics, 2014
151 Jennifer Schultens, Introduction to 3-Manifolds, 2014
150 Joe Diestel and Angela Spalsbury, The Joys of Haar Measure, 2013
149 Daniel W. Stroock, Mathematics of Probability, 2013
148 Luis Barreira and Yakov Pesin, Introduction to Smooth Ergodic Theory, 2013
147 Xingzhi Zhan, Matrix Theory, 2013
146 Aaron N. Siegel, Combinatorial Game Theory, 2013
145 Charles A. Weibel, The K-book, 2013
144 Shun-Jen Cheng and Weiqiang Wang, Dualities and Representations of Lie Superalgebras, 2012
143 Alberto Bressan, Lecture Notes on Functional Analysis, 2013
142 Terence Tao, Higher Order Fourier Analysis, 2012
141 John B. Conway, A Course in Abstract Analysis, 2012
140 Gerald Teschl, Ordinary Differential Equations and Dynamical Systems, 2012
139 John B. Walsh, Knowing the Odds, 2012
138 Maciej Zworski, Semiclassical Analysis, 2012
137 Luis Barreira and Claudia Valls, Ordinary Differential Equations, 2012
136 Arshak Petrosyan, Henrik Shahgholian, and Nina Uraltseva, Regularity of Free Boundaries in Obstacle-Type Problems, 2012
135 Pascal Cherrier and Albert Milani, Linear and Quasi-linear Evolution Equations in Hilbert Spaces, 2012
134 Jean-Marie De Koninck and Florian Luca, Analytic Number Theory, 2012
133 Jeffrey Rauch, Hyperbolic Partial Differential Equations and Geometric Optics, 2012
132 Terence Tao, Topics in Random Matrix Theory, 2012
131 Ian M. Musson, Lie Superalgebras and Enveloping Algebras, 2012
130 Viviana Ene and Jürgen Herzog, Gröbner Bases in Commutative Algebra, 2011
129 Stuart P. Hastings and J. Bryce McLeod, Classical Methods in Ordinary Differential Equations, 2012
128 J. M. Landsberg, Tensors: Geometry and Applications, 2012
127 Jeffrey Strom, Modern Classical Homotopy Theory, 2011
126 Terence Tao, An Introduction to Measure Theory, 2011
125 Dror Varolin, Riemann Surfaces by Way of Complex Analytic Geometry, 2011
124 David A. Cox, John B. Little, and Henry K. Schenck, Toric Varieties, 2011
123 Gregory Eskin, Lectures on Linear Partial Differential Equations, 2011
122 Teresa Crespo and Zbigniew Hajto, Algebraic Groups and Differential Galois Theory, 2011
121 Tobias Holck Colding and William P. Minicozzi II, A Course in Minimal Surfaces, 2011
120 Qing Han, A Basic Course in Partial Differential Equations, 2011
119 Alexander Korostelev and Olga Korosteleva, Mathematical Statistics, 2011

For a complete list of titles in this series, visit the
AMS Bookstore at **www.ams.org/bookstore/gsmseries/**.